PLASMA ASTROPHYSICS AND COSMOLOGY

PLASMA ASTROPHYSICS AND COSMOLOGY

The Second IEEE International Workshop, Princeton, New Jersey, May 10–12, 1993

Edited by

ANTHONY L. PERATT
Los Alamos National Laboratory, NM, U.S.A.

Reprinted from *Astrophysics and Space Science*
Volume 227, Nos. 1–2, 1995

SPRINGER-SCIENCE+BUSINESS MEDIA, B.V.

A C.I.P. Catalogue record for this book is available from the Library of Congress.

ISBN 978-94-010-4181-2 ISBN 978-94-011-0405-0 (eBook)
DOI 10.1007/978-94-011-0405-0

Printed on acid-free paper

TABLE OF CONTENTS

PREFACE

This special issue of the international journal of cosmic physics, *Astrophysics and Space Science,* contains invited contributions delivered at the *Second IEEE International Workshop on Plasma Astrophysics and Cosmology*, held from 10 to 12 May 1993 in Princeton, New Jersey. The Workshop was sponsored by the NSF Division of Atmospheric Sciences, NASA Headquarters, Space Physics Division, and the Nuclear and Plasma Sciences Society of the Institute of Electrical and Electronics Engineers.

It was the purpose of the Workshop to update topics in Plasma Astrophysics and Cosmology presented at the *First IEEE International Workshop on Plasma Cosmology*, La Jolla, California, 20–22 February 1989, and to again bring together observers and theorists to discuss the related links between plasma theory and observation.

Another goal of the Workshop and these proceedings was to highlight the Centennial Celebration (1896–1996) of the founding of Plasma Astrophysics and Cosmology and several papers are devoted to the history of this field of science.

The Workshop topics included:

- Large Scale Structure and Filaments
- Formation of Galaxies and Electromagnetic Fields
- Active Galactic Nuclei and Quasars, Cosmic Rays
- Origin and Abundance of Light Elements
- Star Formation, the Evolution of Solar Systems
- Redshift Periodicities and Anomalous Redshifts
- The Cosmic Microwave Background
- The Evidence For and Against Dark Matter
- General Relativity and Electric Fields
- The History of Plasma Cosmology

The proceedings follows the format of the meeting, which was organized along a plasma–size hierarchy, ranging from well–documented plasma astrophysical phenomena to speculative issues in cosmology.

Astrophysics and Space Science **227**: 1–2, 1995.

We are grateful to Professor Carl-Gunne Fälthammar for his advice and valuable cooperation in bringing out these proceedings. We also wish to thank Princeton University for providing an outstanding location and excellent service for the conference banquet and to the members of the Department of Astrophysical Sciences and the Princeton University Observatory for their heavy participation in spirited debate.

The Scientific Organizing Committee consisted of H. Arp, E. Lerner, and A. Peratt, who wish to express their gratitude to NSF, NASA, and the IEEE for making possible this Workshop.

Anthony L. Peratt

INTRODUCTION TO PLASMA ASTROPHYSICS AND COSMOLOGY

Proceedings of the Second IEEE International Workshop on Plasma Astrophysics and Cosmology, held in Princeton, New Jersey, 10–12 May 1993

ANTHONY L. PERATT
Physics Division
Los Alamos National Laboratory
Los Alamos, New Mexico 87545, USA

Abstract. The year 1996 marks the Centennial Celebration of the founding of Plasma Astrophysics and Cosmology; its origins may be traced to the seminal research first published by Kristian Birkeland in 1896. This special workshop issue reports on advances in issues of importance to the plasma universe; topics as timely as when first raised a century ago.

Key words: Radio Astronomy, Plasma Astrophysics, Large Scale Structure, Filaments, Formation of Galaxies, Magnetic Fields, Active Galactic Nuclei, Quasars, Abundance of Light Elements, Redshifts, Microwave Background

1 Introduction

The year 1996 marks the Centennial Celebration of the founding of Plasma Astrophysics and Cosmology; its origins may be traced to the seminal research of Kristian Birkeland published in 1896 that began his life–long study of laboratory produced cathodic rays and corpuscles[1] and their analogies to astrophysical and cosmological phenomena. This work was presented in two papers: "Sur un spectre des rayons catodiques" in *Comptes Rendus*, 28 September 1896, and a paper in *Archives des Sciences Physiques et Naturelles*, Geneva, 4th period, vol. I, 1896, that announced his discovery of magneto–cathode rays. It was in this work that, according to Birkeland (1908):

> ...I expressed for the first time my belief that the northern lights are formed by corpuscular rays drawn in from space, and coming from the sun.

In addition to his solving the mystery of the Aurora with his now–famous terrella experiments; electron beams in vacuum from magnetized copper globe cathodes, Birkeland utilized his data to formulate a theory about a plasma–filled universe populated with *systems* of nebula (galaxies).

We quote from Birkeland (1908, Volume 1, Section 131.):

[1] The term 'plasma' was not to be coined by I. Langmuir until 1923.

Astrophysics and Space Science **227**: 3–11, 1995.

The Worlds in the Universe. From the conceptions to which our experimental analogies lead us, it is possible to form, in a natural manner an interesting theory of the origin of the worlds. This theory differs from all earlier theories in that it assumes the existence of a universal directing force of electro–magnetic origin in addition to the force of gravitation, in order to explain the formation round the sun of planets—which have almost circular orbits and are almost in the same plane—of moons and rings about the planets, and of spiral and annular nebulæ.

...Poincaré, at the conclusion of the preface to his book, 'Hyptheèses Cosmogoniques', says[2]:

"Un fait qui frappe tout le monde, c'est la forme spirale de certaines nébuleuses; elle se rencontre beaucoup trop souvent pour qu'on puisse penser qu'elle est due au hasard. On comprend combien est incomplète toute théorie cosmogonique qui en fait abstraction. Or aucune d'elles n'en rend compte d'une manière satisfaisante, et l'explication que j'ai donné moi-mème un jour, par mainière de passe–temps, ne vaux pas mieux que les autres. Nous ne pouvons donc terminer que par un point d'interrogation."

Much of Birkeland's work was rediscovered in the 1980s with renewed interest about the role of large scale magnetic fields and currents in explaining astrophysical, galactic, and cosmological scale phenomena, including the origin and structure or galaxies and the containment of intergalactic gas 'clouds' and filaments, formerly attributed to dark matter gravitational binding energy. While the study of plasma phenomena in astrophysics and cosmology is now a century old, plasma–based theories for the explanations of light element abundances, the cosmic microwave background, and the origin of large–scale structures are relatively recent topics but have rapidly developed since their introduction at the first IEEE International Conference on Plasma Cosmology, La Jolla, California, USA, 20–22 February 1989 (IEEE, 1990).

An important element of the Workshop was the history of the temperature of the universe whose study, incidentally, is also at least a century old.

When Birkeland developed his cosmogony and cosmology, it was known that the temperature of space was 5–6K blackbody, the value of which was a topic of considerable interest to both lay– and professional scientists of that era. We quote from his contemporary, Ch.-Ed. Guillaume in the article 'La Température de L'Espace', *La Nature*, vol.24, series 2, pp.210–211, 234

[2] "One fact that strikes everyone is the spiral shape of some nebulae; it is encountered much too often for us to believe that it is due to chance. It is easy to understand how incomplete any theory of cosmogony which ignores this fact must be. None of the theories accounts for it satisfactorily, and the explanation I myself once gave, in a kind of toy theory, is no better than the others. Consequently, we come up against a big question mark."

Fig. 1. Kristian Birkeland.

(1896)[3]:

[3] Captain Abney has recently determined the ratio of the light from the starry sky to that of the full Moon. It turns out to be 1/44, after reductions for the obliqueness of the rays relative to the surface, and for atmospheric absorption. Doubling this for both hemispheres, and adopting 1/600,000 as the ratio of the light intensity of the Moon to that of the Sun (a rough average of the measurements by Wollaston, Douguer and Zöllner), we find that the Sun showers us with 15,200,000 time more vibratory energy than all the stars combined. The increase in temperature of an isolated body in space subject only to the

Le capitaine Abney a déterminéé récemment le rapport de la lumière du ciel étoilé à celle de la pleine Lune; il le trouve égal à 1/44, toutes réductions faites pour l'obliquitée des rayons par rapport à la plaque; et pour l'absorption atmosphérique. En doublant pour les deux hemisphères, et en adoptant 1/600 000 comme rapport de l'intensité lumineuse de la Lune à celle du Soleil (moyenne grossière des mesures de Wollaston, de Douguer et de Zöllner), on trouvera que le Soleil nous envoie 15 200 000 fois plus d'énergie vibratoire que l'ensemble des étoiles. L'élévation de la température d'un corps isolé dans l'espace, et soumis seulement à l'action des étoiles, sera égale au quotient de l'élévation de température due au Soleil sur l'orbite de la Terre par la racine quatrième de 15 200 000, soit enviorn 60. Ce nombre doit, du reste, être considéré comme un minimum, les mesures du capitaine Abney, faits à South Kensington, ayant pu être faussées par quelque source de lumière étrangère. Nous en conclurons que la radiation des étoiles seule maintiendrait à la termpérature de 338/60 = 5,6 abs = –207°,4 centigrades l'éprouvette que nous avons supposée placée en divers points du ciel.

Il ne faudrait pas en conclure que la radiation des étoiles élève de 5 ou 6 degrés la température des corps célestes. Si l'astre en question possède déjà une température très différente du zéro absolu, sa perte de chaleur est beaucoup plus forte; nous trouverons l'élévation de la température due à la radiation des étoiles en calculant la perie par la loi de Stefan. On trouve ainse que, pour la Terre, l'élvévation de température due à la radiation des étoiles est inférieure à un cent-milliéme de degré. Encore devons–nous considérer ce nombre comme une limite supérieure de l'action nous cherchons à évaluer.

The history of the cosmic background temperature during a crucial period from 1926 to the present is reviewed in this journal by Assis and Neves who show how the tired light models predicted the correct value of 2.7K prior to the epic discovery by Penzias and Wilson in 1965. To this paper we might add the 1953 2.3K Finlay-Freundlich's prediction (*Nachrichten*, N.7,

---

action of the stars will be equal to the quotient of the increase of temperature due to the Sun on the Earth's orbit divided by the fourth root of 15,200,000, or about 60. Moreover, this number should be regarded as a minimum, as the measurements of Captain Abney taken in South Kensington may have been distorted by some foreign sources of light. We conclude that the radiation of the stars alone would maintain the test particle we suppose might have been placed at different points in the sky at a temperature of 338/60 = 5.6 abs. = −207°.4 centigrade.

We must not conclude that the radiation of the stars raises the temperature of the celestial bodies to 5 or 6 degrees. If the star in question already has a temperature that is very different from absolute zero, its loss of heat is much greater. We will find the increase of temperature due to the radiation of the stars by calculating the loss using Stefan's law. In this way, we find that for the Earth, the temperature increase due to the radiation of the stars is less than one hundred-thousandth of a degree. Furthermore, this figure should be regarded as an upper limit on the effect we seek to evaluate.

pp. 95–102, 1953):

Auf die hier für die B-Sterne benutzten Einheiten umgerechnet resultiert die Gleichung:

$$T_R^4 \cdot l_R = T_*^4 \cdot l_*$$

wo T_R = Temperatur des Weltraumes, $l_R = 6 \cdot 10^{22}$ cm, entspr. d. beob. Exp. $T_* = 20000°$ für B-Sterne, $l_* \approx 10^7$ cm. Das Ergebnis ist:

$$T_R = +2°, 3\text{K},$$

ein durchaus plausibler Wert.

The Workshop benefited from the attendance of Emil Wolf, the person responsible for bringing Finlay-Freundlich's work[4] to the attention of his teacher, Max Born.[5]

Max Born (*Nachrichten*, N.7, pp.102–108, 1953) realized the seriousness of Finlay–Freundlich's few–degree temperature prediction and suggested radio astronomy as an arbitrator between expanding and infinite cosmologies as they differed orders of magnitude in energy density. It is noteworthy that Born's manuscript was printed 12 years before the Penzias–Wilson measurement. We quote from the opening of Born's paper:

Freundlich glaubt zeigen zu können, daß die übliche relativistische Deutung der R.-V. durch die Beobachtungen nicht bestätigt wird, wogegen die Formel $[\Delta\nu/\nu = -AlT^4,\ A = 2 \cdot 10^{-29}\ cm^{-1} grad^{-4}]$ mit allen bekannten Tatsachen in Einklang ist, einschließlich der Nebelflucht (Hubble-Effekt), sofern man dem Weltenraum eine Temperatur von wenigen Graden beilegt. Ein solcher Widerspruch gegen die auf einfachsten Überlegungen beruhende relativistische Erklärung ist natürlich eine sehr bedenkliche Sache. Trotzdem schien es mir angebracht, die Freundlichsche Formel ein wenig zu analysieren; dabei bin ich zu dem Schluß gelangt, daß die Formel eine einfache wenn auch seltsame Deutung erlaubt, bei der überdies Zusammenhänge mit einer anderen Gruppe von Erscheinungen, nämlich der Radioastronomie, nahegelegt werden.

Today, plasma cosmology is a vigourous field of study supported by new observations such as filamentation, large–scale structure, the 2.7K blackbody background, non-Doppler redshifts, the existance of charged particle beam

[4] E. Finlay–Freundlich befriended the younger Wolf, both post–war refugees, during their stay at the University Observatory of St. Andrews. Finlay–Freundlich, long noted for his careful observations, was unable to accept the recessional velocity interpretation as the source of galaxy redshifts. He often paced the Observatory's halls pondering the nature of redshifts.

[5] Born at first refused, viewing cosmology as a form of philosophy or art, or something less respectable.

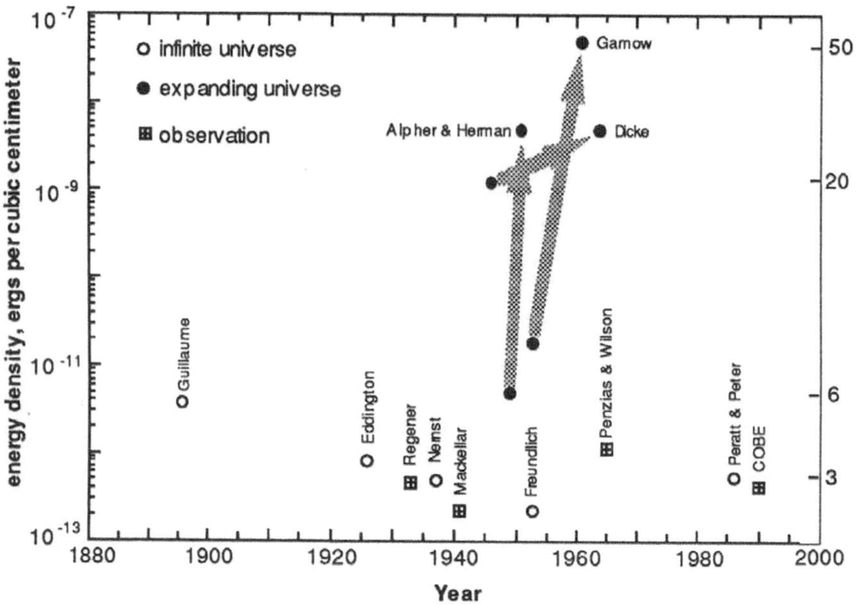

Fig. 2. History of the blackblody cosmic background radiation over 100 years. Note that energy density, *i.e.*, relating to a *physical mechanism*, is a better cosmological indicator than is the relatively insensitive (fourth-root of energy) temperature. (Adapted from Assis & Neves, 1995).

phenomena in cosmic objects, and the morphological observations of the Hubble Space Telescope. This chain of success really started in 1974 when a rocket–launched satellite verified the existence of Birkeland Currents in the earth's magnetosphere/ionosphere (IEEE, 1986, 1989, 1992; Peratt, 1988).

The workshop topics were organized in a size hierarchy, starting from what is now well–documented space and laboratory phenomena and ending with speculative issues in cosmology. T. Chang (MIT) reported on charged particle energization in space physics and astrophysics. K. Nishikawa (University of Iowa) presented three–dimensional, fully electromagnetic particle–in–cell simulations of Birkeland's orginal hypothesis: the interaction of the solar wind with the earth's magnetosphere to drive magnetic–field–aligned currents into the lower atmosphere.

In order, the following topics were discussed: galaxy formation and structure, origin of the light elements, and large scale structure, ending with a presentation by P. Kronberg (University of Toronto) on magnetic field estimates in distant galaxy systems. Active galactic nuclei, cosmic rays, and dark matter were next on the agenda, followed by the two largest sessions, "theoretical explanation of the Hubble relation" and " 'anomalous' redshifts and periodicities". J. Vigier (Université Pierre et Marie Curie) opened the former session while E. Wolf and D. James (University of Rochester) substantiated the latter session with a presentation on spectral invariance from light scattering on anisotropic media (James & Wolf, 1994), outlining quite possibly the non–recessional redshift mechanism sought by E. Finlay–Freundlich.

Table I delineates the topics covered over three days of sessions.

Periodically throughout the Workshop, issues now a hundred years old were debated including plasma cosmology's traditional refusal to claim any knowledge about an 'origin' of the universe (*e.g.*, Alfvén, 1988). Addressing this lively issue, the noted plasma theoretician and cosmologist David Bohm (Buckley & Peat, 1979) was quoted[6]:

> ...the universe is an unending transformation in flux whose previous states we are not privileged to know...

with plasma physicists generally agreeing that their work would simply lead to more questions about the nature of the universe. The Workshop ended with a lively debate "Is the Big Bang Theory Supported by Observation?" between E. Lerner (Lawrenceville Plasma Physics) and D. Spergel (Princeton University), both laudable in their respective cosmologies presentations.

In closing the Workshop, it is worthwhile to repeat Birkeland again:

> *According to our manner of looking at the matter, every star in the universe would be the seat and field of activity of electric forces of a strength that no one could imagine.*
>
> *We have no certain opinion as to how the assumed enormous electric currents with enormous tension are produced, but it is certainly not in accordance with the principles we employ in technics on the earth at the present time. One may well believe, however, that a knowledge in the future of the electrotechnics of the heavens would be of great practical value to our electrical engineers.*
>
> *It seems to be a natural consequence of our points of view to assume that the whole of space is filled with electrons and flying electric ions of all kinds. We have assumed that each stellar system in evolutions throws off electric corpuscles into space. It does not seem unreasonble therefore to think that the greater part of the material masses in the universe is found, not in the solar systems or nebulæ, but in "empty" space.*

It is to this point of view that the Second IEEE International Workshop on Plasma Astrophysics and Cosmology was devoted.

---

[6] According to Bohm: "It means a lot to these people [big bang advocates] that they are explaining their own origin with the origin of the universe. That gives them tremendous impetus to do the work."

TABLE I
Plasma Astrophysics & Cosmology Workshop Paper Topics.

author	size					phenomena							
	inter-planetary	inter-stellar	quasars/ galaxies	clusters	cosmological	B field	E field	Bennett Pinch	filaments	cmb	red–shifts	dark matter	relativity
Assis & Neves					X					X	X		
Tifft					X						X		
Ghosh				X							X	X	
Miller & Miller			X		X						X		X
Lerner				X	X	X	X	X	X	X			
Healy					X								
Reber			X										
Peratt			X	X	X	X	X	X	X	X			
Kanipe			X	X	X	X	X	X	X	X			
Roscoe													X
Brandenburg					X	X	X			X			
Lerner					X					X			
Wszolek			X									X	
Venturi & Giovanni				X		X							
Snell & Peratt			X			X	X	X	X				
Whitney			X			X	X						X
Verschuur		X				X	X	X	X				
Chargeishvili & Sakai		X				X			X				
Goldstein	X	X				X			X				
Healy & Peratt			X			X	X	X	X				
Wells	X					X	X						
Nishikawa et al.	X					X	X						

2 Documentation

Updates and further information related to the plasma universe and plasma cosmology will be posted on the Plasma Science & Technology homepage on the World Wide Web (http:\\www-plasma.umd.edu).

References

Alfvén, H.: 1988, *Has the Universe an Origin?*, unpublished
Assis, A. K. T., Neves, M. C. D.: 1995, "The Redshift Revisited", *Astrophys. & Space Sci. this edition*
Birkeland, Kr. : 1908, 1913 *The Norwegian Aurora Polaris Expedition 1902–1903*, Christiana, Norway, Aschehoug, Secs. 1 and 2
Buckley, P., Peat, F. D.: 1979, "David Joseph Bohm" in *A Question of Physics: Conversations in Physics and Biology*, University of Toronto Press, Toronto, p.124
IEEE: 1986, *Transactions on Plasma Science* **Vol. 14**, Special Issue on Space and Cosmic Plasma
IEEE: 1989, *Transactions on Plasma Science* **Vol. 17**, Special Issue on Space and Cosmic Plasma
IEEE: 1990, *Transactions on Plasma Science* **Vol. 18**, Special Issue on Plasma Cosmology
IEEE: 1992, *Transactions on Plasma Science* **Vol. 20**, Special Issue on Space and Cosmic Plasma
James, D. F. V., Wolf, E.: 1994, "Doppler-Like Frequency Shifts Generated by Dynamic Scattering", *Physics Lett. A* **Vol. 146**, pp. 167–171
Peratt, A. L.: 1988, *Laser and Particle Beams* **Vol. 6**, part. 3, Special Issue on Particle Beams and Basic Phenomena in the Plasma Universe

THE REDSHIFT REVISITED

A. K. T. ASSIS*
Instituto de Física 'Gleb Wataghin'
Universidade Estadual de Campinas, Unicamp
13083-970, Campinas, São Paulo, Brasil

and

M. C. D. NEVES
Physics Department
State University of Maringá
87020-900 Maringá, PR, Brazil

Abstract. We analyse the history of modern cosmology based on the redshift phenomenon and on the cosmic background radiation (CBR). We show the models of different authors for the interpretation of the redshift and how the tired light models predicted the correct value of 2.7 K temperature previous to Gamow and collaborators.

Key words: Cosmology, Cosmological Redshift, History of Cosmology, Cosmic Background Radiation

1 Introduction

The origin of the redshift of stellar sources, galaxies and quasars has been discussed for a long time. The majority of the work on this subject interpreted the redshift phenomenon as a Doppler effect associated with the recession of the sources (mainly galaxies). This interpretation leads directly to the idea of the big bang as most galaxies present a redshift and only a few of them located nearby present a blueshift.

In this article we discuss ideas presented in some recent work (Reber, 1986; Arp, 1987; Assis, 1992) which show an alternative interpretation that is also consistent with the data. Moreover, we make a historical analysis of the subject presenting discordant voices of this paradigm of big bang.

2 Different Views of the History of Modern Cosmology

In his interesting article, "How cosmology became a science," Stephen G. Brush makes a historical analysis of two models of modern cosmology: the big bang one and the steady state theory of Hoyle, Narlikar and Gould, (Brush, 1992). According to him, the discovery of the cosmic background radiation, the CBR, in 1965, was the decisive factor in favour of the standard cosmological model of the big bang against the steady state theory. The CBR spectrum has been found to be equivalent to the spectrum of a blackbody

* Also Collaborating Professor at the Department of Applied Mathematics, IMECC, State University of Campinas, 13081-970 Campinas, SP, Brazil

Astrophysics and Space Science **227**: 13–24, 1995.

with a characteristic temperature of 2.7 K. As the steady state theory did not predict such a temperature, while the big bang did, this discovery settled the question in favour of the big bang, according to Brush.

The main characters in Brush's history are Gamow and his collaborators, Alpher and Herman, who had predicted the correct value of the temperature of cosmic space prior to the discovery of Penzias and Wilson. He mentions briefly the work of A. Eddington, written in 1926, where he evaluated the temperature of interstellar space as 3.2 K, (Eddington, 1988a). Although Eddington's work arose much earlier than Gamow's estimates in the period 1949 to 1961, it has a problem, at least according to Brush: "Eddington did not propose a specific procedure for testing his prediction." Later on we will return to this point.

Beyond Eddington and the works of Gamow and collaborators, and of Dicke and Hoyle, Brush only considers the work of Andrew Mackellar, who utilized the levels of excitation of the cyanogen molecule (CN) in intergalactic space to evaluate the temperature of the intergalactic medium. In this remarkable work he obtained the value of 2.3 K, in 1941, without the big bang conjecture.

We emphasize Brush's paper here, not only because of its importance in the field of the history of modern cosmology, but also because of its impact on popular perceptions; *e.g.*, his articles published in *Scientific American*. But his work is similar to statements found in almost all textbooks on this subject. This paper discusses a line of development of the history of the cosmic background radiation that is usually neglected by most authors.

3 A Steady State Theory Without Expansion and Without Continuous Creation of Matter

Brush's paper and the work of most cosmologists usually compare only two models of the universe: the big bang model and the steady state theory of Hoyle, Bondi and Gold. These two models have an important aspect in common: both accept the interpretation of the cosmological redshift as being due to a Doppler effect. So both theories accept the expansion of the universe without further question. But there is a third model of the universe, developed in this century by scientists like Regener, Nernst (the father of the third law of thermodynamics), Finlay-Freundlich, and the Nobel laureates Max Born and Louis de Broglie. Unfortunately, this third model is always nearly neglected in the textbooks and is virtually unknown to todays physicists and astrophysicists. It is the work of the above notables that we wish to rescue.

The model developed by these authors has in common an interpretation of the cosmological redshift as being due to some kind of interaction of the photon in its journey from a distant galaxy to the earth. These explanations are usually called "tired light" theories. What is generally unappreciated is that these authors predicted the correct value of the temperature characteristic of the CBR prior to the works of Gamow and his collaborators. This means that the discovery of Penzias and Wilson can not be considered decisive in favour of the big bang because an alternative model had also predicted the correct value of the temperature.

First, let us go back to Eddington's book of 1926, (Eddington, 1988a). The remarkable aspect of the temperature of interstellar space as 3.2 K is that it was due, according to him, to the total radiation field emitted by the stellar sources being counterbalanced by the incident radiation over them and being absorbed by them. This is typical of an equilibrium situation. Moreover, he utilized Stephan-Boltzmann's law, according to which the total flux F emitted by a black body is given by

$$F = \sigma T^4 \tag{1}$$

where σ is Stephan-Boltzmann's constant ($\sigma = 5.67 \times 10^{-8} W m^{-2} K^{-4}$).

Later on, Eddington changed his cosmological views and accepted the idea of an expanding universe (he even wrote the book *The Expanding Universe*, in 1933, (Eddington, 1988b)). But at least his prediction of 1926, of a temperature of 3.2 K, was not based on an expanding universe.

This utilization of Stephan-Boltzmann's law, characteristic of a black body spectrum, is an extremely important element in the works of Regener, Nernst, and Finlay-Freundlich.

In 1933 Regener, (Regener, 1933), analysing the energy of cosmic rays arriving on earth, wrote:

> Ein Himmelskörper, der die zur Absorption der Ultrastrahlung notwendige Dimension hat (...) wird sich durch die Ultrastrahlung erwärmen. Die Erwärmung wird proportional der zugestrahlten Ultrastrahlungsenergie S_U und der Oberfläche O sein. Er wird sich so lange erwärmen, bis die emittierte Wärmestrahlung, bei schwarzer Strahlung also $= \sigma T^4 O$, ebensogroβ geworden ist. Es ergibt sich die Endtemperatur $T = \sqrt[4]{S_U/\sigma}$. Das gibt nach Einsetzung der Zahlenwerte 2.8 K.[1]

Following this work Nernst presented a remarkable paper in 1937, (Nernst, 1937). Nernst believed in a stationary universe. Making reference to Regener's work he commented:

---

[1] "A celestial body which has the sufficient dimensions to absorb the cosmic radiation (...) is heated by means of this cosmic radiation. The heating results to be proportional to the energy S_U from the cosmic radiation and to the surface O [of the body]. It is heated until it emits the same amount of radiation, so that the black radiation becomes equal to $\sigma T^4 O$. The final temperature is determined by $T = \sqrt[4]{S_U/\sigma}$ and is found to be 2.8 K" (free translation).

> In der soeben erwähnten wichtigen Arbeit von Regener findet sich die Angabe, daβ im Universum ein die Kosmiche Strahlung absorbierender Körper sich bis auf 2.8^o abs. erwärmen müβte.[2]

Nernst, utilizing Regener's work, advocated a model of a boundless universe, homogeneous in large scale and without expansion. He suggested an equation to explain the light absorption by cosmic dust or something similar, due to a decrease of the luminous quantum of energy, resulting in the reddening of the photon:

$$-d(h\nu) = H(h\nu)dt \qquad (2)$$

where h is Planck's constant ($h = 6.6 \times 10^{-34} Js$), ν is the light frequency and H is Hubble's constant. One of his conclusions in this article is that the cosmological redshift is not due to a Doppler effect.

In 1954 Finlay-Freundlich discussed the redshift of the spectral lines of B-stars and O-stars belonging to the Orion Nebula group, (Finlay-Freundlich, 1954). He analysed the influence of the gravitational potential over the results of observed redshifts. He summarized his results of the B-stars stating: "The B-stars in Orion nebula show a systematic redshift relative to the lines in the nebula amounting to at least $+10km/s$. This value is, by a factor of the order of ten, larger than the redshift predicted by the theory of relativity." Freundlich found, for O-stars, that the redshifts result to be about $+18km/s$. Analysing binary systems of stars he found redshifts larger by a factor of 10 to 20 than the predicted by general relativity (gravitational redshift). He says about this fact:

> It is quite improbable that they are produced by a systematic motion of the stars in the Orion Nebula group relative to the nebula itself, or by a systematic motion of the O-stars relative to the B-stars in the same cluster. (...) We see thus that the large redshifts reveal a physical effect which cannot be interpreted either as a gravitational displacement or as a true recession effect.

Trying to explain the observed redshifts, Freundlich suggests an interesting hypothesis:

> I propose to introduce as an additional hypothesis that light passing through deep layers of intense radiation field, loses energy — perhaps due to photon-photon interaction — and that the energy loss is proportional both to the density of the radiation field and to the length of path of the light through the radiation field.

Freundlich writes thus an empirical formula to explain these redshifts:

---

[2] "In Regener's important work cited above it is found the fact that, in the universe, a celestial body that absorbs cosmic radiation must be heated until 2.8^o K" (free translation).

$$\frac{\Delta\nu}{\nu} = -AT^4 l \tag{3}$$

where $\Delta\nu$ is the change of frequency of the line, ν is the original frequency, A is a constant, T is the temperature of the radiation field and l is the length of path traversed by light through the radiation field. The constant A is obtained when we have $l = 10^7 cm$, $\Delta\nu/\nu = -3.3 \times 10^{-5}$ and $T = 20000K$ for a B-star temperature. Therefore the value of A is $2 \times 10^{-29} K^{-4} cm^{-1}$.

Freundlich applies his formula to the explanation of the redshift of the sun, A-stars, supergiant M-stars, Wolf-Rayet stars, and white dwarfs, with great success.

With these results, Freundlich compared the cosmological redshift and stellar redshifts (for instance, B-stars). He then applied his formula to the cosmological redshift. In his analysis, Freundlich derived a blackbody temperature for intergalactic space. The two extremum values obtained by Freundlich's formula for the mean temperature of intergalactic space were $T = 1.9K$ and $T = 6.0K$.

Finlay-Freundlich concluded his article writing:

> We may have, therefore, to envisage that the cosmological redshift is not due to an expanding universe, but to a loss of energy which light suffers in the immense lengths of space it has to traverse coming from the most distant star systems. That intergalactic space is not completely empty is indicated by Stebbins and Whitford's discovery (1948) that the cosmological redshift is accompanied by a parallel unaccountable excess reddening. Thus the light must be exposed to some kind of interaction with matter and radiation in intergalactic space.

TABLE I

Predictions of the temperature of the CBR according to different models of the universe and different authors.

year	stationary universe	big bang	temperature
1926	Eddington		$3.2\ K$
1933	Regener		$2.8\ K$
1937	Nernst		$2.8\ K$
1949		Alpher and Hermann	$T \geq 5\ K$
1953		Gamow	$7\ K$
1954	Finlay-Freundlich		$1.9\ K \leq T \leq 6.0\ K$
1961		Gamow	$50\ K$

4 Gamow's Different Predictions of the Temperature of the CBR

In his paper, Finlay-Freundlich references Gamow's 1953 temperature of $7K$, a value obtained from thermodynamical considerations, for the mean temperature of intergalactic space. Freundlich did not mention Alpher and Hermann's paper of 1949, (Alpher and Hermann, 1949). These authors, collaborators of Gamow, wrote:

> (the present density of radiation, $\rho_{r''} \cong 10^{-32} g/cm^3$) corresponds to a temperature now of the order of $5^o K$. This mean temperature for the universe is to be interpreted as the background temperature which would result from the universal expansion alone. However, the thermal energy resulting from the nuclear energy production in stars would increase this value.

So, according to these authors, the temperature characteristic of this radiation should be **at least 5 K**.

In 1961 Gamow published a revised edition of his popular book *The Creation of the Universe*, (Gamow, 1961). This is the last work of Gamow known to us, where he discusses the temperature of interstellar space prior to the discovery of the CBR by Penzias and Wilson in 1965. There is only one place in the book where he mentions the temperature of the CBR. Let us quote in full these important paragraphs (our emphasis):

> The relation previously stated between the value of Hubble's constant and the mean density of the universe permits us to derive a simple expression giving us the temperature during the early stages of expansion as the function of the time counted from the moment of maximum compression. Expressing that time in seconds and the temperature in degrees (see Appendix, pages 142-43), we have:
>
> $$\text{temperature} = 1.5 \times 10^{10}/[\text{time}]^{1/2}$$
>
> Thus when the universe was 1 second old, 1 year old, and 1 million years old, its temperatures were 15 billion, 3 million, and 3 thousand degrees absolute, respectively. Inserting the present age of the universe ($t = 10^{17}$ sec) into that formula, we find
>
> $$T_{\text{present}} = 50 \text{ degrees absolute}$$
>
> *which is in reasonable agreement with the actual temperature of interstellar space.* Yes, our universe took some time to cool from the blistering heat of its early days to the freezing cold of today!

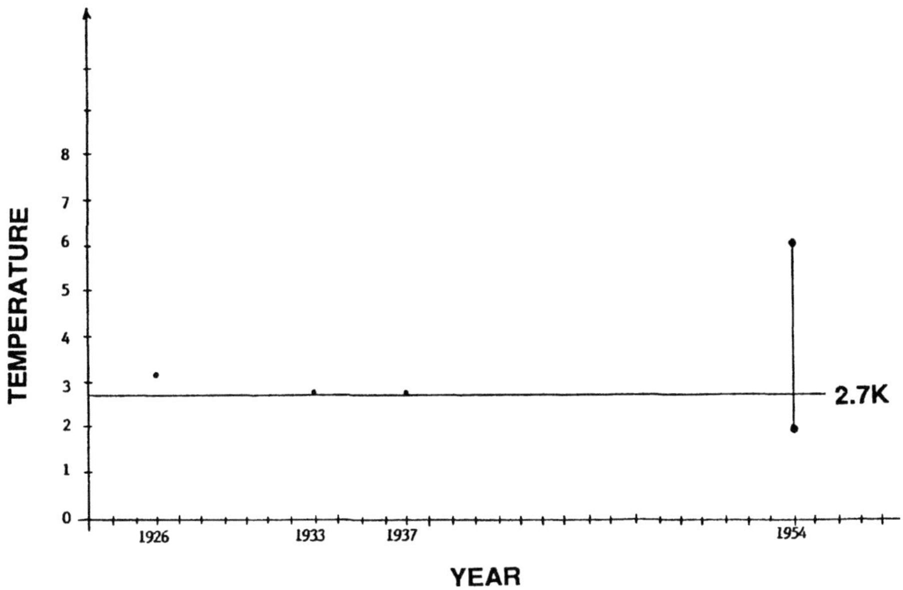

Fig. 1. Predictions of the temperature of the CBR based on a universe in dynamical equilibrium without expansion (for short called stationary universe): Eddington, 1933 [see Eddington 1988] ($T = 3.2\ K$); Regener, 1933 ($T = 2.8\ K$); Nernst, 1937 ($T = 2.8\ K$); and Finlay, 1954 ($1.9\ K \leq T \leq 6.0\ K$). The known observational value of 2.7 K is also shown.

While the theory provides an exact expression for the temperature in the expanding universe, it leads only to an expression with an unknown factor for the density of matter, in fact, one can prove (see Appendix) that

$$[\text{density of matter}] = \text{constant}/[\text{time}]^{3/2}$$

We see in Chapter III that the value of that constant may be obtained from the theory of the origin of atomic species.

This value of 50 K is, obviously, very different from that obtained by Penzias and Wilson in 1965, namely, $T = 3.5 \pm 1.0\ K$, (Penzias and Wilson, 1965). A trajectory of the estimates of the temperature of cosmic space year by year, by Gamow and collaborators, diverges away from the value finally measured in 1965. In Figure 1 we plot, in chronological order, the predictions based on a non-expanding universe according to Eddington, Regener, Nernst

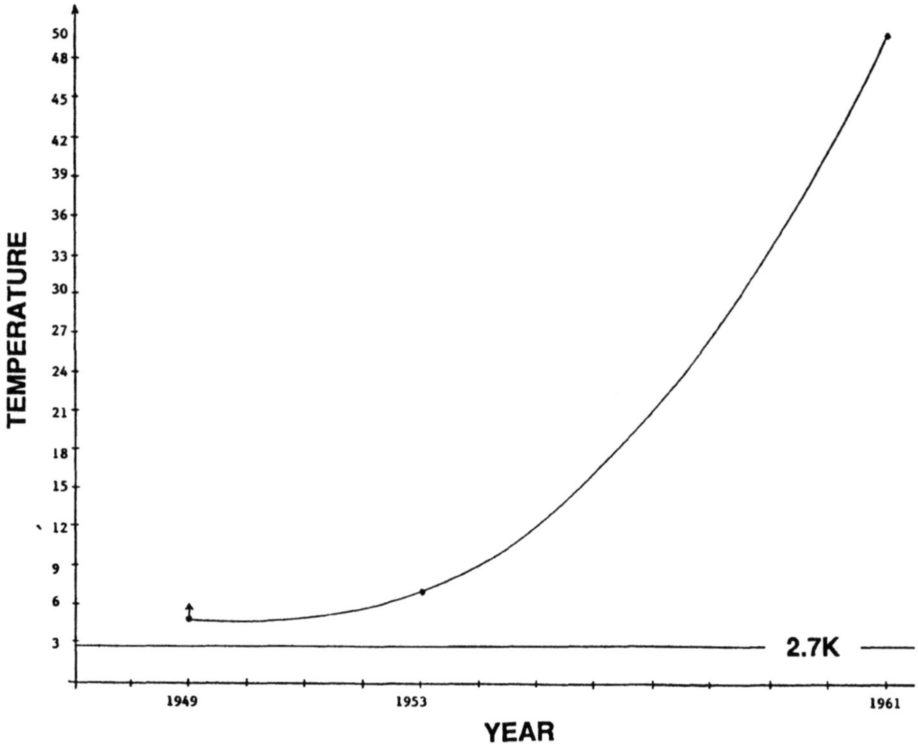

Fig. 2. Predictions of the present value of the temperature of the CBR according to Gamow and collaborators (1949: $T \geq 5\ K$, 1953: $T = 7\ K$, 1961: $T = 50\ K$). It is also presented the known observational value of 2.7 K.

and Finlay-Freundlich. In Figure 2 we present, in a chronological order, the predictions of the temperature of the CBR according to Gamow and collaborators. This data is collected in Table 1.

It is relevant to remember here a letter sent by Gamow to Arno Penzias, in 1965 (which was curiously dated 1963). This letter was reproduced in Penzias's article, entitled "Cosmology and microwave astronomy," (Penzias, 1972). We reproduce it here again:

"Sept 29th 1963

Dear Dr. Penzias,

Thank you for sending me your paper on 3 K radiation. It is very nicely written except that "early history" is not "quite complete". The theory of, what is now known, as, "primeval fireball", was first developed by me in 1946 (Phys. Rev. 70, 572, 1946; 74, 505, 1948; Nature 162, 680, 1948). The prediction of the numerical value of the present (residual) temperature could be found in Alpher & Hermann's paper (Phys. Rev. 75, 1093, 1949) who estimate it as 5 K, and in my paper (KongDansk. Ved. Sels 27 n° 10, 1953) with the estimate of 7 K. Even in my popular book *Creation of the Universe* (Viking 1952) you can find (on p. 42) the formula $T = 1.5 \times 10^{10}/t^{1/2}$ K, and the upper limit of 50 K. Thus, you see the world did not start with almighty Dicke.

Sincerely,

G. Gamow"

This letter, as we have seen, does not correspond to the true facts. Gamow, in the revised edition of his book of 1952, published in 1961, calculated a temperature **equal to 50 K**. Thus, Gamow did not estimate in this work an **upper limit** of 50 K.

The chronology of predictions of the temperature of the CBR suggests a different history than that presented in cosmological textbooks and in articles written about cosmology.

In this regard, we quote another part of Penzias paper, (Penzias, 1972):

> It is beyond the scope of this contribution to weigh the various theoretical explanations of the 3° K. Still the unique claim of the hot evolving universe theory is that it predicted the background radiation before the fact. At the 4th "Texas" Symposium on Relativistic Astrophysics, George Gamow was the chairman of the session on Microwave Background Radiation. He ended his remarks with a comment which, to the best of my recollection, went, "If I lose a nickel, and someone finds a nickel, I can't prove that it's my nickel. Still, I lost a nickel just where they found one." The applause was loud and long.

As a matter of fact Gamow did not lose a single coin, but many of them. Moreover, these coins had different values, in a divergent series relative to the correct value found at a latter date. It is even more remarkable that other people had lost nickels much closer to where they were later found to be, and at an earlier date than Gamow!

5 Discussion and Conclusion

Two other important authors in the subject of a non-expanding universe are Max Born, (Born, 1954), and Louis de Broglie, (de Broglie, 1966). Max Born showed that the theory of Finlay-Freundlich (photon-photon collisions as the cause of redshift) was scientifically sound. When discussing the cosmological redshift in this paper, Max Born made a remarkable prediction: "Thus the redshift is linked to radio-astronomy." This was written eleven years before the discovery of the CBR by Penzias and Wilson, (Penzias and Wilson, 1965), utilizing a horn reflector antenna built to study radio astronomy.

Despite this fact Max Born never stated, to the best of our knowledge, that he did not believe in the expanding universe, he did not feel at all confortable with the big bang theory, as indicated by the following quotations from his book *Einstein's Theory of Relativity*, (Born, 1962):

> The reader may get the impression that modern cosmology has strayed from the sound empirical road into a wilderness where statements can be made without fear of observational check. Indeed, this can be said of the theories just sketched, particularly as the mixed feeling of admiration and slight disgust which they produce is enhanced by the almost fanatical assurance with which they are advertised by their authors. Unfortunately but rather naturally, this state of affairs has been used by different ideologies to claim one of these theories as a confirmation of their dogma and to anathematize the other. (Born, 1962), p. 369

> Views of this kind, preached as dogma, are foreign to the spirit of science, and each of them can be refuted by showing that it does not take all aspects into account. Those who welcome the idea of a "beginning" forget that all one can assuredly say is that this is a state of high density of matter quite distinct from the distribution of isolated stars known to us; one may doubt that in this state the notions of space and time are applicable, because these notions are intimately related to the dispersed system of stars. The "beginning" refers only to our ability to describe the state of things in terms of accustomed concepts. Whether there was a creation from nothing is not a scientific question, but a matter of belief and beyond experience, as the old philosophers and theologians like Thomas Aquinas knew. (Born, 1962), p. 369

Louis de Broglie states a "photon aging" due to a continuous loss of energy by the photon. A more detailed discussion about these two authors can be found in (Assis, 1992) and (Assis, 1993).

In this paper, we have presented another view of the history of cosmology; one that is very different from that presented by Brush. We have emphasized the fact that their exists a large body of research, from a number of notable physicists, that is critical of the Doppler effect as the explanation of cosmic redshifts. It should also be emphasized that there are numerous papers on the topic of anomalous redshift observations (see, for instance, (Reboul, 1981), for a list of 772 untrivial redshifts). Anomolous redshifts cannot easily be explained by the Doppler model. But if the redshift is not due to a Doppler effect, what is its origin?

Finlay-Freundlich believed in a photon-photon interaction in the intense radiation fields of the stars. Marmet believes in a redshift produced by inelastic collisions of photons on atoms and molecules. Reber and Kierein pointed out the Compton effect (interaction photon-electron). Vigier and Monti proposed the resistivity of intergalactic medium. Arp believes in an effect due to the age of celestial bodies. For further discussion of these models and for the references, see (Assis, 1992) and (Assis, 1993).

With respect to the Compton effect (scattering of photons by free electrons), it is known that the variation of wavelength is given by:

$$\lambda - \lambda_o = \frac{h}{mc}(1 - \cos\theta) , \tag{4}$$

where λ is the scattered wavelength of the photon, λ_o is the incident wavelength, h is Planck's constant, m is the electron's mass, c the speed of light and θ is the angle between incident and scattered photon.

Therefore, if the cosmological redshift is due to a Compton effect, we would have for each interaction the following contribution to the redshift:

$$\frac{\Delta\lambda}{\lambda} = \frac{1}{\lambda}\frac{h}{mc}(1 - \cos\theta) . \tag{5}$$

The constants h, m and c do not depend on λ. Then we are led to conclude that the cosmological redshift should scale as $1/\lambda$, but this is not observed. As a matter of fact Hubble's constant seems to be independent of wavelength. It thus seems improbable that the cosmological redshift is due to a Compton effect.

What is the real mechanism that produces the observed values for the redshift? This question continues to be a great mystery. A possible answer could arise in the future from a stationary model of the universe.

We close this article with three quotations by Hubble, as given by Reber, (Reber, 1986):

> "Light *may* lose energy during its journey through space, but if so, we do not yet know how the energy loss can be explained."

> "The disturbing features are all introduced by the recession factor, by the assumption that red-shifts are velocity-shifts. The departure from linear law of red-shifts, the departure from uniform distribution, the curvature necessary to restore homogeneity, the excess material demanded by the curvature; each of these is merely the recession factor in another form. These elements identify a unique model among the array of possible expanding worlds, and, in this model, the restriction in time-scale, the limitation of spatial dimensions, the amount of unobserved material, is each equivalent to the recession factor.

> On the other hand, if the recession factor is dropped, if redshifts are not primarily velocity-shifts, the picture is simple and plausible. There is no evidence of expansion and no restriction of the time-scale, no trace of spatial curvature and no limitations of spatial dimensions."

"We seem to face, as once before in the days of Copernicus, a choice between a small, finite universe, and a universe indefinitely large plus a new principle of nature."

6 Acknowledgments

A. K. T. A. wishes to thank FAPESP, FAEP and CNPq (Brazil) for financial support during the past few years. M. C. D. N. wishes to thank DFI, PPG (State University of Maringá) and the Organizing Committee of the Second IEEE International Workshop on Astrophysics and Cosmology for financial support. The authors thank Prof. Emil Wolf for important suggestions regarding the first version of this paper.

References

Alpher, R. A. and Hermann, R. C.: 1949, "Remarks on the evolution of the expanding universe", *Phys. Rev.*, **Vol. 75**, pp. 1089-1095.

Arp, H.: 1987, *Quasars, Redshifts and Controversies*, Interstellar Media, Berkeley.

Assis, A. K. T.: 1992, "On Hubble's law of redshift, Olbers' paradox and the cosmic background radiation", *Apeiron*, **Vol. 12**, pp. 10-16.

Assis, A. K. T.: 1993, "A steady-state cosmology", in: *Progress in New Cosmologies: Beyond the Big Bang*, H. C. Arp, C. R. Keys and K. Rudnicki (eds.), Plenum Press, New York, pp. 153-167.

Born, M.: 1954, "On the interpretation of Freundlich's red-shift formula", *Proc. Phys. Soc. A*, **Vol. 67**, pp. 193-194.

Born, M.: 1962, *Einstein's Theory of Relativity*, revised ed., Dover Publications, New York.

Brush, S. G.: 1992, "How cosmology became a science", *Scientific American*, **Vol. 267**, pp. 62-70.

de Broglie, L.: 1966, "Sur le déplacement des raies émises par un objet astronomique lointain", *Compt. Rendues l'Acad. Sci. Paris*, **Vol. 263**, pp. 589-592.

Eddington, A. S.: 1988a, *The Internal Constitution of the Stars*, Cambridge University Press, Cambridge, Chapter 13, p. 371. Reprint of 1926 edition.

Eddington, A. S.: 1988b, *The Expanding Universe*, Cambridge University Press, Cambridge. Reprint of 1933 edition.

Finlay-Freundlich, E.: 1954, "Red shifts in the spectra of celestial bodies", *Phil. Mag.*, **Vol. 45**, pp. 303-319.

Gamow, G.: 1961, *The Creation of the Universe*, Viking Press, New York, revised edition, pp. 42-43.

Penzias, A. A.: 1972, "Cosmology and microwave astronomy", in: *Cosmology, Fusion & Other Matters*, F. Reines (ed.), Colorado Associated University Press, Boulder, pp. 29-47.

Penzias, A. A. and Wilson, R. W.: 1965, "A measurement of excess antenna temperature at 4080 Mc/s", *Astrophys. J.* **Vol. 142**, pp. 419-421.

Nernst, W.: 1937, "Weitere prüfung der annahme lines stationären zustandes im weltall", *Zeit. Phys.*, **Vol. 106**, pp. 633-661.

Reber, G.: 1986, "Intergalactic plasma", *IEEE Trans. on Plasma Sci.*, **Vol. PS-14**, pp. 678-682.

Reboul, K. J.: 1981, "Untrivial red shifts: A bibliographical catalogue", *Astron. Astrophys. Supp.*, **ser. 45**, pp. 129-144.

Regener, E.: 1933, "Der energiestrom der ultrastrahlung", *Zeit. Phys.*, **Vol. 80**, pp. 666-669.

REDSHIFT QUANTIZATION – A REVIEW

W. G. TIFFT
Steward Observatory, University of Arizona
Tucson, Arizona 85721 USA

Abstract. Redshift quantization has three main facets: 1) the internal organization of galaxies, 2) differential effects between galaxies in physical systems, 3) global effects linking all galaxies and cosmology. The subject originated as an outgrowth of redshift correlation studies including studies of internal kinematics of galaxies. While possibly central to understanding redshift quantization, this aspect is complex and largely undeveloped. The bulk of the evidence for redshift quantization comes from differential and global periodicity testing. More recently redshift variability has been associated with the phenomenon. Early work in theory led to a determination of q_0 close to 0.5. A new association between quantization and the Cosmic Background Radiation further links redshift quantization with basic cosmology.

Key words: Redshifts, Periodicities, Cosmic Background Radiation, Galaxies

1 Introduction

The redshift is normally assumed to be an extrinsic quantity, unrelated in most respects to intrinsic properties of individual galaxies. Within the standard framework the redshift can be decomposed into three parts, the Hubble flow due to expansion of the Universe, local peculiar motion governed by the mass distribution on various scales, and internal motions within galaxies. This viewpoint is the logical extension of theories of mechanics developed within a *very* tiny part of one galaxy, and while seeming to explain many aspects of the Universe it has created some well known problems. One problem which has attracted much attention is the 'missing mass' or 'dark matter' problem. When interpreted as standard Doppler shifts, redshift measurements imply the presence of far more mass than is obviously visible in ordinary barionic form. One of the more recent versions of this problem is the apparent requirement for large chaotic motions on very large scales. Our motion with respect to larger and larger reference frames of galaxies does not appear to be related to the cosmic background dipole anisotropy.

Few people have seriously questioned the basic assumptions which underlie the standard model of the Universe. Numerous tests are possible, however, and beginning in 1970 a series of experiments on the 'nature of the redshift' was initiated by the author at Steward Observatory. The work has developed within three broad categories, beginning with, (1), correlations between the redshift and intrinsic properties of galaxies. These studies quickly led to the discovery of apparent redshift quantization which provides the second type of test. (2) Is the redshift a discrete or continuous quantity? The need for high precision and a thorough understanding of uncertainties led more recently to

Astrophysics and Space Science **227**: 25–39, 1995.

the detection of apparent redshift variability which provides a third form of testing. (3) Is the redshift rapidly variable? Throughout the development of the program it has seemed increasingly clear that the redshift has properties inconsistent with a simple velocity and/or cosmic scale change interpretation. Various implications have been pointed out from time to time, but basically the work is observationally driven. It seems that the more we look, the less we can presume we know.

2 The Discovery of Redshift Quantization

Some of the original correlation studies are relevant here. For broader discussions see IAU Symposium 58 (Tifft 1974) and IAU Colloquium 37 (Tifft 1977c). Redshift was found to correlate in unexpected ways with galaxy luminosity (redshift-magnitude bands), morphology, and activity (radio, optical emission). The work suggested that at least part of the redshift was intrinsic to the galaxies or their constituent matter. All or part of the Hubble Law could be interpreted to show redshift/galaxy evolution displayed in lookback time. Redshift spread could arise from a dispersion in rates or stages of evolution applicable to all galaxies, not just certain types. It was in this correlation-induced galaxy-centered evolutionary atmosphere that redshift quantization was discovered as redshift precision was progressively improved. This work was done using clusters of galaxies. The Coma cluster provided the first calibration of the quantization interval.

The quantization concept was formally presented in a series of three Astrophysical Journal papers in 1976-77. The papers examined internal properties of individual galaxies (Tifft 1976), systems of galaxies (Tifft 1977a), and stars along with peculiar galaxies (Tifft 1977b). Two basic principles are introduced in the 1976 paper. the first states that "the redshift occurs in discrete steps with a step size near 70-75 km s^{-1} and/or multiples thereof". The second states that "all major galaxies (and presumably related objects) contain two states of redshift which appear on opposite sides ... of the nucleus. Lesser amounts of other states may be present." In later usage this second concept is more generally stated as "galaxies consist of a superimposition of specific redshift states". The correlation studies had already made it likely that galaxies (and redshifts) evolved from within. The quantization picture reinforced that idea and gave it structure.

Aside from the definition paper, and some commentary at the 1987 Venice Symposium (Tifft 1988), there has been little discussion of quantized structure for individual galaxies. A follow-up paper exists as Steward Observatory Preprint 513 but it did not survive the intense review process that attends radical new approaches. This is, nonetheless, where the quantization concept and the basic number 72 km s^{-1} were born. They originated there because good differential measurements in rotation curves showed character-

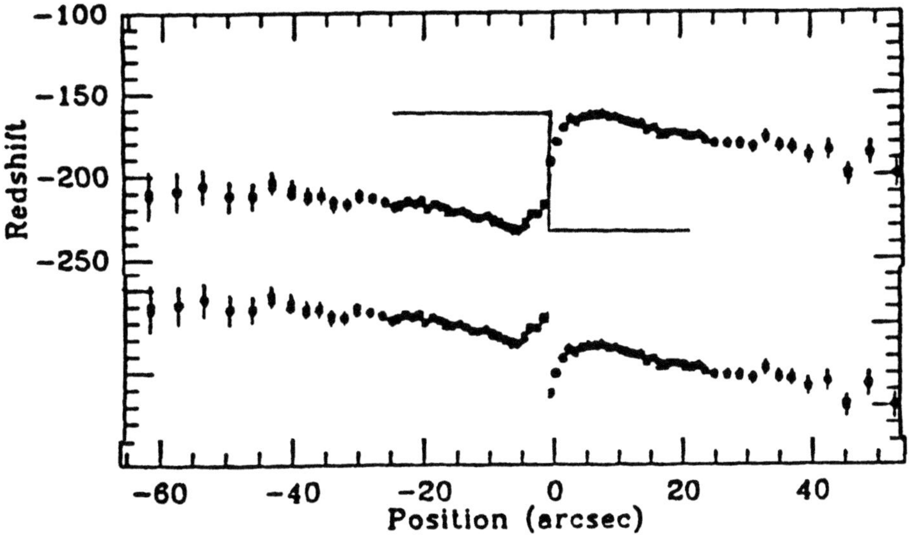

Fig. 1. The rotation curve for M32 from Tonry (1984). The upper curve shows rotation velocity as a function of angular distance from the nucleus. A step function of 72 km s^{-1} is shown for comparison. The lower curve removes a 72 km s^{-1} step from the rotation curve.

istic offsets in a regular pattern consistent with the presence of *two* velocity distributions differing by a constant. That constant was ≈ 72km s^{-1}. Most redshifts were then too uncertain to make many precise comparisons between galaxies.

Figure 1 shows a rotation curve for M32, from Tonry (1984), which provides an example of internal offsets. The original curve shows a sharp transition across the nucleus with peak rotation occurring close to the nucleus. The curve then drops back to essentially the nuclear value. The transition amplitude is indistinguishable from 72 km s^{-1} as shown with a step function. When the curve is cut and realigned shifted by 72 km s^{-1} the second curve is produced. A smooth rotation curve with 'flat' wings, well known in more rapidly rotating galaxies, results. Note that if the outer rotation curve drops back below the nuclear value a situation is produced where the nucleus and outer parts rotate in opposite directions. Such 'counter-rotation' is observed in the cores of several galaxies. A study of the frequency of offsets of specific sizes would be interesting.

The original explanation of offsets envisioned outflowing streams of intrinsically different material. The two 'streams' define a dipole or 'multiple state' structure of a galaxy. The third initial paper (Tifft 1977b) examined stars and interstellar material for evidence that the velocity pattern could be caused by a variation in electron massor some other process at the atomic physics level. Several velocity correspondences were found but there has been no follow-up work. Whatever the structure is, intrinsically different material, quantum-mechanically constrained velocity 'wave-functions', or something else, is not known. Nuclei do seem to be an origin point for the phenomenon. A multiple state structure does not deny effects of normal kinematics. States seem to be wound up and intermingled by ordinary motion; they are easily seen only in kinematically simple systems, like ellipticals, or in galaxies carefully modeled to remove rotation and/or radial flow. Galaxies need not be presumed to be static systems and could evolve by radial outflow. This is consistent with correlations which suggest that luminosity and/or morphological evolution could be occurring (Tifft 1979).

The second introductory paper (Tifft 1977a) extended quantization constraints to differentials between galaxies, at least physically associated ones. This constraint was already implied by earlier work suggesting that redshift-magnitude correlations could be linked together into some absolute global structure (Tifft 1973). Such a linkage is obviously at variance with conventional extrapolated gravitational theory. It either eliminates large scale motion or constrains it to quantized jumps. Quantization was apparently present, continuity was not. Quite explicit predictions could be made. The quantized redshift problem was defined.

3 Differential Redshift Testing

The second definition paper discussed expected patterns for redshift differentials in binary galaxies. Barring extreme biases or selection, conventional dynamics predicts a smooth monotonic distribution of differentials peaking at zero. The quantized redshift requires peaks near multiples of 72 km s^{-1} with clear gaps between. The first sample of pairs for which this test could be carried out became available when Peterson (1979) published 21 cm radio data for widely spaced pairs. The Peterson data were shown to conform closely to the quantization prediction (Tifft 1980). A revision of the original study gave the same result (Tifft 1982a). A program to measure accurate differentials for more than 200 close pairs using optical techniques further confirmed quantization (Tifft 1982b).

The significance of the original double galaxy quantization tests has been actively debated. Sharp (1984,1990) concluded that the periodicity was present but dismissed its significance from a dynamical viewpoint. Despite the fact that quite precise predictions preceded the testing, Newman et al (1989) argued that this did not matter and that the findings were not significant. Cocke & Tifft (1991) reaffirmed that proper statistical procedures were followed and that the significance was quite high. Disagreement remains concerning the accuracy of some optical differentials; compare Sharp (1990) with the Appendix in Tifft & Cocke (1989).

Periodicity analysis was quickly extended beyond simple pairs although isolated pairs retain a special significance. W. J. Cocke joined with Tifft in an analysis of compact groups (Cocke & Tifft 1983) which again confirmed the presence of specific multiples of 72 km s^{-1}. Efforts were also begun to formulate a quantum mechanical basis for redshift periodicities where redshift retains a velocity interpretation subject to quantum constraints (Cocke 1983, 1985). Observations of companion galaxies by Arp (1982, 1986), Arp & Sulentic (1985) and Sulentic (1984) also verified quantization effects in the early 1980 period. Many of the companion studies refer to groups, and include data from varied sources, which complicates comparisons with isolated pairs. Redshift accuracy and sample homogeneity is of paramount importance in discussing quantization (Tifft 1982b).

Emission activity in the large study of close isolated pairs (Tifft 1982b) has been discussed (Tifft 1985). Activity increases as differential redshift decreases, but does not seem to associate with a specific quantum level as Arp & Sulentic (1985) have suggested. The morphology of an interacting pair is generally related to emission activity, but all clearly interacting pairs show quantization independent of interaction morphology or the intensity of emission (Tifft 1982b).

The next stage of differential redshift studies was marked by the introduction of several new samples or sample upgrades. Schneider et al (1986) compiled a large sample of 21 cm data, Schweizer (1987) provided an accurate optical sample of Southern Hemisphere pairs, and Tifft & Cocke (1988) provided improved 21 cm redshifts for a large number of galaxies, including many in pairs and groups. Schneider et. al. found the sharp drop in frequency of differentials below 72 km s^{-1} predicted by the quantization model, but argued against the presence of larger preferred intervals. This group has gone on to propose a dynamical explanation based upon radial infall and strong selection biases in defining pairs (Schneider & Salpeter 1992). Models show that a small peak can be produced; however, the dip inside 72 km s^{-1} is shallow and no additional peaks exist at larger differentials. The Schneider pairs were selected using a very weak isolation criterion; when more stringent criteria are applied the distribution is consistent with the multiple peaks seen previously. Cocke (1992) finds, by both likelihood and Baysian analysis, that a quantized distribution is by far the best fit to existing samples.

The precision 21 cm redshift survey by Tifft & Cocke (1988) removed most doubts about the accuracy of 21 cm data being used. It also indicated that the 'zero' differential redshift peak is displaced from zero and centered near 24 km s^{-1}, 1/3 of the basic 72 km s^{-1}. Identical redshifts do not seem to occur in close proximity to one another. This suggested that some type of exclusion principle could be operating. Tifft & Cocke (1989) introduced the zero deviation and applied a revised model to the new Schweizer (1987) sample, finding excellent agreement. Figure 2 shows differentials for isolated pairs. The first panel shows 21 cm data. The initial sample and later additions are distinguished. The other panels add the Schweizer pairs and the

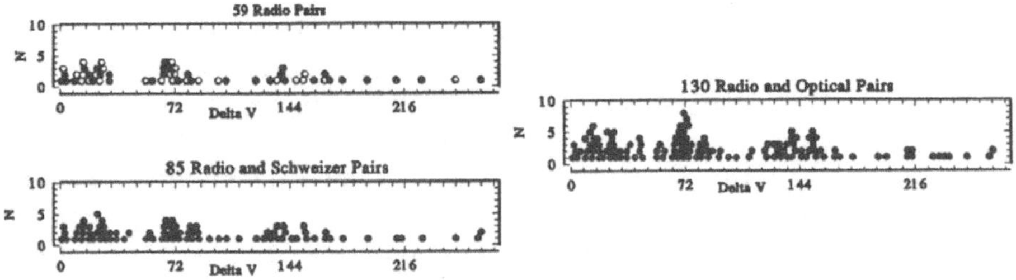

Fig. 2. The upper left panel shows differential redshifts for galaxy pairs measured at 21 cm. Filled circles refer to pairs from Tifft (1982a); open circles refer to additional 21 cm pairs from Schneider et. al. (1986) which meet the same standards. The lower left panel adds pairs measured optically by Schweizer (1987), as described in Tifft and Cocke (1989). The right panel adds the most accurate subset of close pairs measured optically by Tifft (1982b).

best data on close optical pairs from Tifft (1982b).

Principal areas of disagreement between workers concern the definition of isolation and the use of wide pairs. In discussing 21 cm data Tifft & Cocke require more stringent isolation than Schneider et al. The widest pairs were also excluded from the Schweizer sample in order to generate a sample comparable to close optical pairs known to be well isolated. The importance of such restrictions was demonstrated by Tifft & Cocke (1989) by means of a 'triplets' test. The test shows that potential pairs, which have a third galaxy closer to one of the objects than the purported companion, show systematic offsets from the standard pattern of differentials. The combination rules are presumably more complex for complex systems; homogeneous samples are essential for testing. To cite an example, Schneider et. al. consider UGC12808 to be a pair with UGC12815 which is 5.4 arc minutes away. UGC12815, however, has a companion, UGC12813, only 1.4 magnitudes fainter and 1.1 arc minutes distant, with which it shows direct signs of interaction. While the wide 'pair' may be suitable for some dynamical modeling it is not acceptable in a quantization test of isolated pairs.

The risk of contamination increases and the strength of interaction drops with separation. Large differentials can be expected only in relatively close pairs. Tifft (1988) has suggested that the range over which a given differential can be traced corresponds to the range over which the standard gravitational potential could generate the motion conventionally. Large differentials at larger separations could involve a group potential rather than the pair potential. Group members provide the primary source of contamination in studies of pairs. It is well known that galaxies occur in groups and rarely in isolation. Crossing times in groups are significantly less than a Hubble time, hence it is unlikely that many unperturbed very wide pairs exist. The effective testing ground for quantization in pairs is necessarily concentrated in the close pairs. Resolution limits at 21 cm and lack of neutral hydrogen in many galaxies place most such work in the optical domain.

One test involving very wide pairings is of interest. If the range of a quantization level is set by the potential field, and if an exclusion condition prevents identical redshifts, then the range over which a zero differential does not occur is a measure of the potential range. Tifft (1988) and Tifft & Cocke (1989) found that the zero deficiency extends out to roughly a Megaparsec. One possible consequence of an exclusion condition is that at high galaxy density galaxies could be forced into higher and higher levels. Large 'velocity' dispersions could be a consequence of degeneracy, not large masses. This suggests that the degree of ordered periodicity on the large scale could relate to galaxy density. Tifft (1988) has found that high density regions around the Coma cluster are strongly periodic in redshift. At low density the periodicity fades. When the distribution of galaxies is examined as a function of phase (fractional position in a periodic cycle, V/P, P = period, V = redshift) structural elements associated with superclusters are suggested. This study is at variance, however, with work by Guthrie & Napier (1990) who found quantization in the lower density parts of the Virgo cluster.

4 Global Redshift Testing

Studies in the early 1970s suggested a globally linked pattern in redshift correlations (Tifft 1973). Quantization definition studies (Tifft 1977a) reached the same conclusion. Associations on progressively larger scales implied that an absolute frame of reference of quantum levels existed. Quantization is a property of galaxies; galaxy redshifts are presumed to be means over specific states. If the pattern of states is linked into a global hierarchy one must be in the quantum frame to see global effects. Specifically one must remove 'ordinary' solar motion with respect to the frame. Differential redshifts for closely spaced centers are insensitive to global corrections, hence quantization is seen locally without global reference. To go further 'solar motion' corrections were required. Presumably galactocentric redshifts, not heliocentric ones, should reflect global quantization.

Tifft (1977a) demonstrated that Local Group galaxies were periodically distributed in redshift at 72 km s^{-1} when galactocentric corrections were applied. Arp (1986) confirmed and extended the work using an independent solar motion determination. Napier *et al.* (1988) further verified the periodicity. Since the Local Group is a physical group, quantization at this level is a three-dimensional generalization of the differential work in pairs and groups. The analysis shows the significance of the transformation to the galactic center. To extend the quantization linkage between groups, Tifft (1977a) went on to show that the redshift pattern in the external M101 group was apparently in phase with the Local Group. The intergroup phasing was then extended out to the Coma cluster to provide an accurate estimate of the periodicity, (72.46±0.5 km s^{-1}), assuming a global scale for linked redshift states. General studies followed.

Global quantization studies are much more complex than differential work. First, the redshifts must be accurate; only 21 cm data easily meet this condition. Second, one should work with homogeneous classes of objects. If galaxies are superimpositions of states one should compare similar combinations to minimize phase or period differences. Sorting by the width of the 21 cm profile is one way this is done. Third, the solar motion is not accurately known. A global quantization study is, in effect, a solar motion study; can one find a reasonable solar motion which reveals quantization? A sizeable amount of accurate homogeneous data is required. The first global solution (Tifft 1978a,b) used the original Fisher-Tully (1975) 21 cm study of dwarf galaxies. In retrospect, the sample was too small for the task but it verified a galactocentric reference and indicated a dependence upon profile width and perhaps shape. Periods were simple fractions of 72 km s^{-1}. Narrow profile galaxies (the simplest galaxies?) were clearly distinguished. The transition from dwarf irregulars (with narrow single peaked 21 cm profiles) to rotating disk systems (with double horned profiles) occurs for profile widths just sufficient to accommodate a 72 km s^{-1} interval along with a minimum natural width. An effort was made to construct galaxies from simple combinations of states. Various combinations are required to match different types of galaxies. Wide and narrow profile galaxies must be distinguished in most work.

Tifft & Cocke (1984) used the extended Fisher-Tully (1981) 21 cm study of more than 1000 galaxies to show that narrow 21 cm profile galaxies were strongly periodic in redshift at $72.45/3 = 24.15$ km s^{-1} for a well defined solar motion vector $(\theta, \pi, z) = (231 \pm 2, -35 \pm 3, 1 \pm 2)$ km s^{-1}. θ, π, and z are the tangential, radial, and perpendicular components of galactic rotation, positive in the direction of rotation, inward, and toward the north galactic pole. The significance of the narrow profile finding is clouded by the number of parameters fit; however, for the same motion vector the widest profile galaxies were shown to be strongly periodic at $72.45/2 = 36.2$ km s^{-1}. The agreement of the solar motion with standard values (see Arp (1986) for a discussion), and the fitting of galaxies at both extremes of profile width with 72 km s^{-1} related periods, is very unlikely if accidental. Redshifts appear to be globally periodic. At intermediate profile widths it is complex; at the extremes it seems to be simpler.

Several independent investigations followed. M. Croasdale (1989) examined 21 cm data for wide profile galaxies in Arecibo data. Although most of the Arecibo data is in the unexplored domain of higher redshifts, he generally confirmed the presence of a 36 km s^{-1} periodicity. Guthrie & Napier (1991) examined a sample of very nearby galaxies and found a strong periodicity at 37.1 km s^{-1}. This study, and a newer version extending the redshift range (Napier & Guthrie 1993), have recently heightened interest in the quantization phenomenon. Guthrie & Napier do not restrict the type of galaxy studied. Their sample is dominated by bright galaxies in nearby groups with profiles between 100 and 300 km s^{-1} wide. The result is not readily connected to less luminous galaxies with similar profiles in the Fisher-Tully data. The quantization phenomenon depends upon properties of galaxies. There seem to be various ways to find compatible samples, but much work remains to understand the factors involved.

If the redshift is a velocity, regardless of how it is constrained, it should show relativistic and/or cosmological effects. The global quantization study by Tifft & Cocke (1984) included a correction in the form $V_{corr} = c\ln(1 + V/c)$. As attempts to find a theoretical approach to quantization developed (see Cocke & Tifft (1989) for details), various models suggested that the quantization interval should scale as the square root of the Hubble constant. Within a standard Friedman cosmology $H(t)$ can be expressed as a function of H_0 and q_0, hence the quantum interval can be written as a function of z and q_0. Including a transformation to the external rest frame observed intervals should scale as

$$\Delta V = (1 + \bar{z})^{-3/2}(2q_0\bar{z} + 1)^{-1/4}\delta V_{obs}.$$

This relationship was used to explore quantization in the Lyman alpha forest of quasars. On the hypothesis that Lyman alpha clouds could be associated with dwarf galaxies or protogalaxies, Cocke & Tifft (1989) predicted that the 24 km s^{-1} global periodicity associated with dwarf galaxies could be reflected in the spacings of Lyman alpha lines. They found that a strong periodicity did occur for q_0 close to 1/2, the value associated with inflationary cosmologies. For q_0 of exactly 1/2, observed and corrected redshifts can be directly related. For small deviations of q_0 from 1/2 the expression for ΔV can be integrated in a Taylor expansion to give redshift corrections (Tifft 1991b). This correction is included in recent work, but is of little consequence locally. It seems that quantization observations have the potential to precisely determine q_0. The association of predicted periods with a unique value of q_0 would be a remarkable accident if quantization had no cosmological significance. New work, discussed in Section 6, has now associated quantization with the 3 degree Cosmic Background radiation rest frame, further strengthening a cosmological connection.

5 Redshift Time Dependence

Commentary on the accuracy of 21 cm redshifts is useful here. Tifft & Cocke (1988) reobserved many galaxies with older 21 cm redshifts to determine

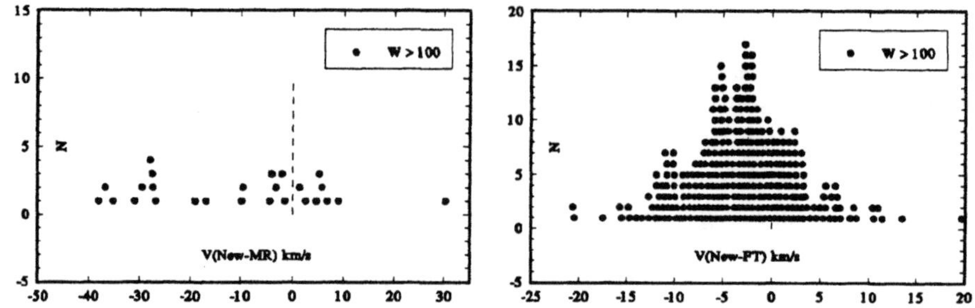

Fig. 3. Deviations between new and old redshifts. Data from Roberts (1968) are shown at the left. The Fisher-Tully (1981) comparison is shown at the right. The dashed line marks zero deviation. There appear to be multiple components and, on average, a significant negative displacement.

uncertainties and get accurate values. Tifft (1990) and Tifft & Huchtmeier (1990) made multi-telescope comparisons and established redshift standards. 21 cm redshifts involve precise frequency measurements.

After eliminating a few system or software errors, systematic error in 21 cm data was found to be vanishingly small. Beam size differences can produce systematic effects in resolved disk galaxies; inner regions seem to show a slight blueshift not seen in non-rotating dwarf galaxies (Tifft, 1991a). Precision comparisons require data sets from the same telescope. Pointing ultimately limits precision at high signal-to-noise levels. With consistent pointing random error is related to the signal-to-noise ratio. The most common method of redshift measurement uses the mean of two frequencies measured at specific levels on each wing of the profile. Observers differ in how the points are defined, but at high S/N the accuracy of each measure depends only on the bandwidth employed. The repeatability of high S/N data (S/N $>$ 10) varies with the square root of the bandwidth used and is basically independent of profile width. Single observations made with a 5 MHz bandwidth can repeat within about 0.8 km s^{-1}.

The interpretation of a redshift is different problem. From a conventional viewpoint random fluctuations in cloud motions or hydrogen distribution limit the degree to which a redshift measure, no matter how precise, can represent the center-of-mass motion of a galaxy. This limit is assumed to increase with profile width. If, however, galaxies are quantized structures it is not obvious that this uncertainty applies. The ultimate test of what level of regularity can be seen in redshifts has to be determined empirically, it cannot be determined from preconceived concepts of dynamics which may not apply. Random fluctuations cannot consistently produce predicted periodicities. Sizeable systematic errors in 21 cm data are difficult to generate, or when present, difficult to casually dismiss.

It was therefore interesting to find that comparisons of new and old 21 cm data contained *periodic systematic deviations.* It had been known previously (Rood 1982), that Fisher-Tully redshifts were, on average, slightly too large. Tifft (1991b) demonstrated this offset conclusively (9 sigma), and showed that other studies from the same era also deviated. Still older data from the 1960s by Roberts (1968) contain larger displacements, some exceeding 30 km s^{-1}. Figure 3 shows the comparison with Roberts (left) and Fisher-Tully (right). Despite the regular trend, one would be hesitant to suggest real changes based purely on systematic deviations in older data. However, the deviations correlate with global redshift phase. There is a direct connection with quantization, very unlikely if accidental. The effect was first reported at the Venice Symposium (Tifft 1988), where new redshifts were compared with Fisher-Tully values. Deviations were subsequently confirmed (Tifft & Cocke 1990) using entirely new redshifts spaced 2-3 years apart. The initial work is summarized in Tifft (1991b) where the temporal variation concept was formally developed. The 1991 temporal variation paper also examined the effect of rapid variation on the determinations of the period and solar motion. Slightly modified values were derived by optimizing periodicities in a two dimensional phase-deviation diagram. Figure 4 illustrates phase-deviation diagrams comparing new observations with Fisher-Tully data for galaxies with narrow 21 cm profiles. The upper left ordinate is phase at the recent epoch, the lower left panel shows phase at the initial epoch. The amount of 'drift', down and to the left or straight to the left, depending upon phase epoch, is a function of phase. The upper right panel contains a schematic six-level phase-deviation diagram showing how redshift might cascade between narrowly spaced levels to produce the observed correlations. A level-and-transition model was proposed which views the redshift as occurring in specific relatively stable levels. During brief periods, perhaps several years apart, transitions bridging levels must occur. The population within levels may be modulated over a range of levels to give rise to longer periodicities. The frequency and/or times of transition must differ between levels. The dominant change is down in redshift. Surrealistic though this model sounds, continuing monitoring of standards supports the concept. The lower right panel in the figure shows an enlarged, less homogeneous sample; the phase dependent drift pattern is consistent.

A primary objection to the level-and-transition model is that the finer levels are spaced only a few km s^{-1} apart. While compatible with measurement error the visibility of this fine structure requires models for galaxies at variance with popular opinion. Newer work permits demonstration of changes at longer periods although finer levels remain. Monitoring is continuing.

The problem, of course, is the rate of change. Even if the Hubble Law was entirely due to a lookback effect in redshift evolution, changes should be unobservable if transit times across galaxies or intergalactic distances are considered. If the variation is real, the cause must be very local or a

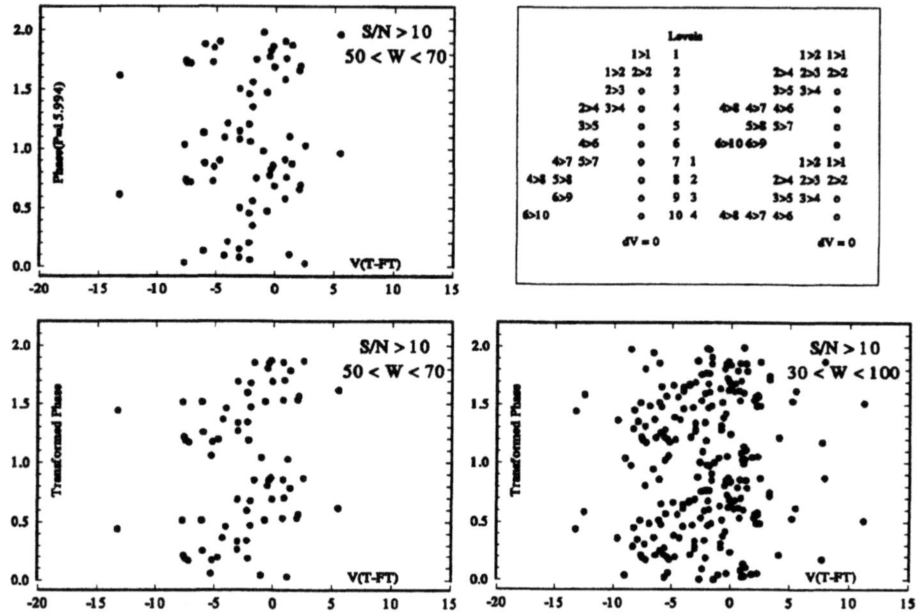

Fig. 4. Phase-deviation diagrams comparing recent observations with Fisher-Tully redshifts for galaxies with narrow 21 cm profiles. The ordinate is redshift phase, plotted as a double cycle to illustrate periodicity. The abscissa is redshift difference (new−old) in km s^{-1}. At upper left, phase is calculated at the new epoch; the figure at lower left shows the same data with phase figured at the older epoch. The diagram at upper right indicates how jumps between levels in a multi-level phase diagram can produce the observed drift patterns. Numbers refer to phase levels, arrows indicate transitions taking place between the two epochs. This example contains six levels drifting at different rates. Points are displaced diagonally or horizontally depending on the epoch used to determine phase. Some duplicate points are omitted for clarity. At lower right an extended sample shows the general nature of the effect.

completely different view of galaxies is required. If galaxies are described by a purely quantum mechanical analog with particle physics then there may be a mechanism which could produce rapid oscillations. Time scale estimates for an analog of the Zitterbewegung or 'jitter' phenomenon in relativistic quantum mechanics indicate changes might occur over a few years (Tifft & Cocke 1990).

6 New Directions

A question often asked when quantized redshifts are discussed is 'is there any connection with the Cosmic Background Radiation?' The answer to this is now yes. When redshifts are transformed to the CBR rest frame using the COBE cosmic dipole velocity and apex, widespread quantization is seen at the original 72 km s^{-1} period. Optimum tuning occurs at or within the error box for the COBE vector. The original demonstration (Tifft & Cocke

1993, Cocke & Tifft 1993), utilizes three independent data samples, each of which shows the CBR association with likelihoods of accidental coincidence at or below the 0.001 level. Periodic clumping of redshifts is consistently enhanced when viewed from the CBR rest frame. A large velocity space search around the CBR coincidence finds only chance fluctuations which do not reinforce from sample to sample. Figure 5 shows the velocity space correspondence with the COBE dipole vertex. The upper left panel is a phase-deviation diagram showing the phase concentration which occurs when redshifts are referred to the CBR rest frame. The lower left panel shows how solar motion choices which produce strong periodicities cluster at the COBE dipole vertex. The diagrams show the $\theta - \pi$ velocity plane for z velocities near 275 km s^{-1}. A Fisher-Tully galaxy sample was used in the left figures; Arecibo data on Perseus supercluster galaxies was used in the right panel. The CBR solar motion vector is $(-242, -31, +275)$ km s^{-1}, which compares with $(-245, -23, +275)$ km s^{-1} for COBE. Uncertainties are in the 5-10 km s^{-1} range.

A direct link between global and differential quantization seems apparent. Differential quantization has consistently shown 72 km s^{-1} intervals. Since redshift differentials are largely unaffected by any reasonable rest frame transformation, they might be expected to resemble the view from any fundamental frame which existed. This is what occurs in the CBR rest frame. Quantization has apparently achieved the CBR link which conventional large scale motion studies find very elusive. On the other hand the connection poses some interesting problems. There are now *two* frames which produce quantization, both involving apparent velocity transformations. There are indications that the CBR reference is widely applicable. The galactic reference may be more local.

In addition to the recognition of a possible fundamental rest frame recent work gives new insight into the hierarchy of periods present. Since this work is in development it is not discussed here. It appears possible to demonstrate variability using periods which are large compared with uncertainties, and some insight into what could cause changes now exists. We do not know what galaxies or systems of galaxies are, but it seems unlikely that conventional dynamics with or without dark matter will be sufficient to explain them.

References

Arp. H.: 1982, *ApJ* **256**, 54
Arp. H.: 1986, *A&A* **156**, 207
Arp. H.,& Sulentic, J. W.: 1985, *ApJ* **291**, 88
Cocke, W. J.: 1983, *Astroph. Lett.* **23**, 239
Cocke, W. J.: 1985, *ApJ* **288**, 22
Cocke, W. J.: 1992, *ApJ* **393**, 59
Cocke, W. J.,& Tifft, W. G.: 1983, *ApJ* **268**, 56
Cocke, W. J.,& Tifft, W. G.: 1989, *ApJ* **346**, 613

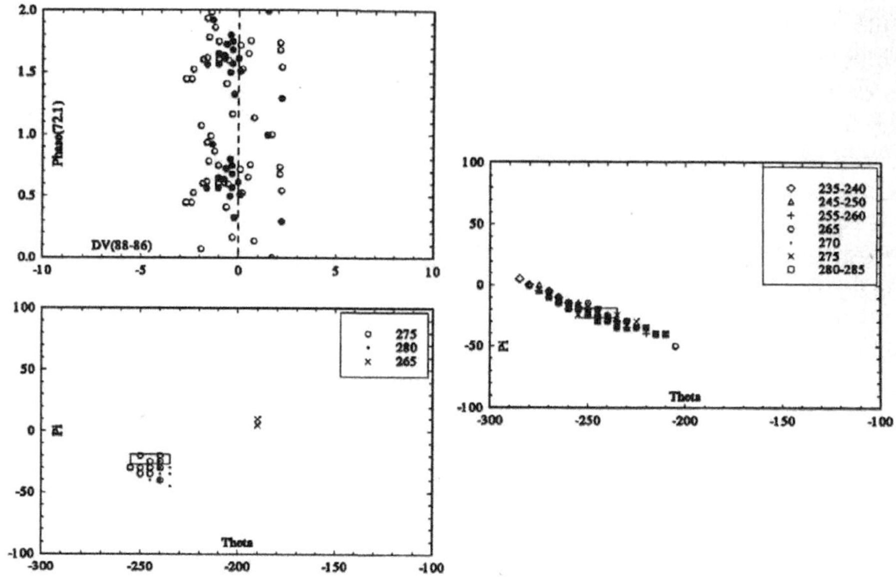

Fig. 5. Diagrams showing the association of redshift quantization with the Cosmic Background Radiation dipole vector. The upper left panel is a phase-deviation diagram for one sample. Filled circles show points with the highest signal-to-noise. The ordinate is a double phase cycle for a period of 72.1 km s^{-1} after referral of redshifts to the CBR rest frame. The abscissa is the redshift difference between 1986 and 1988. The dashed line denotes zero deviation. The other panels show the $\theta - \pi$ velocity plane for a range of z velocities. The small rectangle is the error box for the vertex of the CBR dipole moment centered at $z = 275$. The points mark the velocity coordinates of maximum phasing of redshifts. The sample at lower left is based upon Fisher-Tully galaxies; the right panel contains Arecibo data.

Cocke, W. J.,& Tifft, W. G.: 1991, *ApJ* **368**, 383
Cocke, W. J.,& Tifft, W. G.: 1993, *ApJ* submitted
Croasdale, M. R.: 1989, *ApJ* **345**, 72
Fisher, J. R.,& Tully, R. B.: 1975, *A&A* **44**, 151
Fisher, J. R.,& Tully, R. B.: 1981, *ApJS* **47**, 139
Guthrie, B. N. G.,& Napier, W. N.: 1990, *MNRAS* **243**, 431
Guthrie, B. N. G.,& Napier, W. N.: 1991, *MNRAS* **253**, 533
Napier, W. N., Guthrie, B. N. G., & Napier, B.: 1988, **New Ideas in Astronomy**, ed. F. Bertola, J. Sulentic, B. Madore (Cambridge: Cambridge Univ. Press), 173
Napier, W. N.,& Guthrie, B. N. G.: 1993, **Progress in New Cosmologies**, ed. Arp, Keys, Rudnicki (Plenum)
Newman, W. I., Haynes, M. P., & Terzian, Y.: 1989, *ApJ* **344**, 111
Peterson, S. D.: 1979, *ApJS* **40**, 527
Roberts, M. S.: 1968, *AJ* **73**, 945
Rood, H. J.: 1982, *ApJS* **49**, 111
Schweizer, L. Y.: 1987, *ApJS* **64**, 427
Sharp, N. A.: 1984, *ApJ* **286**, 437
Sharp, N. A.: 1990, *ApJ* **354**, 418
Schneider, S. E.,& Salpeter, E. E.: 1986, personal communication
Schneider, S. E.,& Salpeter, E. E.: 1992, *ApJ* **385**, 32

Sulentic, J. W.: 1984, *ApJ* **286**, 442
Tifft, W. G.: 1973, *ApJ* **179**, 29
Tifft, W. G.: 1974, IAU Symp. 58, **The Formation and Dynamics of Galaxies**, ed. J. R. Shakeshaft (Dordrecht: Reidel), 239
Tifft, W. G.: 1976, *ApJ* **206**, 38
Tifft, W. G.: 1977a, *ApJ* **211**, 31
Tifft, W. G.: 1977b, *ApJ* **211**, 377
Tifft, W. G.: 1977c, IAU Colloq. 37, **Decalages Vers Le Rouge et Expansion De L'Univers**, (Centre National de la Recherche Scientifique: Paris), 159
Tifft, W. G.: 1978a, *ApJ* **221**, 449
Tifft, W. G.: 1978b, *ApJ* **221**, 756
Tifft, W. G.: 1979, *ApJ* **233**, 799
Tifft, W. G.: 1980, *ApJ* **236**, 70
Tifft, W. G.: 1982a, *ApJ* **257**, 442
Tifft, W. G.: 1982b, *ApJ* **262**, 44
Tifft, W. G.: 1985, *ApJ* **288**, 65
Tifft, W. G.: 1988, **New Ideas in Astronomy**, ed. F. Bertola, J. Sulentic, B. Madore (Cambridge: Cambridge Univ. Press), 173
Tifft, W. G.: 1990, *ApJS* **73**, 603
Tifft, W. G.: 1991a, *ApJ* **368**, 105
Tifft, W. G.: 1991b, *ApJ* **382**, 396
Tifft, W. G.,& Cocke, W. J.: 1984, *ApJ* **287**, 492
Tifft, W. G.,& Cocke, W. J.: 1988, *ApJS* **67**, 1
Tifft, W. G.,& Cocke, W. J.: 1989, *ApJ* **336**, 128
Tifft, W. G.,& Cocke, W. J.: 1990, IAU Colloq. 124, **Paired and Interacting Galaxies**, ed. J. Sulentic, W. Keel, C. Telesco, (NASA Conf. Pub. 3098), 479
Tifft, W. G.,& Cocke, W. J.: 1993, *BAAS* **25**, 796
Tifft, W. G.,& Huchtmeier, W. K.: 1990, *A&AS* **84**, 47
Tonry, J. L.: 1984, *ApJ* **283**, L27

DETERMINATION OF TRUE VELOCITY DISPERSION AND THE DARK MATTER PROBLEM IN CLUSTERS OF GALAXIES

AMITABHA GHOSH
Indian Institute of Technology
Kanpur, India 208016

Abstract. According to the convention normally followed the redshifts of the galaxies in a cluster are assumed to be of purely dopplerian origin. The resulting velocity dispersion, when used in the virial theorem, leads to a very large proportion of dark matter to be present in the galaxy clusters. However, the recently proposed model of velocity dependent 'cosmic' drag cause redshifts of photons and it is necessary to develop a procedure to determine the true velocity dispersion from the gross redshift data. A method for this has been presented in the paper. Coma and Perseus clusters have been investigated using this procedure and the M/L ratios for both were found to be approximately of the order of 30, i.e., approximately the order of M/L ratios for individual galaxies. A study of the $m - z$ relation indicates that the galaxies with higher redshifts have fainter magnitudes. Distortion of the redshift plots and the typical elongation of the core regions along the line–of–sight is also explained.

Key words: Galaxy Clusters, Dark Matter, Redshifts

1 Introduction

Dynamical study of galaxy clusters, assuming the observed redshifts of the individual galaxies to be of purely dopplerian origin, results in the presence of a disproportionately large amount of gravitating matter as compared to the luminosity and the number count of galaxies. This has led to the interpretation that most of the universe is composed of dark matter (DM). It is also found (Ostriker et al 1973 and Trimble 1987) that the ratio of the dark matter to the luminous matter increases with the size of the system considered. This raises some doubt about the nature of the redshifts. If a major fraction of the observed redshift is an indicator (in the presence of the proposed cosmic drag machanism to cause the redshifting of the photons)of the distance rather than the recessional velocity of a galaxy along the line–of–sight then it is obvious that the redshift dispersion indicates the diameter of the system of galaxies under consideration. On the other hand, increase of the ratio of the dark matter to the luminous matter with the size of the system is difficult to explain whatever may be the nature of dark matter distribution. Furthermore, the $m - z$ relationship shows that the galaxies with fainter magnitudes posses higher redshifts. Thus, it is reasonable to assume that a major fraction of the redshift is a distance indicator rather than the

Astrophysics and Space Science **227**: 41–52, 1995.

velocity. Jaakkola (Jaakkola 1983) and Arp (Arp 1993) also express similar suspicions.

In this paper a method has been proposed to extract the true velocity dispersion from the gross redshift data in case of spherically symmetric rich galaxy clusters. The author considers the velocity dependent inertial induction, which has yielded consistent results in a number of cases (Ghosh 1984, 1986a, 1986b, 1988a, 1988b, 1993), for producing redshifts approximately proportional to the distance. Coma and Perseus clusters have been investigated using the proposed procedure.

2 Determination of True Velocity Dispersion

The method for separting the true velocity dispersion from the gross redshift data presented below assumes rich clusters to be spherically symmetrical. The model of velocity dependent inertial induction (Appendix) applied to a quasistatic, infinite, homogeneous universe shows that a photon is subjected to a cosmic drag $h\nu\sqrt{\pi G_o \rho}/c$ where, h is the Planck's constant, ν is the frequency of the photon, G_o is the constant of gravitation, ρ is the mean density of the universe and c is the speed of light. It can be easily shown that this drag leads to a redshift approximately proportional to the distance as follows (Ghosh 1984, 1986a, 1992):

$$z = \frac{\Delta\lambda}{\lambda} \approx \frac{\sqrt{\pi G_o \rho}}{c} x = Kx \quad (\text{for} z << 1) \tag{1}$$

where z is the fractional redshift, λ is the photon wavelength at the source, $\Delta\lambda$ is the increase in the wavelength, x is the distance of the source and K is the proportionality constant (which is equivalent to the Hubble constant H_o when multiplied by c, the speed of light). Since this redshift is generated by some interactive mechanism it is reasonable to assume that the intensity of the mechanism (and, therefore, the value of K) is dependent on the matter density in the path of a photon.However to startwith let us neglect the effect of local variation of K. The redshift of a galaxy in a cluster will be primarily due to two reasons (the magnitudes of the intrinsic redshifts are comparatively small and, therefore, neglected) as follows:

(i) Cosmological redshift proportional to the distance of the galaxy, z_c.

(ii) Doppler red (or, blue) shift due to the line–of–sight component of the velocity of the galaxy, z_v.

Thus, z_c can be taken as a distance indicator and when z_c (to some scale representing the distance) is plotted against the distance from the core centre (with the same scale) the diagram will be of approximately semicircular shape as shown in Fig.1. This is because it will be nothing but the diagram

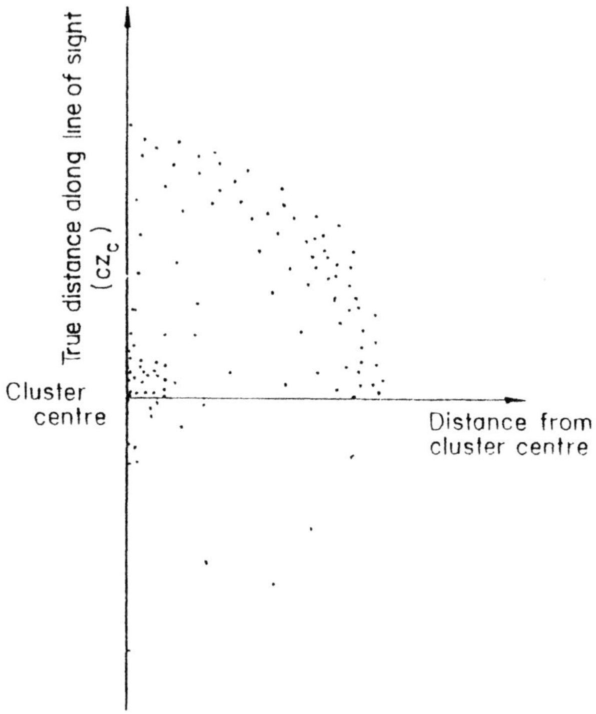

Fig. 1. Redshift - distance from centre plot of a galaxy cluster.

obtained by folding the circle (representing the projection of the spherical cluster on the plane containing the line–of–sight) about the diameter along the line of sight. The total redshift of a galaxy

$$z = z_c + z_v \tag{2}$$

Since z_v is due to the velocity component along the line of sight, when z is plotted against the distance from the cluster centre the diagram will be elongated along the line–of–sight. This is because many galaxies near the boundary of the far side of the cluster have velocities away from the observer causing the upper quarter circle (of the semicircle shown in Fig. 1) to move up by an amount corresponding to $z_v(= v/c$, where v is the velocity of such galaxies and c is the speed of light). Similarly many galaxies near the cluster boundary at the near end move with velocity v towards the observer. Thus a blueshift of magnitude z_v will result in a lowering of the bottom quarter circle of the semicircle shown in Fig. 1. Thus, the plot of the total redshift (in a suitable distance scale) against the distance from the cluster centre will have the appearence of two split quarter - circles as shown in Fig. 2. Since

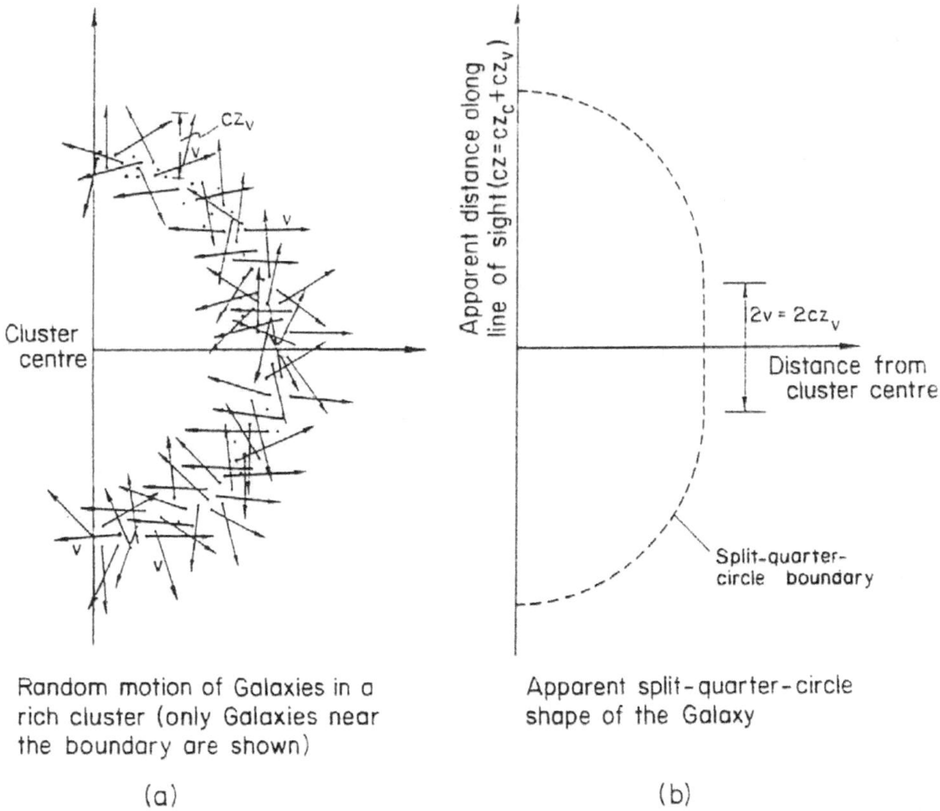

Fig. 2. Split quarter–circle appearance due to velocity dispersion

the redshift - distance relation given by (1) is used, the order of magnitude of the true velocity v is given by cz_v where z_v is determined from the amount of the split of the quarter circles (as shown in Fig. 2).

3 Effect of Local Fluctuation in K and Shape Distortion

The density at the core of a rich cluster is much more than the average density of the universe. So, if the cosmological redshift z_c is due to velocity dependent inertial induction then the local value of $K(= H_o/c)$ inside the core region will be much higher introducing substantial distortion in Fig. 1 because of intensified local interaction as the matter density in this region is more. Figure 3 shows two diametrically opposite galaxies A and B on the outer boundary of a cluster along the line–of–sight. The distance between

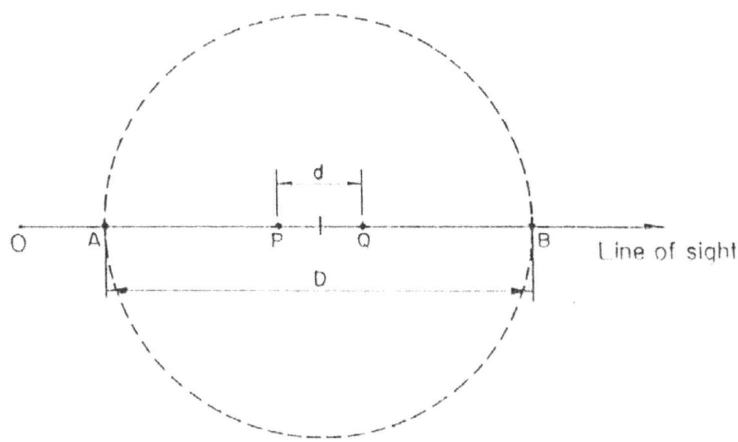

Fig. 3. True Nature of distribution of galaxies in a spherical rich cluster

these two (i.e., the diameter of the cluster), D, depends on the redshift difference as follows:

$$z_{c_B} - z_{c_A} = K_{BA}.D \tag{3}$$

where K_{BA} is the local average value of K when a photon travels from B to A, respectively. Similarly, for the distance, d, between two galaxies P and Q at the core region (Fig. 3) we get the following relation:

$$z_{c_Q} - z_{c_P} = K_{QP}.d \tag{4}$$

where z_{c_P} and z_{c_Q} are the cosmological redshifts of P and Q, respectively, and K_{QP} is the local average value of K for a photon going from Q to P.

Taking the orders of magnitude of the values from the data for Coma it is seen that $K_{BA} \approx K$ whereas $K_{QP} > K$ using the model of inertial induction (Ghosh 1986a, 1993). As a result the shpherically symmetric core, when plotted using z_c as the distance indicator with K as the average universal value of the proportionality constant, appears to be elongated along the line of sight. However, the outer shape of the cluster remains undistorted once the redshift due to the true velocity dispersion is determined (in a manner discussed in the previous section) and subtracted from the gross redshift data. This is so because

$$d_{ap} = \frac{z_{c_Q} - z_{c_P}}{K} > \frac{z_{c_Q} - z_{c_P}}{K_{QP}} = d$$

and

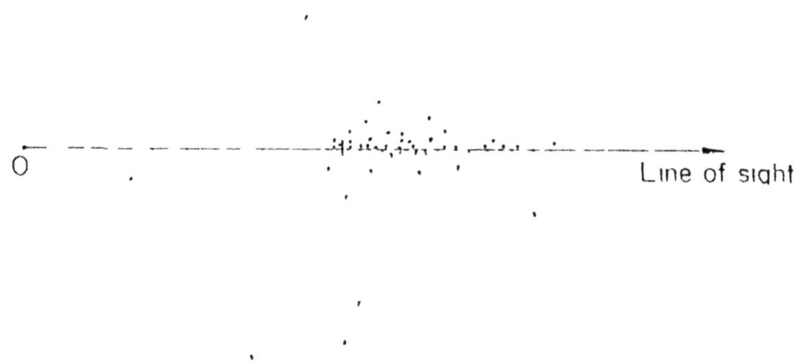

Fig. 4. Apparent nature of distribution of galaxies from the redshift data

$$D_{ap} = \frac{z_{cB} - z_{cA}}{K} \approx \frac{z_{cB} - z_{cA}}{K_{BA}} = D$$

where d_{ap} and D_{ap} represent the apparent diameters of the core and the cluster, respectively. Figure 4 shows the apparent shape from the cosmological redshift plot.

The mean position of the elongated core also gets shifted away from the observer and the geometric centre of the cluster as indicated in Fig. 4. This happens because of the systematic excess redshifts of the galaxies behind the cluster centre.

4 Dependence of Apparent Magnitude on Redshift

Since a major fraction of the observed redshift is proposed to be due to velocity dependent inertial induction, a larger redshift, in general, also implies a larger distance. Assuming the average intrinsic luminosity of the galaxies of a particular type to be constant the apparent magnitude should increase with redshift. However, it is difficult to observe this effect distinctly because of the smallness of the effect and a large scatter in the intrinsic luminosities of the galaxies of a particular type in a cluster. Spiral galaxies are most suitable for this study because of the relatively smaller scatter in their luminosities.

If two galaxies A and B with equal intrinsic brightness be at distances x_A and x_B, $(x_A > x_B)$ respectively, their apparent magnitudes, m_A and m_B, should satisfy the relation

$$m_A - m_B = 5 \log_{10}(\frac{X_B}{X_A}) \tag{5}$$

5 Analysis of Coma and Persus Clusters

Both Coma and Perseus clusters are rich clusters and have been studied quite extensively (Kent and Gunn 1982, Kent and Sargent 1983). Though Coma is a very clean system with spherical symmetry, the extent of Perseus is not that well identified. Figure 5 shows a plot of the line - of - sight distance against the distance from the core centre taking a value of $K(\equiv H_o)$ equal to 50 km s^{-1} Mpc^{-1}. It should be noted that the result will not depend on this choice as both the line - of - sight distance scale and the scale for the distance from the cluster centre depend on this choice. A numerical value has been adopted only for the convenience of representation. It appears that if Coma is spherically symmetric then certain galaxies in the foreground and in the background do not truely belong to the cluster though so far they have been treated as members of the cluster because of their small angular distance from the centre.

The split - quarter - circle configuration is quite obvious from the figure. Following the procedure developed earlier in this paper the order of magnitude of the overall true velocity dispersion is found to be about 350 km s^{-1} (Fig.6). Application of the virial theorem yields an M/L ratio of about 30 (instead of about 180 as estimated in the conventional manner) which appears to be more realistic and closer to the value of M/L ratios for individual galaxies.

Figure 7 shows the $m - z$ relation for the spiral galaxies in Coma which agrees well with the relationship predicted in (5). Arp (Arp 1988) has also obtained a similar dependence of m on z. Figure 6 also shows the elongation of the core region as per the expectation because K in the core region is a few times more than that used for converting the redshift into distance. It is not only the shape which is elongated, the mean position of the core is also shifted away from the geometric centre of the cluster as predicted.

An analysis of the data from Kent and Sargent 1983 indicates the possibility that Perseus could be in reality two smaller clusters one behind the other along the line–of–sight. Converting the redshift data into the apparent distances and plotting these against the distance from the centre (to the same scale) results in the diagram shown in Fig.8. There are two split - quarter - circles representing two spherical clusters. The velocity dispersions of these clusters are of the order of 380 km s^{-1} and 120 km s^{-1} as shown. If the unusually high conventional value of 1200 km s^{-1} is replaced by 380 km s^{-1} M/L drops from a conventional value of 300 to about 30. Am $m - z$ plot for the galaxies in Perseus was not attempted as adequate data on spiral galaxies were not available.

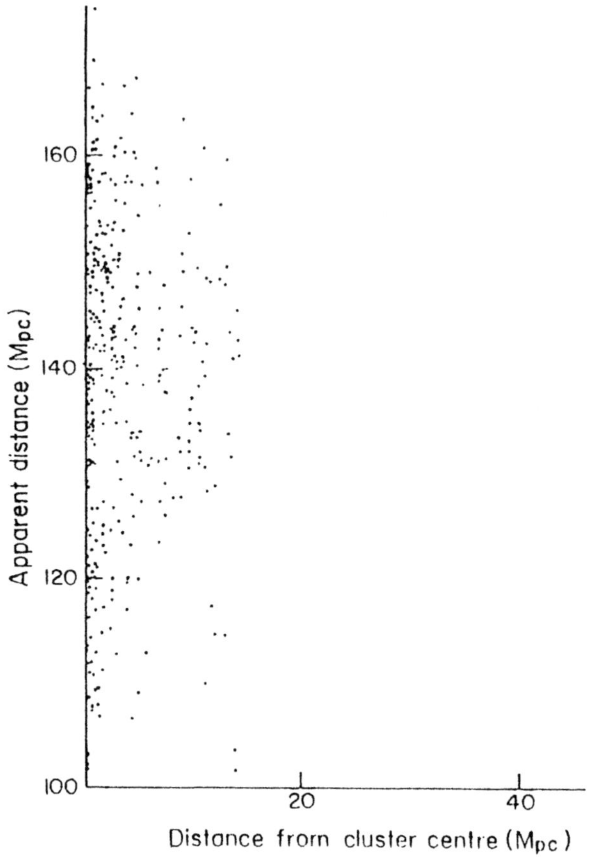

Fig. 5. Plot of Coma cluster

6 Concluding Remarks

If the redshifts of the galaxies in clusters are assumed to represent primarily the distance (rather than velocity) because of some interactive mechanism (like velocity dependent inertial induction) causing the photons to undergo redshifts, it is possible to limit the M/L ratio to the values obtained for individual galaxies. A method for extracting the true velocity dispersion of the clusters from their gross redshift data has been developed which yields cosistent results. The typical elongation along the line–of–sight is explained. Analyses of Coma and Perseus yield realistic results with a much lower value of M/L. It is also found that Perseus may be in reality two smaller clusters side by side along the line–of–sight.

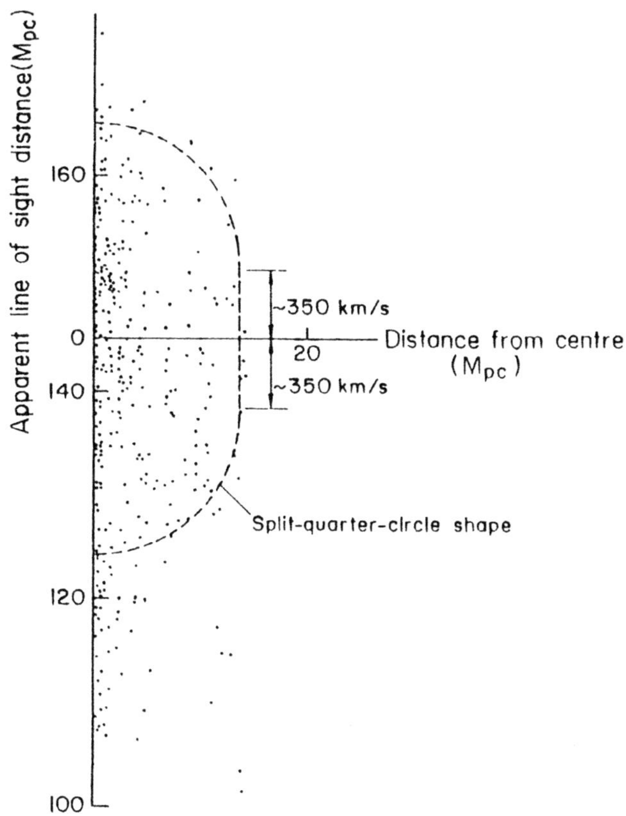

Fig. 6. Determination of true velocity dispersion for Coma cluster

7 APPENDIX

According to the phenomenological model of inertial induction the gravitational force on a body A due to another body B is given by (Ghosh 1986a, 1993).

$$\boldsymbol{F}_{AB} = -\frac{Gm_A m_B}{r^2}\boldsymbol{u}_r - \frac{Gm_A m_B}{c^2 r^2} v^2 \boldsymbol{u}_r f(\theta) - \frac{Gm_A m_B}{c^2 r} a \boldsymbol{u}_r f(\phi)$$

where $\boldsymbol{r}(= r\boldsymbol{u}_r)$, $\boldsymbol{v}(= v\boldsymbol{u}_v)$ and $\boldsymbol{a}(= a\boldsymbol{v}_a)$ are the position, velocity and acceleration of body A with respect to B ($\boldsymbol{u}_r, \boldsymbol{u}_vand\boldsymbol{u}_a$ being the unit vectors). $f(\theta)$ and $f(\phi)$ (with $\cos\theta = \boldsymbol{u}_r.\boldsymbol{u}_v$ and $\cos\phi = \boldsymbol{u}_r.\boldsymbol{u}_a$) represent the inclination effects, m_A and m_B are the relativistic gravitational masses of A and B, respectively. In this work we have assumed $f(\theta) = \cos\theta.|\cos\theta|$ and $f(\phi) = \cos\phi.|\cos\phi|$ Using the above model the force on a particle of

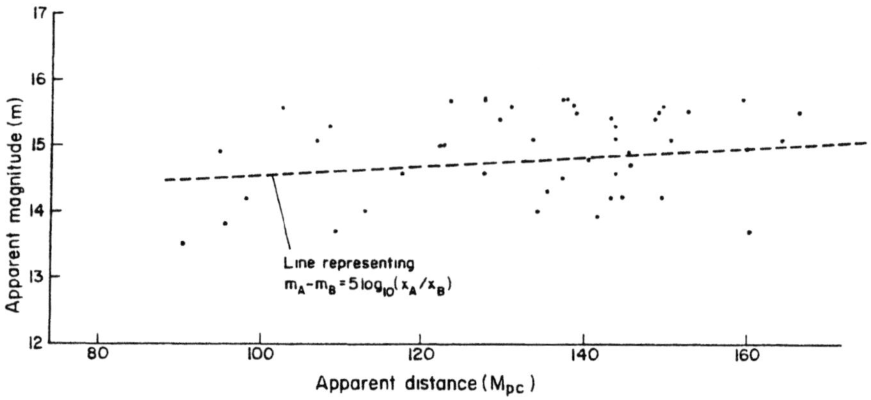

Fig. 7. Apparent magnitude-redshift relation for the spiral galaxies in Coma cluster

relativistic gravitational mass m due to matter in an infinite, quasistatic non expanding universe can be shown to be (Ghosh 1984, 1986a, 1993)

$$F = -\frac{k}{c} m v^2 \boldsymbol{u}_v - m\boldsymbol{a}$$

where $k = \sqrt{\pi G_o \rho}$ with G_o as the local value of gravitational constant, ρ is the mean density of mass-energy in the universe, boldmath$v (= v\boldsymbol{u}_v)$ and $\boldsymbol{a}$ are the velocity and acceleration of the particle with respect to the mean rest frame of the universe. The first term indicates that a cosmic drag acts on a body even if it moves with a constant velocity and the second term implies Newton's second law of motion alongwith the equivalence of gravitational and inertial masses.

Local interactions according to the proposed model lead to a number of interesting observable effects and in all cases the observations have supported the theoretical prediction. A consolidated discussion of all these phenomena is presented in (Ghosh 1993).

8 Acknowledgment

The author gratefully acknowledges the help and encouragement received from C. Roy Keys during the work. The suggestions and corrections pointed out by H.Arp are also thankfully acknowledged.

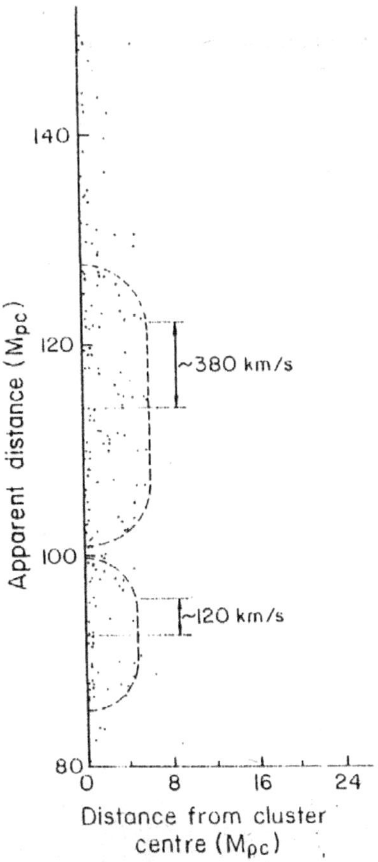

Fig. 8. Determination of true velocity dispersion for Perseus cluster

References

Arp, H., 1988, *Astronomy and Astrophysics*, 202 : 70.

Arp, H., 1993, "Fitting Theory to Observation" in *Progress in New Cosmologies : Beyond the Big Bang*, ed. H.Arp and C. Roy Keys, Plenum Press, New York

Ghosh, A., 1984, *Pramana - Jr. of Physics*, 23:L671.

Ghosh, A., 1986a, *Pramana - Jr. of Physics*, 26:1.

Ghosh, A., 1986b, *Pramana - Jr. of Physics*, 27:725.

Ghosh, A., Rai, S. and Gupta, A., 1988, *Astrophysics and Space Science*, 141:1.

Ghosh, A., 1988, *Earth, Moon and Planets*, 42:169.

Ghosh, A., 1991, *Apeiron*, 9-10:35.

Ghosh, A., 1993, "Astrophysical and Cosmological Consequences of Velocity Dependent Inertial Induction" in *Progress in New Cosmologies : Beyond the Big Bang* ed. H Arp and C. Roy Keys, Plenum Press, New York

Jaakkola, T., 1983, *Astronomische Nachrichten*, 304 :103.

Kent, S.M. and Gunn, J.E., 1982, *The Astronomical Journal*, 87:945.

Kent, S.M. and Sargent, W.L.W., 1983, *The Astronomical Journal*,88 : 697.

Ostriker, J.P. and Peebles, P.J.E., 1973, *Astrophysical Journal*,186 : 467.
Trimble, V., 1987, "Existence and Nature of Dark Matter in the Universe" in *Annual Review of Astronomy and Astrophysics*, 25 :425.

LOW CORRELATION BETWEEN REDSHIFT AND INTRINSIC BRIGHTNESS FOR QUASARS AND GALAXIES IN A DE SITTER UNIVERSE

JOHN B. MILLER
6 Benham Street, Danbury, Connecticut 06811 USA

and

THOMAS E. MILLER
Columbia University, Department of Radiology
One Atwell Road, Cooperstown, New York 13326 USA

Abstract. Since the discovery of quasars, it has been recognized that these objects must have either an extraordinary intrinsic brightness or a nonlinear redshift. The most widely accepted current belief incorporates a linear (Hubble) redshift- distance relation and time evolution within a Big Bang model. We reconsider the possibility of a nonlinear (de Sitter) redshift-distance relation and find quasar intrinsic brightness to be not at all extraordinary. Given a de Sitter law, intrinsic brightness is found to be independent of redshift over five orders of magnitude.

Key words: de Sitter, Nonlinear Redshifts, Quasars

1 Introduction

The first relativistic gravitational redshift was predicted by Schwarzchild (1916); his solution for an isolated mass implies a metrical gravitational redshift. Shortly thereafter, in 1917, de Sitter obtained a similar, but "cosmological" metrical redshift, one that becomes increasingly apparent at large (cosmological) distances. This was the first-known cosmological redshift and became known as the de Sitter redshift. De Sitter wrote, in the last of his three papers that constitute the first English language version of Einstein's theory of general relativity (de Sitter, 1917),

> Consequently the frequency of light-vibrations diminishes with increasing distance from the origin of co-ordinates. The lines in the spectra of very distant stars or nebulæ must therefore be systematically displaced towards the red, giving rise to a spurious positive radial velocity.

This prediction of a redshift was apparently made in ignorance of the contemporary observational work by Slipher demonstrating such a redshift. "Slipher's list of 13 velocities, although published in 1914, had not reached de Sitter, probably as a result of the disruption of communications during the war" (Hubble, 1936). Ehrenfest seems to have had the basic idea behind the de Sitter solution, suggesting that if space is curved, time should be curved as well (Kerzberg, 1989). Of course, this idea is also contained

Astrophysics and Space Science **227**: 53–60, 1995.

in the Schwarzchild solution. It is the time curvature which produces the Schwarzchild and the de Sitter metrical gravitational redshifts.

There are three possibilities for a quasi-static, homogeneous, isotropic universe: the "flat" spacetime of Minkowski, Einstein's "cylindrical" spacetime in which space is curved but time is flat, and de Sitter's "spherical" spacetime in which both space and time are curved. Einstein's "cylindrical" world and de Sitter's "spherical" world were the dominant theories throughout the 1920's, although they have now slipped largely into obscurity. (Both the Einstein solution and the de Sitter solution are distinct from the Einstein-de Sitter solution.) Edwin Hubble wrote in his vintage paper demonstrating a correlation between redshift and distance (Hubble, 1929),

> The outstanding feature, however, is the possibility that the velocity-distance relation may represent the de Sitter effect .... In the de Sitter cosmology, displacements of the spectra arise from two sources, an apparent slowing down of atomic vibrations and a general tendency of material particles to scatter.

Notice that two types of de Sitter redshift were being discussed in 1929: metrical and kinematic. The kinematic de Sitter redshift led to the linear Hubble relation that gave rise to the the expanding universe theory, while the original, metrical de Sitter redshift was nearly forgotten.

During the 1930's, the time-independent theories of Einstein and de Sitter faded in favor of the time-dependent solutions of Friedmann and Lemaître. The Einstein world was generally thought to be unstable, and the de Sitter world was thought to be empty. An equation arises in the de Sitter solution, $p + \rho = 0$, where p is pressure and ρ is density. This implies that one must have either negative pressure, negative density, or an "empty" universe in which $p = \rho = 0$. It seems that there was a general dislike for "negative pressure" and even more dislike for "negative density." Thus the de Sitter world was thought to be empty and unrealistic. Modern theorists, imbued with esoteric concepts such as time reversal and confronted with particles like the positron, are less complacent about immediately dismissing "negative pressure" or "negative density."

During the war of 1939–45, scientists were engaged with more-or-less practical problems and cosmology languished. In the early 1960's, a few researchers were reconsidering the de Sitter solution. For instance, Browne wrote, "In many respects the most satisfactory of the cosmological models based on solutions of Einstein's gravitational field equations is the steady-state de Sitter model" (Browne, 1962). Hawkins at one time advocated a de Sitter metrical redshift (Hawkins, 1960), but he apparently abandoned the de Sitter redshift with the discovery of quasars (Hawkins, 1969):

> Actually both the steady state theory and the static universe were dealt a severe blow by the discovery of *quasars*. Neither theory can explain a

concentration of these objects at the boundary of the universe, existing at an epoch of 3 billion years in the past.

It is not clear why Hawkins abandoned the de Sitter redshift with the discovery of quasars when he might have portrayed quasars as a manifestation of the metrical de Sitter redshift which he had been advocating. Despite Hawkins's capitulation, we reconsider below the metrical gravitational de Sitter redshift.

2 Theory

Given the relativistic spacetime metric

$$ds^2 = -\gamma^{-1}\, dr^2 - r^2\, d\theta^2 - r^2 \sin^2\theta\, d\phi^2 + \gamma\, dt^2, \tag{1}$$

the de Sitter solution to the Einstein field equations can be written as

$$\gamma = 1 - (r/R)^2, \tag{2}$$

with $R^2 = 3/(8\pi\rho)$, where the constant R is the radius of spacetime curvature and ρ is the mean mass density of the universe (assuming simplified units so that $G = c = 1$). There are many different formulations of the de Sitter solution, depending on which coordinate transformations are employed (Eddington, 1924). In order to use the inverse-square law, we choose "pseudo-Euclidean" coordinates (r, θ, ϕ, t) that preserve the Euclidean formula for surface area: $A = 4\pi r^2$. Other coordinate transformations entail unfamiliar surface area formulae and do not obey inverse-square-law dimming. Redshift z is defined as

$$z = \lambda_r/\lambda_o - 1 = \gamma^{-1/2} - 1, \tag{3}$$

where λ_o is the wavelength of the unshifted photon and λ_r is the wavelength of the photon observed at a distance r. Distance can then be given as a function of redshift,

$$r = R\left[1 - (z+1)^{-2}\right]^{1/2}. \tag{4}$$

Equation (4) is the de Sitter redshift-distance law—the de Sitter equivalent of the Hubble law (Fig. 1).

Assuming inverse-square-law dimming, absolute magnitude M and apparent magnitude m are related to distance r by

$$M + C = m - 5\log_{10}(r) \tag{5}$$

where C is a constant determined by observation. According to the Hubble law, redshift is directly proportional to distance, thus

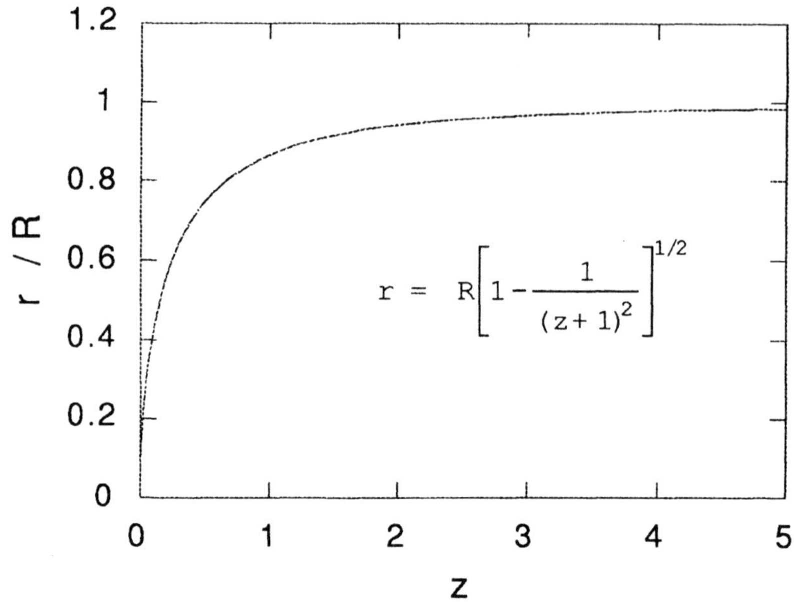

Fig. 1. Distance (r/R) plotted *versus* redshift z in a de Sitter universe (Equation 4).

$$M_H + C_H = m - 5\log_{10}(z), \tag{6}$$

where M_H is Hubble absolute magnitude and C_H is an observational constant that incorporates the Hubble constant. Combining equation (4) and equation (5) gives the de Sitter version of the magnitude-redshift relation

$$M_D + C_D = m - 2.5\log_{10}\left[1 - (z+1)^{-2}\right], \tag{7}$$

where M_D is de Sitter absolute magnitude and C_D is an observational constant that incorporates the de Sitter radius R.

3 Methods

To compare the Hubble and the de Sitter redshift-distance laws in the context of recent observations, we examined large catalogs of galaxies and quasars with published measurements of both redshift and apparent magnitude (Huchra, 1990; Huchra and Clemens, 1991; Hewitt and Burbidge, 1989). We transformed apparent magnitude to absolute magnitude according to the Hubble and the de Sitter laws using equation (6) and equation (7) respectively. The

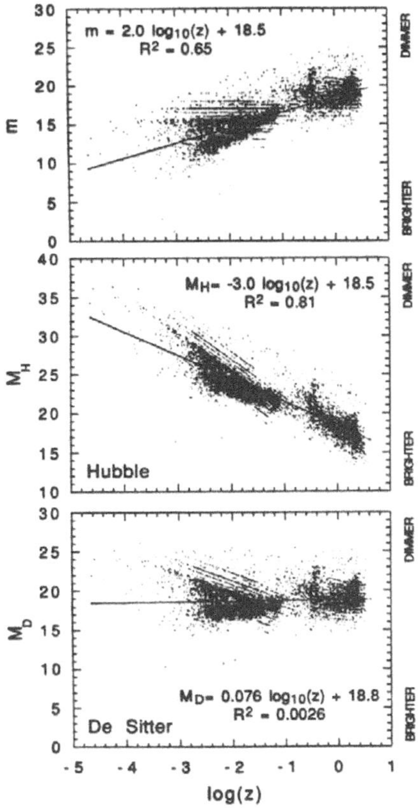

Fig. 2. *(a)* Apparent magnitudes of galaxies and quasars *versus* $\log_{10}(z)$. The corduroy texture represents lines of iso-apparent magnitude. *(b)* Absolute magnitudes obtained by assuming a linear Hubble redshift- distance relation and applying Equation 6 to the data set in *2(a)*. The actual values are arbitrary, since the constant C_H was set equal to zero. This affects the intercept, but not the slope nor the correlation coefficient. *(c)* Absolute magnitudes obtained by assuming a nonlinear de Sitter redshift-distance relation and applying Equation 7 to the data set in *2(a)*. The actual values are arbitrary, since the constant C_D was set equal to zero.

following scatter plots show magnitude plotted versus redshift: first the raw data or apparent magnitude, second the absolute magnitude assuming a Hubble law, and third the absolute magnitude assuming a de Sitter law (Fig. 2). As a statistical assessment, a least-squares-fit line is drawn for each set of data in the plots.

TABLE I

A comparison of the dependence of absolute magnitude on redshift for the Hubble and de Sitter laws. Slope refers to the slope of a least-squares linear regression curve for a scatter plot of absolute magnitude *versus* $\log_{10}(z)$, as in Figure 2. Intercept gives the $\log_{10}$ value of this line in the limit of $z = 0$ and with $C_H = C_D = 0$. R^2 is the correlation coefficient (which is distinct from the de Sitter radius R). Galaxies are taken from (Huchra, 1990) ($n = 15597$, $z \leq 0.33$) and (Huchra and Clemens, 1991) ($n = 521$, $z > 0.33$). Quasars are taken from (Hewitt and Burbidge, 1989). Only objects with published measurements of both redshift and apparent magnitude are included. A few blue-shifted galaxies have been excluded.

		Hubble			deSitter		
	Number	Slope	Intercept	R^2	Slope	Intercept	R^2
Galaxies	16118	-3.0	18.4	0.60	-0.27	18.2	0.011
Quasars	4197	-3.6	18.6	0.52	0.33	19.0	0.0089
Total	20315	-3.0	18.5	0.81	0.076	18.8	0.0026

4 Results

Almost all objects fall within a band spanning about ten magnitudes vertically (for a given redshift). This entire band is down-sloping assuming a Hubble law, but is roughly horizontal assuming a de Sitter law. The line in the Hubble plot (Fig. 2b) is down-sloping, but the line in the de Sitter plot (Fig. 2c) is nearly horizontal. De Sitter absolute magnitude is practically uncorrelated with redshift, while Hubble absolute magnitude is fairly strongly correlated with redshift. This is found to be the case not only for the combined catalogs, but also for galaxies and quasars considered separately (Table I).

Recognizing that Eq. 6 is strictly accurate only for a deceleration parameter $q_o = 1$, we also analyzed the data substituting Hubble luminosity distance D_L for coordinate distance r, where D_L is given by

$$D_L = \left(c/H_o q_o^2\right)\left[1 - q_o + q_o z + (q_o - 1)\left(2q_o z + 1\right)^{1/2}\right] \quad (8)$$

where H_o is the Hubble constant (Zombeck, 1990). The data were reanalyzed assuming $q_o = \{1, 0.5, 0\}$, but no statistically meaningful difference was obtained (Table II). This reflects the well-known situation that, with currently available observations, it has not been possible to determine the value of q_o.

TABLE II

A statistical comparison of a Friedmann universe (using three different values for the deceleration parameter q_o) with a de Sitter universe (q_o is not applicable).

	Friedmann			de Sitter
	$q_o = 1$	$q_o = 0.5$	$q_o = 0$	(q_o n/a)
R^2	0.81	0.83	0.85	0.0026
Slope	-3.0	-3.2	-3.6	0.076
Intercept	18.5	18.1	17.5	18.8

5 Discussion

Some increase in intrinsic brightness is to be expected as a result of the well-known Malmquist bias: more distant objects will tend to be brighter because we will only be able to see distant objects if they are intrinsically brighter than closer objects. However, Malmquist bias is generally thought to be inadequate to explain quasar intrinsic brightness. Of course, the downward slope of the least- squares-fit line in the Hubble plot (Fig. 2b) could still be explained by luminosity evolution, as is widely accepted. That is, as we look out into space, we look back to a time when objects existed that were intrinsically much brighter than anything currently (or locally) extant.

Given the asymptotic character of the de Sitter redshift at large distances, quasars with high redshifts may not be much more distant than are intermediate-redshift objects. Absolute magnitudes are roughly the same at all redshifts assuming a de Sitter law. Quasars may be considered in the context of a nonlinear de Sitter redshift-distance law without invoking luminosity evolution.

Quasars and their redshifts have been the subject of considerable controversy since their discovery thirty years ago. Assuming a linear redshift-distance law, high-redshift quasars must be extraordinarily powerful, since they are assumed to be much farther away than are objects of intermediate redshift, and yet appear to be nearly as bright. One might think that all possible quasar redshift mechanisms would have been exhausted by now, but the de Sitter redshift has been somewhat neglected in the search for quasar mechanisms. The asymptotic character of the metrical de Sitter redshift may provide a reasonable explanation for the quasar phenomenon.

References

Browne, P. F.: 1962, *Nature*, **vol. 193**, p. 1019.
de Sitter, W.: 1917 *MNRAS*, **vol. 78**,p. 3.

Eddington, A. S.: 1924, *The Mathematical Theory of Relativity, Second Edition* (University, Cambridge).

Hawkins, G. S.: 1960, *Astron. J.*, **vol. 65**, p. 52.

Hawkins, G. S.: 1960, *Splendor in the Sky, Rev. Ed.*, (New York, Harper and Row), p. 282.

Hewitt, A. and Burbidge, G.: 1989, *A New Optical Catalog of Quasi-Stellar Objects, Selected Astronomical Catalogs, Volume 1, CD ROM* (Astronomical Data Center, NASA Goddard Space Flight Center, Greenbelt, Maryland, USA).

Hubble, E.: 1936, *The Realm of the Nebulae*, (Yale University Press, New Haven, Connecticut), p. 109.

Hubble, E.: 1929, *Proc. Nat. Acad. Sci.* **vol. 15**, p. 168.

Huchra, J.: 1990, *CfA Redshift Catalogue, Selected Astronomical Catalogs, Volume 1, CD ROM* (Astronomical Data Center, NASA Goddard Space Flight Center, Greenbelt, Maryland, USA).

Huchra, J. and Clemens, C.: 1991, *ZBIG Catalog*, personal communication.

Kerzberg, P.: 1989, *The Invented Universe: The Einstein-de Sitter Controversy (1916–17) and the Rise of relativistic Cosmology*, Clarendon Press, p. 178.

Schwarzchild, K.: 1916, *Sitzber. Deut. Akad. Wiss. Berlin, Kl. Math.-Phys. Tech.*, pp. 189–196.

Zombeck, M. V.: 1990, *Handbook of Space Astronomy and Astrophysics, Second Edition*, Cambridge, Cambridge University Press, pp. 250–252.

INTERGALACTIC RADIO ABSORPTION AND THE COBE DATA

ERIC J. LERNER
Lawrenceville Plasma Physics
Lawrenceville, New Jersey, USA

Abstract. The COBE data on cosmic Background radiation (CBR) isotropy and spectrum are generally considered to be explicable only in the context of the Big Bang theory and to be confirmation of that theory. However, this data can also be explained by an alternative, non-Big Bang model which hypothesizes an intergalactic radio-absorbing and scattering medium. A simple, inhomogenous model of such an absorbing medium can reproduce both the isotropy and spectrum of the CBR within the limits observed by COBE, and in fact gives a better to fit to the spectrum observations than does a pure blackbody. Such a model does not contradict any other observations, such as the existence of distant radio sources.

Key words: Plasma Cosmology, Cosmic Microwave Background, COBE

1 Introduction

The Planck spectrum and high isotropy of the CBR has long been widely considered as the key observational support for the Big Bang. However, in 1988 the author proposed that the CBR was isotropized and thermalized by synchrotron absorption and reemission of electrons trapped in dense magnetically pinched filaments in the IGM (Lerner, 1988). A similar theory was independently put forward by Peter and Peratt (1990). In the author's theory, the energy for the CBR was supplied not by the Big Bang, but by hot, intermediate mass stars in early galaxies, which also produced the observed "primordial" helium.

Observational evidence for the existence of such an absorbing IGM came from studies of radio and infrared brightness of galaxies at various distance from Earth (Lerner, 1990, 1993). For a given infrared brightness, more distant galaxies were systematically dimmer in the radio, indicating a selective absorption of radio frequency radiation.

However, the new data provided by COBE (Bennett *et al.*, 1994) measuring the anisotropy of the CBR and the detailed spectrum (Mather *et al.*, 1993), present a challenge for any theory. In particular, with an absorbing, emitting medium, subject to the Hubble relation, radiation from more distant parts of the medium will be redshifted, so the sum of the radiation from the medium will not be a precise blackbody. Can a realistic model of this sort produce a spectrum as close to a blackbody as that observed by COBE? In addition, parts of the medium that are denser than others, or slightly

Astrophysics and Space Science **227**: 61–81, 1995.

hotter, should produce anisotropies. Can these approximate those detected in the COBE data?

The present paper is intended to give a preliminary answer to these questions. However, it should be noted that the Big Bang itself does not provide a very satisfactory explanation of the Planckian spectrum and isotropy observed As has been known for more than a decade, the original version of the theory does not predict even rough isotropy of the CBR, since portions of the sky separated by more than about 10 degrees would not have had time to be in contact with each other and so could not have reached thermal equilibrium with each other. Neither isotropy nor a Planck spectrum would thus be expected on large angular scales, in gross contradiction to observation (This is the so-called horizon problem). The widely accepted solution to this problem is the additional hypothesis of an inflationary stage that expands a small portion of the universe that has reached thermal equilibrium. This is a wholly ad hoc hypothesis, since it is derived from grand unified theories that are themselves without any experimental justification. Indeed, the only unequivocal predictions of these theories, the instability of the proton, has been contradicted by experiment (Becker-Szend, 1990).

If inflation is accepted, it predicts an omega of unity for the density of the universe. Since the directly observed density leads to an omega of only 0.02 at most, this is another violent contradiction with observation. Indeed, an omega of unity with ordinary matter is inconsistent with Big Bang nucleosynthesis predictions which allow at most a density of 0.04 (Olive *et al.*, 1990). To overcome this inconsistency and contradiction with observation, a third ad hoc hypothesis has been widely proposed that a new and wholly unobserved type of matter, dark matter, constitutes 98% of the universe. Again, the existence of such dark matter in the required amounts is utterly without experimental or observational support.

So the Big Bang does not provide a good explanation of the COBE results, despite claims to the contrary. Does an absorption model?

2 Theoretical Model

To calculate the spectrum that results from the hypothesized IGM, three elements are required. First we need an explicit, theoretically derived expression for the absorptivity of a given density of filaments for all wavelengths observed by COBE. Second, we need an hypothesized model of the variation of the density of the IGM with distance. Third, we need a formula to relate the radiation received from a filament to the distance to the given filament.

The absorbent medium has been described analytically previously (Lerner, 1992). The wavelength that the filaments are opaque to depends of the radii of the individual filaments. The ensemble of filaments act as essentially a perfect blackbody in a wavelength range extending from about 2000 μ

to 200 m. They become slightly transparent by 800 μ and then increasingly transparent up to 200 μ, where they are completely transparent to radiation. The model hypothesizes that the density of the filaments in space is approximately proportional to the energy density of the intergalactic magnetic field and to the smoothed density of matter.

To summarize this model, we begin by hypothesizing force free filaments which are electrically neutral, and have magnetic field energy equal to ion kinetic energy. Electrons and ions travel along field lines, absorbing and emitting synchrotron radiation. In equilibrium with a background radiation field, electrons have a temperature T_p perpendicular to the field lines. Following Lerner (1992) we find that the optical depth A at the synchrotron wavelength λ of each individual filament with radius R (cm) and magnetic field B (gauss) is

$$A = \lambda^2 J \;/ 2\pi k T_p = \lambda^2 n_e B^2 \gamma_e h r_e^2 R f \;/ 3mkT_p c \tag{1}$$

where J is emissivity, n_e is electron density, r_e is the classical electron radius, m is the electron mass and λ is the synchrotron wavelength and $f = abs(\nu dR \;/ R d\nu)$, where $\nu = c \;/ \lambda$.

This equation is simply Kirchoff's law relating absorptivity and emissivity, with J being the standard synchrotron radiation emissivity. (This formulation is valid when $kT_p < h\nu$, so that electrons are predominantly at the zero energy state relative to the magnetic lines of force, and "jump" off them to the energy $h\nu$ when they absorb a photon. In the long wavelength limit $kT_p >> h\nu$, the formula converges on the classical equation which is larger by a factor of $kT \;/ h\nu$, since then the average electron spends most of its time near an energy of kT). For $kT_p << mc^2$, which we'll see is the case for these electrons traveling along field lines, $\nu = eB\gamma_e \;/ 2\pi mc$ (γ_e is the relativistic factor). This is smaller than the normal formula $eB\gamma_e^2 \;/ 2\pi mc$ for synchrotron radiation from randomly oriented electrons and applies to radiation emitted in the forward direction, as nearly all the radiation is. Thus,

$$\lambda = 2\pi mc^2 / eB\gamma_e \tag{2}$$

With magnetic field energy equal to ion energy, $n_p = n_e = B^2 / 8\pi\gamma_p Mc^2$ where M is the proton mass.Substituting these values of n_e and λ into (1) we get

$$A = \pi h r_e B^2 R f \;/ 6kT_p Mc\gamma_e\gamma_p \tag{3}$$

To find the values in (3), we must look at the formation of the filaments. As detailed in (Lerner, 1992) these filaments are generated in quasars, active galactic nuclei and Herbig Haro objects when beams are accelerated by high electric fields. If a proton beam is to accelerate to energy $\gamma_p Mc^2$ in a length

r, energy loss by synchrotron radiation will balance energy gained by acceleration if

$$r/c = 24Mc\ /r_p^2 B^2 \gamma_p \tag{4}$$

or

$$I^3/r = 3(20)^3 e^3\ /2r_p^4 = 2.95 \times 10^{39}\mathrm{Amp}^3/\mathrm{cm} \tag{5}$$

Where I is beam current and r_p is the classical proton radius. This is equivalent to beam power $P_c = 2.8 \times 10^{34} r^{2/3}$ erg/sec. For quasars with $r = 10^{18}$ cm, $P_c = 2.8 \times 10^{46}$ erg/sec. For AGN with $r = 10^{15}$ cm, $P_c = 2.8 \times 10^{44}$ erg/sec.

For any object with $I^3\ /r$ larger than that defined by (4), the beam will reach a steady state in which it is losing energy to synchrotron radiation as fast as it is gaining it by the accelerating field. The beam will be unstable against filamentation since for all inhomogeneties in B, the energy gain per unit time is $\gamma_p Mc^3\ /r$ and the energy loss is $B'^2 \gamma_p'^2 r_p^2 c/24$, where B' is the local magnetic field and $\gamma_p' Mc^2$ is the local energy of the protons. Equating these quantities, we have,

$$B'^2 \gamma_p'^2 = 24\gamma_p Mc^2/r_p^2 r \tag{6}$$

For these filaments, which contain only current carriers, $n_e' = B'/2\pi Re$, where R is the filament radius. In addition, particle energy and magnetic field energy must be equal for a steady state, so $n_e' = B'^2/8\pi\gamma_p' Mc^2$ and thus $\gamma_p' = B'Re/4Mc^2$ so,

$$B'^2 R = 2^{7/2} 3^{1/2} r_p^{-3/2} Mc^2 (\gamma_p/r) \tag{7}$$

Physically, any local increase in B' will cause protons to lose energy to synchrotron radiation, shrinking their gyroradius thus concentrating the local currents and further increasing B'. A hierarchy of filaments will thus form, with the same $B'^2 R$ and $B'\gamma_p'$.

(Filamentation will not be inhibited by electron or proton thermal motion. In this situation, the temperature T_p perpendicular to the field will rapidly decrease for both electrons and protons and the filaments will assume a force-free form. The characteristic cooling time of the electrons is a factor of $(M/m)^3$ shorter than for the protons, so they will cool very rapidly. Electron temperature will thus not prevent filamentation. For protons, the cooling rate increases as B with a given jet, and as I^3/r comparing different jets. For jets with I exceeding criterion (5) by a factor of two, T_p for protons will fall by 50-fold halfway through the accelerating region even for those in the lowest B fields, again allowing rapid filamentation.)

Electrons and protons will be accelerated in the opposite direction by the accelerating fields. It is easy to see that, for the accelerating beams $\gamma_p' = \gamma_e'$. This is the case because electrons traveling along the field lines radiate at the same rate as protons, given the same γ. So, if accelerating power and radiated power is balanced for protons at a given γ it is for electrons at the same γ.

Once such filamentary beams leave the accelerating region, the electrons and protons will lose energy rapidly through synchrotron radiation. However, such cooling will stop when the proton synchrotron frequency $\nu_p = eB\gamma_p^2/8\pi Mc = \nu_{pe}$, the electron plasma frequency that is, the plasma becomes opaque to proton synchrotron radiation. Since the electrons and protons are counterstreaming, we have

$$(4\pi n_e e^2\ /m)^{1\ /2} = eB\gamma_{p^2}\gamma_e\ /4Mc \tag{8}$$

since $B'^2/n_e = 8\pi\gamma_p Mc^2$,

$$\gamma_p'^5\gamma_e'^2 = 8M/m \tag{9}$$

And, since $\gamma_e' = \gamma_p', \gamma_p' = (8M/m)^{1/7} = 3.94$.

With γ_e being a constant the synchrotron frequency of a filament simply depends on B'. The filamentary hierarchy with B'^2R=constant has the property that the total area of sky covered by all filaments of a given radius R and thus a given B' and n, is independent of R. To demonstrate this, we note that if an ensemble of N filaments each of radius R and current I is spread through a cylindrical region of radius R', then the average spacing between filaments, L is defined by

$$NL^2 = R'^2 \tag{10}$$

But the field B'' at a distance L from each filament must equal the average field b in the whole region, so

$$B'' = I'/5L = b \tag{11}$$

and thus $L = I'/5b$. So, from (10), $NI'^2 = 25b^2R'^2$.

But since $B'^2R = I'^2/25R = b^2R', NR = R'$, for all ensembles of radius R, regardless of what R is. So that each ensemble with a given radius R covers the same area of sky.

The relation $B'^2R = K$, a constant, will continue to hold as the filamentary assemblage expands and K decreases, so long as the filaments retain their internal structure. Eventually, the expansion ceases when B'^2R decreases to the level of the background field configuration.

Theoretical considerations of large-scale supercluster filaments (Lerner, 1986) lead to the conclusions that in such filaments $nR = 1.0 \times 10^{-19}\text{cm}^{-2}$ and $B'^2 = 6.08 \times 10^{-7}n$. B'^2R therefore is expected to be $6 \times 10^{12}G^2$ cm.

$$B^2R = 6 \times 10^{12} G^2 \text{cm}. \tag{12}$$

For comparison, in a typical supercluster filament, observations (Kim, *et al.*, 1989) indicate $B' = 2\mu G, R' = 3 \times 10^{24}$cm (1Mpc), $B^2R = 1.2 \times 10^{13} G^2$ cm, a reasonable agreement.

Returning to eq.(3), using the results of (9) and (12) we find that the optical depth of individual filaments is

$$A = \pi h r_e B^2 R f / 6kT_p Mc(8M/m)^{2/7} = 9.1 \times 10^{-12} B^2 R f / T_p = 54.6 f / T_p \tag{13}$$

With $B^2R = K$, $f = 2$ so for equal to the microwave background temperature, 2.75K, A=40. The opacity is thus a constant independent of radius and thus of frequency and is quite high.

We can find the high frequency cut off to this opacity by finding the B of the smallest filaments in the assemblage. These will be the filaments with $\gamma_p' = (8M/m)^{1/7}$ in the initial assemblage, before synchrotron cooling. At this point, $\gamma_p' = B'Re/4Mc^2$ so, $B'_{\text{max}} = 4Mc^2(8M/m)^{1/7} R_{\text{min}} e$. But $R_{\text{min}} = K/B^2_{\text{max}}$, so $B'_{\text{max}} = Ke/4Mc^2(8M/m)^{1/7}$. From (2)

$$\lambda = 8\pi Mc^2/Kr_e = 1.34 \times 10^{11}/K \tag{14}$$

or about 220 μ. Such filaments will have B of about 12 T and R of 4 m. However, this simple picture is complicated by the infant mortality some of the filaments suffer from. To understand this, we must look at collisional processes in the filaments. During acceleration of the beam, collisions in the denser filaments will lead to electrojet formation as protons and electrons collect into single charged streams. That is, electrons will suffer more collisions in regions of local proton over density and will collect in regions of proton underdensity with the same process taking place, although more slowly, with protons. Taking account of these electrojets, the collision time, t_c is determined by the ratio of the interaction time of particles in the counter streaming jets to the reciprocal of the electron cyclotron frequency:

$$t_c = \gamma_e' \gamma_p' (M/m)(d/D)(e^{2D/d})/\pi^2 r_e^2 cn \tag{15}$$

(Jackson, 1963) where $d = \gamma_e \gamma_p c/\nu_c$, $\nu_c = eB/mc$ and D is electrojet radius. Since $B' = 4\gamma_p' Mc^2/Re$, and $n_e = B'^2/8\pi\gamma_p' Mc^2 = 2\gamma_p'/\pi r_p R^2$

$$t_c = \gamma_e'^2 \gamma_p' (R/D) e^{8MD/\gamma_e' mR} / 4\pi^2 r_e^2 cn \tag{16}$$

By the end of acceleration $t_c = L/c$, where L is the length of the accelerating region and since $\gamma_p = \gamma_e$, we thus have:

$$\begin{aligned} D/R &= (m\gamma_p'/8M)\ln[(D/R)L(4\pi^2 r_{e^2} n/\gamma_{p^3})] \\ &= (m\gamma_p'/8M)\ln[(D/R))(L/R^2)(8\pi r_e M/m\gamma_{p^2}] \end{aligned} \tag{17}$$

Assuming $L = r$, the beam radius during acceleration, we know that $L/I^2 = R/I^2$, where $I' = 20Mc^2\gamma_p/e$ is the filament current. So we have

$$R = (L/I^2)(400\gamma_p'^2 Mc^2/r_p) \tag{18}$$

Using eq. (18) and (5), we have

$$D/R = (m\gamma_p'/8M)\ln((D/R)I(3\pi r_e M/5me\gamma_{p^6})) \tag{19}$$

Once the filaments leave the acceleration region and begin to expand along with the beam, the ratio D/R remains constant, as the electrojets expand as well. Since γ_p' drops rapidly to 3.94 t_c increases rapidly beacuse of the exponential term in (16) and then continues to increase as n decreases with R^2. Since the distance traveled increases only with R, for a roughly constant opening angle, t_c will remain greater than the travel time t_t. Thus collisional losses will remain small.

However, as the filaments expand, their collisional behavior will change, since at a certain point the filaments will "crystallize", since the particles would not have sufficient energy to move past each other. At this point, (16) will no longer be valid, because the field felt by an individual particle will no longer vary in a time determined by the distance between the electrojets, but will rather vary rapidly with a time $n^{1/3}/c$. This is due to the coherent effect of the structured beams once crystallization has occurred. (This crystallization phenomenon is observed in laboratory beams). Crystallization of the beam occurs when the thermal energy of the particle perpendicular to the field lines is less than the electrostatic energy required to move by a nearest neighbor,

$$T_c = e^2 n^{1/3}/2k \tag{20}$$

where T_c is the temperature of crystallization. Initially, early in the expansion of the beam, cooling is limited since the particles cannot radiate at frequencies below $n_p e$. Therefore, the thermal energy of particles $kT_p > h\nu_{pe}$. For a filament to condense:

$$T_c < h\nu_{pe}/k \tag{21}$$

or $h/k(4\pi n_e e^2/m)^{1/2} < e^2 n^{1/3}/2k$

This is equivalent to $n < 1/(16\pi a_0)^3$ where a_0 is the Bohr radius. Using $I'^2/R = I^2/r$ for all filaments at any given instant we find that since $B = (8\pi\gamma_p' Mc^2 n)^{1/2}$

$$I^2/r < 5(\gamma_p'/a_0)^{3/2}(Mc)^{1/2}I'/2^{9/2}\pi \tag{22}$$

So, condensation of a filament with I' occurs when the beam radius expands to the value r defined by (22), taken as an equality. For filaments broken up into electrojets of radius D, after condensation the time for collisional decay is

$$t_c = \gamma_e'\gamma_p' M D^3/2\pi m r_{e2} c \tag{23}$$

If a filament is to survive the transition to a condensed state, the collision time must exceed the time it takes to expand significantly

$$t_c > r_{bc}/\theta c \tag{24}$$

where r_{bc} is the beam radius at condensation and θ is the opening angle at that point. Using (23), (19) and (22) taken as an equality we can find a limit on I, total beam current, as a function of I', filament current, that defines which filaments of given I' will survive the condensation transition,

$$I^2 < (\theta\pi\Delta^3/\gamma_p)(r_e/5e)^5(m/\alpha M)^6(I'/2)^7 \tag{25}$$

where $\Delta = 1n((D/R)I(3\pi(5e/r_e)^5(M/m)^7(4/I)^6$, and a is the fine structure constant. For a given I and I', (19) and (25) can be solved numerically to yield D/R and the current I_c' for the smallest surviving filaments for a beam of given I. Physically, if r_{bc} is too large, and thus I too large when a given filament condenses, the filament will collisionally self-destruct before it has time to expand. Since $P = 1.6 \times 10^8 I^2$ erg/sec, where P is the total power of the beam

$$P < 400(\theta\pi\Delta^3/\gamma_p')(r_e/e)^5(m/\alpha M)^6 I'^7 \tag{26}$$

Since, from (2), the absorbing wavelength of a given filament λ is directly proportional to the current I' in the filament, $\lambda = 2\pi I e/5\gamma_p' r_e K$, (26) implies that for each jet source of power P there is a minimum absorbed wavelength λ_c. Equivalently, for an ensemble of filaments from many different sources, at each absorbing wavelength λ there will be a maximum source power P_c.

For progressively shorter λ, fewer and fewer sources will have power, P_c and thus the total number of absorbing filaments will decreases, and thus the absorption at the wavelength. From the above relation between λ and I',

$$P_c = (5\Delta/2)^3(5h\gamma_p'/2\pi^2 mMc^3)^6(K\lambda)^7\theta = 2.72 \times 10^{-40}\Delta^3(K\lambda)^7\theta \tag{27}$$

TABLE I
Filament absorbency.

$\lambda(\mu)$[1]	I' GA	P_c ergs/sec	Decrease in abs.
490	.564	3.0×10^{44}	6.3×10^{-2}
620	.713	1.0×10^{45}	2.4×10^{-2}
770	.887	3.0×10^{45}	9.0×10^{-3}
832	.959	4.47×10^{45}	6.0×10^{-3}
1000	1.15	1.11×10^{46}	2.6×10^{-3}
1200	1.38	2.73×10^{46}	1.0×10^{-3}
1400	1.61	5.88×10^{46}	4.2×10^{-4}
1600	1.84	1.13×10^{47}	1.9×10^{-4}
1800	2.07	2.0×10^{47}	9.4×10^{-5}
2000	2.30	3.4×10^{47}	4.8×10^{-5}
2400	2.76	8.5×10^{47}	1.1×10^{-5}
2800	3.22	1.8×10^{47}	3.3×10^{-6}

[1] $K = 6 \times 10^{12}$

As we shall see, this relationship indicates a small but measurable variation in absorbency over the range of wavelengths studied by COBE.

We can use observations to determine the opening angle θ. A survey of radio jets by de Rutier*et al.* (1990) indicates that θ, measured at half the observed length of a jet, is defined by

$$\theta = 2.6 \times 10^8 P^{-0.21} \tag{28}$$

where P, the power of the jet is taken to be $3 \times 10^{19} P_t$, the observed 1.4 Ghz power in W/Hz. The same survey indicates that L', the length of the jet is $10^{15} P^{0.19}$. Observation by Walker*et al.* (1987) of QC120 show that θ also varies with distance, d, from the central source, varying as $d^{-1/8}$. Combining these observations, we find

$$\theta = 2 \times 10^{10} d^{-1/8} P^{-0.19} \tag{29}$$

Since $d = r/\theta$ and since, at condensation of a given filament with current I', $r = I^2/5(\gamma_p'/a_0)^{3/2}(Mc^2)^{1/2} I'/2^{9/2}\pi$,

$$\theta = 3 \times 10^{11} I'^{1/7}/I^{0.73} \tag{30}$$

Substituting (30) into (25) we get

$$I^2 = 2.77 \times 10^{-13} I^{5.23} \Delta^{2.2} \text{ or } P = 4.4 \times 10^{-5} I^{5.23} \Delta^{2.2} \tag{31}$$

Solving (19) and (30) for Δ, we can find P_c, the limiting source power for a filament of given I' and thus, given K, for a given absorbing wavelength λ. These results are shown in Table 1.

To obtain the relative absorbency at a given wavelength, we need to know what proportion of filaments are generated by sources above a given power. From the radio luminosity function in Fanti and Perola (1977), we find that the fraction of total power radiated by sources with $P > 3\times10^{44}$ is 6.3×10^{-2}, $P > 3\times10^{45}$ is 9×10^{-3}, $P > 3\times10^{46}$ is 8.8×10^{-4}, $P > 3\times10^{47}$ is 5.4×10^{-5}, and $P > 3 \times 10^{48}$ is 1.36×10^{-6}. Interpolating this data, we can obtain the decrease in relative absorbency shown in the last column of Table I.

We thus have obtained from the theoretical model the absorbency as a function of wavelength. It should be noted that after this initial period of "infant mortality" collisional loses should not be important. When the filaments expand to their steady state size we find from (23) that even the densest filaments have collisional lifetimes of over 1400 Gy.

The second part of the theoretical model is a density function which describes the density of absorbers as a function of distance from the observer. In the general filament model, since the individual filaments are essentially perfect absorbers and scatterers, the absorbency is a function of the portion of the sky covered by filaments, which increases with distance from Earth. A hierarchical bundle of filaments should, according to the model developed above have an optical depth of unity, viewed from the outside.

However, a realistic model must take into account the observed inhomogenity of the universe. Observations of galactic distribution indicate that our galaxy is located inside an extensive filament of galaxies, in relatively dense central portion of that filament. Thus we would expect that the concentration of absorbing filaments, proportional to B^2, should decrease with distance from Earth, at least within this broad supercluster of galaxies. Indeed, observations of the absorption of radio radiation form other galaxies indicates that the density of absorbers declines as $1/D$, with D distance form Earth (Lerner, 1990, 1993).

If we are located within a filaments, or more precisely a roughly symmetrical concentration of matter and fields within such a filament, and if we assume that the absorbing filaments are randomly oriented within the much large supercluster filament, then we expect that total optical depth A' will be defined by

$$dA'/dD = 1/3D \text{ or } A' = \ln D/3 \tag{32}$$

This would lead to an expectation that apparent radio luminosity will fall as $D^{-1/3}$, in excellent agreement with observations of $D^{-0.32}$. Thus one part of our density modeled defines the density of absorbers as $0.32/D$.

Presumably this slope must have inner and outer cutoffs. We will put the inner cutoff at 100 kpc. If B^2R is approximately constant within the supercluster filament, then at 100 kpc, B will be around 6 μG, comparable with galactic field strength, so we can expect no higher field in intergalactic space. Observations of absorbency indicate an outer cutoff, Q, of about 50 Mpc, at which point dA/dD appears to level off to a constant value. In examining the model predictions for CBR spectrum and anisotropy, we will examine the effect of varying this cut-off value.

On a larger scale, observations indicate that inhomogenities exist of scales of several hundred Mpc. We therefore crudely model these inhomgenities by assuming that the density of absorbers remains constant with D from Q out to some outer distance C, and then drops to a "universal' value of F/Mpc. We will also examine the effects of varying C and F. To avoid too great a density contrast between the large-scale concentration and the background universe, we somewhat arbitrarily constrain this contrast to be less than 10.

The optical density model must be observationally constrained so as not to produce too much absorption at high redshift. Comparisons of radio quiet to radio loud quasars indicate that absorption at $z = 1$ can be no more than a factor of 3×10^4. Given a total level of absorbency, F can be determined for any given values of Q and C.

The third element of the model defines the spectrum and intensity of radiation from absorbing filaments at a given distance D. This depends on the actual mechanism that creates the Hubble relation in the first place. In the Big Bang formulation, the spectrum of a blackbody radiation at a redshift z is

$$J = 2\pi h\nu^3/c^2(e^{(h\nu(1+z)/kT)} - 1) \tag{33}$$

Here, the resulting spectrum remains a blackbody, and the total flux received drops as $(1+z)^{-4}$. However, other explanations of the Hubble relation produce different formulae. For example, if the Hubble relation is created by a true expansion in space, then the Doppler formula is used which produce the relation

$$J = 2\pi h\nu^3 (1+z)^2 / c^2 \left(e^{(h\nu(1+z)/kT)} - 1\right) \tag{34}$$

Here the redshifted spectrum is no longer a black body and the total flux drops as $(1+z)^{-2}$.

There are two physical processes expected from the model that must be taken into account. Both are a result of the inhomogenity the model assumes. A concentration of filaments with a higher energy and matter density than the surroundings will expand, driven by radiation pressure on the filaments and the surrounding plasma and will cool with time. This means that (33) must be modified. If a body is expanding uniformly, a Doppler shift is introduced in addition to the Hubble shift. If the body is cooling such that $T = T_0 e^{-bt}$, where time $t = (1-z)H_0$, (H_0 is the Hubble time), then

$T = T_p e^{bH_0 z}$ which can be approximated for low z as $T_p(1 + bH_0 z)$ (T_p is present temperature). Thus we have

$$J = 2\pi h\nu^3(1 + az)^2/c^2(e^{[h\nu(1+z)/kT(1+bH_0 z)]} - 1) \tag{35}$$

We can use the theoretical model to determine a and b, given the cutoff values Q and C. To determine the cooling rate b we calculated the radiation rate from a sphere at uniform temperature with the opacity determined from the model, and subtracted the radiation received from the surroundings, assuming the radiation density to be proportional to the modeled density. We then added to the net energy loss rate the temperature reduction resulting from the expansion of the absorbing cloud, to determine a final loss rate (The details are straightforward, so we omit them).

From the energy loss, we can determine the average mass density of the cloud, which is required to find the expansion rate, a. We assume cooling starts from a Hubble time ago, and we thus know the ratio $\varepsilon = e^{4bH_0}$, of the initial energy to the current CBR energy density. If we assume, following Lerner (1988) that the energy of the CBR is derived from thermonuclear processes in an early intermediate mass population of stars (Population III), which created the pregalactic abundance of Helium, we can determine the mass density needed to produce the given energy density.

$$n = 7.59 \times 10^{-15} T^4 e^{4b'}/E_h H_e \xi \ cm^{-3} = 1.8 \times 10^{-7} e^{4b'}/\xi \ cm^{-3} \tag{36}$$

where E_h is the energy per unit mass released in the production of helium, H_e is the mass percentage of the observable matter that is helium, assumed to be 0.22, $b' = bH_0$,and ξ is a factor to account for the additional energy produced by intermediate mass stars in generating the helium and heavy elements trapped within the collapsed cores of such stars. Based on the model developed in Lerner(1988), we estimate $\xi = 3.2$, with slightly less than half the total mass in collapsed cores-neutron stars. Since observation indicate the total mass density in the local universe to be around 3.2×10^{-7}, with about half that as visible matter we expect $4b'$ to be around 1.7 and less than 2.0.

Given n, the expansion rate can be determined, assuming that the radiation pressure equals the difference between the CBR energy density internal and external to radius C. We then get

$$\begin{aligned} a &= 2.6 \times 10^{-5}[1 - (Fq/0.32)]/n(e^{4b'} - 1/4b')(H_t^2/C^2) \\ &= 462(e^{4b'} - 1)(H_t^2/C^2)[1 - (Fq/.32)]/4b'e^{4b'} \end{aligned} \tag{37}$$

where C is in Mpc and H_t is the Hubble time in Gyr.

3 COBE Data

Analysis of the COBE data by the initial researchers has yielded two conclusions: 1) that the spectrum does not deviate from a blackbody by more than

10^{-4} of peak brightness and 2) that there are anisotropies at a scale of 10 degrees of the order of 10^{-5}. For the purposes of this paper we will accept the second conclusion as valid, even though it is still subject to debate. The first conclusion, however warrants a closer look.

The reduced data reported by Mather *et al.* (1993) does not fully support the conclusion that no deviations were detected. In fact, small deviation were detected, outside the calculated experimental errors. However, Mather *et al.* rather arbitrarily increased these statistical errors by a factor of 2.3 to produce an undeviated fit to a blackbody. Since relatively small deviations in the spectrum will be important in testing the present theoretical model, a re-analysis of the data is presented here.

In Table II, I represent the data from Table I of Mather*et al.* Column I is the frequency in units of kT/h, where T is the background temperature, 2.726K. Column 2 is the ΔT at that frequency, based on the "residuals" in Mather*et al.* Table I. Column 3 is the statistical errors for the residual, without the fudge factor of 2.3 included in the original table. To smooth out the data, every three points are averaged into a single ΔT with a single error in Columns 4 and 5 (The point at 5.66 is omitted because of its very large error bars).

We can obtain an estimate of the actual statistical errors, without making unjustified assumptions, by using the deviations of individual points from the smoothed curve defined by the three-point averages. Using the χ^2 test, we estimate the "true" errors to be 1.5 times the calculated errors, much smaller than the "fudged errors" of Mather *et al.* The "true" errors are given in column 6.

In this new analysis, all the points in the spectrum are compatible with a perfect Planck spectrum, except the first point, which is more than 4.7 s too high. This clearly is a definite deviation. The overall χ^2 is 28.74 which has less than a 0.005 chance of being obtained randomly.

4 Results-Predicted Spectrum

To determine the prediction of the theoretical model for the CBR spectrum, we simply integrate over the hypothesized density distribution, redshifting the radiation. Thus the basic formula is

$$\begin{aligned} J = &[\textstyle\int_0^{.1} 10s(e^{-10sr})(1+aHr)^2\nu^3(e^{\nu(1+(1-b)Hr)}-1)^{-1}dr] \\ &+[0.1^s e^{-s} s \textstyle\int_{.1}^{Q} r^{-1-s}(1+aHr)^2\nu^3(e^{\nu(1+(1-b)Hr)}-1)^{-1}dr] \\ &+[(10Q)^{-s} s \textstyle\int_Q^c Q^{-1}e^{-sr/Q}(1+aHr)^2\nu^3(e^{\nu(1+(1-b)Hr)}-1)^{-1}dr] \\ &+[(10Q)^{-s}e^{c(d-s/Q)} \textstyle\int_c^\infty Fe^{-Fr}(1+aHr)^2\nu^3(e^{\nu(1+(1-b)Hr)}-1)^{-1}dr] \end{aligned} \qquad (38)$$

Here, the parameters s and F define the opacity inside and outside the cutoff radius C, while Q defines the radius at which the density is assumed to stop decreasing. H is the Hubble constant, here defined as $H = z/r$, and

TABLE II
COBE Spectrum data.

Frequency (h/kT)	$\Delta T \times 10^4$	error[1]	3 point avg	3 point error	corr. error
1.195	1.75	.38			
1.489	1.27	.41	1.57	.22	.33
1.789	1.68	.34			
2.084	.40	.24			
2.384	.07	.21	-.08	.12	.18
2.684	-.70	.16			
2.98	.11	.14			
3.28	-.13	.10	-.06	.07	.10
3.58	-.16	.10			
3.87	-.10	.09			
4.17	-.05	.08	0.0	.05	.08
4.47	.16	.08			
4.77	-.06	.10			
5.07	.20	.12	.09	.07	.10
5.36	.12	.14			
5.96	-.33	.24			
6.11			.19	.15	.22
6.26	.70	.25			
6.56	.09	.27			
6.86	.83	.29	.39	.17	.25
7.15	.24	.31			
7.45	.27	.32			
7.75	1.17	.37	.36	.22	.33
8.05	-.36	.47			
8.35	.84	.59			
8.64	-2.11	.80	-.13	.48	.72
8.94	.87	1.07			
9.24	-.17	1.33			
9.54	-2.43	1.67	-.87	1.0	1.5
9.84	0	2.00			

[1] in $\Delta T \times 10^4$

a is a parameter determining the Doppler expansion of the emitting cloud. Frequency, ν is measured here in units of kT/h.

What does the spectrum look like for the simplest model, with just Hubble expansion, no Doppler expansion or cooling and with an absorbency independent of frequency? We initially take $s = 0.32$, $Q = 50$ Mpc, $C = 600$

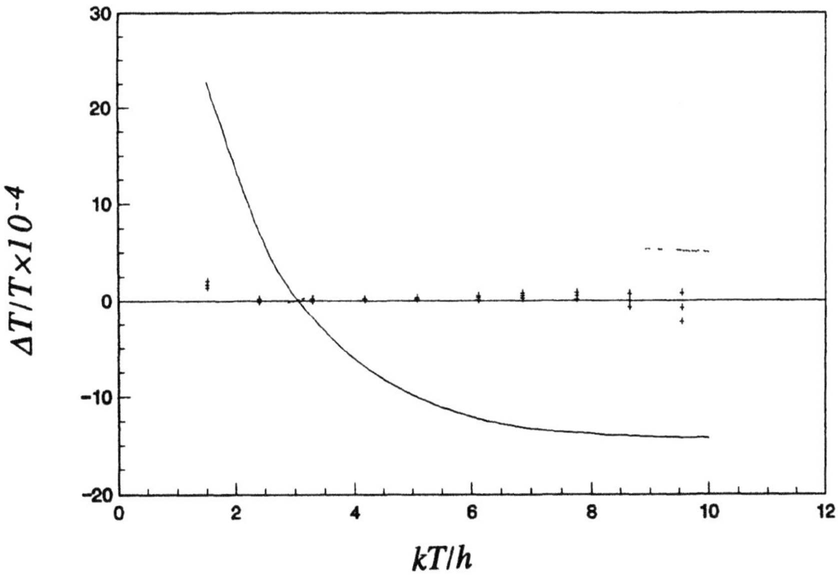

Fig. 1. Predicted $\Delta T/T \times 10^{-4}$ is plotted against frequency, in units of kT/h. Dashed line assumes pure Hubble expansion, dotted line pure Hubble expansion with opacity function predicted by the model, and solid line pure Doppler expansion. Crosses show COBE observations.

Mpc. We find the total absorption to $z = 1$ in this model is

$$A' = (10Q)^s e^{s(C/Q)+F((1/H)-C)} \tag{39}$$

which we set equal to 30,000 for this choice of parameters, giving $F = 0.0013$. We then calculate the spectrum shown as the dashed line in Figure 1. This clearly is outside the limits set by COBE. Similar results are obtained with other Q and C.

In the specific absorber model proposed here, opacity declines with decreasing wavelength, partially counteracting the rise in the dashed line . We find that a decrease in opacity by 10^{-3}, for example, reduces J by 10^{-5}. By including the opacity changes determined from Table I, the dotted line results. This is still a poor fit. So, a model without expansion or cooling will not work.

A pure Doppler expansion ,with $a = 1$, is hypothesized in some cosmologies, such as the Alfvén–Klein (1962) model. A pure Doppler expansion in space would produce rapid cooling due to expansion, since energy density would decrease as t^3 and temperature would decrease as $t^{0.75}$. This has the effect of reducing the effective H by a factor of 4 and produces the solid line in Fig.1, an even worse fit. Here, introduction of changes in opacity will only worsen the fit. It thus seems that a pure Doppler expansion model cannot yield a good correspondence to the COBE spectrum.

However, with the full model proposed here, the situation is quite different. Assuming an underlying redshift determined by the conventional Hubble relation, with the cooling and expansion determined above, all the parameters of the model can be determined by setting only three free parameters-the cutoff distance Q and C and the overall opacity level A'. From these, with s fixed observationally at 0.32, we can determine F from (39), b from the cooling model described above, and a from (37).

We initially set $A' = 30,000$, the maximum value observationally allowed and search the Q, C parameter space for the best fit with the observed spectrum. By requiring that $b < 2.0$, we limit $C > 600$ and by limiting the density contrast between the central mass concentration and the surrounding universe to < 10, we limit $C < 1000$ for $Q = 50$ and < 1300 for $Q = 70$.

Within this allowed parameter space, we find the best fit at approximately $Q = 60$, $C = 700$. This excellent fit is shown in Fig. 2 (solid line) and has a $\xi^2 = 5.06$ for 10 degrees of freedom. By contrast, the Big Bang prediction of a perfect black body has a $\xi^2 = 28.74$. Good fits, with $\xi^2 < 16$ are also found at $(Q, C) = 50, 800(\xi^2 = 9.57)$; $(Q, C) = 50, 850(\xi^2 = 6.96)$; $(Q, C) = 50, 900(\xi^2 = 10.68)$; $(Q, C) = 60, 650(\xi^2 = 12.05)$; $(Q, C) = 60, 750(\xi^2 = 12.67)$; and $(Q, C) = 70, 600(\xi^2 = 7.50)$.

Decreasing A', the total opacity, has a small effect on the results. Thus with $A' = 6,000, Q = 60, \xi = 700$, the dashed line in Fig 2 results with $\xi^2 = 5.52$.

This is an important result. *With no adjustable parameters other than the dimensions of the local matter concentrations, the observed COBE CBR spectrum is excellently fit by a model that assumes the CBR is produced by an intergalactic array of absorbing and emitting magnetic plasma filaments, which obscures distant radio sources by no more than a factor of several thousand, and fits all other observational constraints.*

At frequencies below the range studied by COBE, the model predictions remain well within the observational errors of other experiments. At 0.1 kT/h or 5 cm wavelength the model spectrum is 6.7×10^{-4} above a blackbody, at 50 cm the deviation is 8.2×10^{-4} and at 5 m the deviation is 1.3×10^{-3}.

If a flat opacity function is assumed, rather than the specific curve derived in this model, but the model is otherwise unchanged, the fit becomes unacceptable for all Q and C. For example, the dotted line in Figure 2 shows the prediction for a flat opacity function at 60, 650. The ξ^2 is 43, clearly too high. Thus the modeling indicates that not only does an absorbent IGM model fit the COBE results, but that only one with the opacity vs frequency

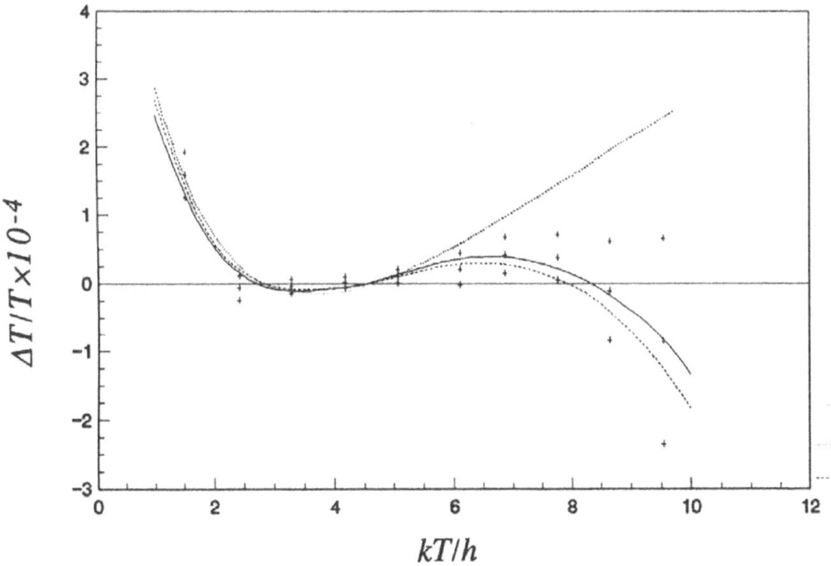

Fig. 2. Predicted $\Delta T/T \times 10^{-4}$ is plotted against frequency, in units of kT/h. Model prediction, including expansion and cooling of large scale concentration. Solid line shows Q= 60 Mpc, C= 700 Mpc, total opacity is 30,000. Dashed line same Q, C, but with total opacity 6,000. Dotted line assumes flat opacity. Crosses show COBE observations.

function derived from the magnetized filament theory provides an adequate fit. This is strong, if indirect, evidence for the validity of this particular mechanism for CBR absorption and scattering.

It is important to test the sensitivity of these results to changes in the model parameters. ξ, which determines the predicted local density of matter and is inversely proportional to the expansion parameter a, can be varied by 10% without altering the spectral predictions. The best fitting value for C decreases by about 14% for a 10% decrease in ξ. A decrease of ξ by 25% to 2.4 yields a worse but still acceptable fit with $\xi^2 < 16$. This is as low as ξ is likely to be. A 25% increase in ξ is also acceptable, although such a high value is quite implausible.

The parameter determining the rate of cooling, b', can also be varied. Indeed more sophisticated models of cooling, taking into account temperature variations, are likely to somewhat decrease b'. Again, we take 25% as the maximum likely change in b'. In the case of Q, C =60,700 with total opacity =6,000, a decrease in b' by 25% will also decrease density by 35%

and increase a by 18%. The optimal C will shift down to 650 Mpc and ξ^2 will still be a very acceptable 9.1.

A third variation in the model is to change the sharp cutoff in density at distance C to a $1/r$ slope. Again, the shape of the spectrum is basically unchanged for $Q = 60$, inner edge of slope at 260 Mpc, outer edge at 940 Mpc.

5 Results-Predicted Anisotropy

The anisotropy of the CBR predicted by this model will be affected by a number of phenomena: the location of our galaxy relative to the center of the mass concentration (the region within the cutoff C), its motion, deviations of the concentration from sphericity and density and temperature variations with the concentration. Since our galaxy will participate in the expansion of the concentration, its velocity relative to the center of motion of the concentration will be aHD, where D is the distance between our galaxy and the center of the concentration. Such a velocity will create a dipole distortion of the CBR. Setting this velocity equal to the observed 600km/sec dipole in the CBR yield D= 80 Mpc, so our galaxy would be in the central 12% by distance of the concentration. Of course, if our galaxy has an additional velocity component relative to the expansion of the concentration, this distance could be greater or less than that calculated.

While the COBE team has not presented a detailed analysis of the limits on the anisotropies of the CBR spectrum, we can ask what changes in the spectrum would be expected in different parts of the sky is the galaxy were 80 Mpc from the central concentration. Using the model parameters connected with line 2 of Fig. 2. (Q, C = 60,700, opacity =6,000), we find that the spectra in the direction of the dipole maximum and minimum are both good fits to the observed spectrum. The variation in T due to this variation in spectrum is only 4% that due to the Doppler dipole so is quite insignificant.

Deviations from sphericity of the IGM concentration will result in deviations in T that would have been observed by COBE. The observed rms anisotropy of 10^{-5} would be produced by sinusoidal deviations from sphericity with a deviation of 20 Mpc or 2.8% of the radius of the concentration, C. Such a relatively good sphere is to be expected for a plasma body that was initially gravitationally bound, and then subject to a uniform expansion pressure that has increased its radius by about 5%. For comparison, the Sun is spherical to within 0.1%.

Similar limits of about 18% can be placed on deviations from sphericity of the local concentration, within the inner cutoff radius Q. Thus the major axis of the inner concentration could be as much as 43% more than the minor axis without exceeding COBE anisotropy limits. The galaxy can be located anywhere within the inner 30% of this local concentration without affecting the observed spectrum of the CBR. Dipole contributions from the galaxy's location would simply affect the dipole anisotropy direction, but would have an insignificant effect on its magnitude.

Another contribution to anisotropy is from inhomogenities in the local density of magnetic filaments, which is assumed to be proportional to the local magnetic energy density. An accurate model of this would require a detailed knowledge of magnetic field strength throughout the local universe, which is not available. However, we can make good estimates of the maximum anisotropies to be expected from this source.

Magnetic field enhancements are to be expected along the roughly cylindrical filaments that contain large concentrations of galaxies. From galactic atlases, such as Tully and Fisher (1987) it can be determined that the angular radii of such filament on the sky does not exceed 0.1 radians. Thus the concentrations are modeled as having physical radii of 0.1 D, where D is the distance to the concentration. The magnetic energy density within the concentration is assumed to be 50% more than the average density for distance D. The concentrations are assumed to be surrounded by compensating depletions of magnetic energy so that the average level remains unchanged.

For concentrations for 20 to 60 Mpc distant, the increase in physical size, from 2 to 6 Mpc radius is compensated by the decrease in average density. At these densities the maximum deviations of CBR temperature T (not average rms deviations) is 3.2×10^{-5}. At larger distances, the large physical size of the concentrations is partially compensated by the increased absorption between the concentrations and the observer. The maximum ΔT at 120 Mpc is 5.3×10^{-5}, at 300 Mpc 5.6×10^{-5} and at 600 Mpc (for concentrations 60 Mpc in radius) 4.6×10^{-5}. These are the maxima that would be observed with a 4 degree diameter beam. A 7 degree beam, like that on COBE, would see ΔT 0.75 times those above.

The ΔT calculated here represent deviations of less than 2.6 σ for the COBE 7 degree anisotropy of 44 mK. For the Tenerife observations at 4 degrees (Hancock *et al.*, 1994), these deviations also represent less than 2.6 σ compared with a measured anisotropy of 2.1×10^{-5}. Nearby concentrations, with 60 Mpc of earth are clearly not a problem, having peak ΔT of only 1.5 σ.

The total rms anisotropy predicted by the model depends on the rate of occurrence of very large concentrations in space, which is not clear from observation. However, the QDOT survey (Saunders*et al.*, 1991) found seven concentrations of approximately 25 Mpc radius within 300 Mpc of earth. Even three times this many concentrations would not create an overall distribution that would fall outside the anisotropy limits set by the COBE or Tenerife observations.

A final source of anisotropy is variations in the intrinsic temperature of the various parts of the large scale concentration within cutoff distance C

(around 700 Mpc). Such temperature variations would be expected only over large physical distances, since equilibriation would be quite efficient within the concentration during the time since the generation of the CBR energy. Indeed even for the largest subconcentrations with radius 60 Mpc, temperature deviations would be of the order of e^{-10} or 5×10^{-5}. However, the effect of absorption is such that even much large temperature variations will not create large anisotropies. For example, for a concentrations with radius 60 Mpc at a distance of 600 Mpc, a 1% temperature difference, or 4% radiation energy difference, will only produce an observe ΔT of 7.7×10^{-6}. *Thus the absorptive model does not produce anisotropies that contradict the COBE or other CBR observations.*

6 Conclusions

A model that hypothesizes the isotropization of the CBR by magnetized plasma filament in the IGM thus provides a better fit to the spectrum of the CBR than Big Bang models and fits as well the small anisotropy observed in the CBR. It does not require a total absorption of radio radiation form high z objects that is beyond that allowed by observation. Indeed, observations of radio to optical flux ratios among quasars (Kellerman*et al.*, 1989) indicate a drop of relative radio luminosity by more than an order of magnitude between a z of 0.25 and 1.25, completely compatible with model predictions.

The model does require a very large scale inhomgenity about 700 Mpc in radius and with a mass of about 8×10^{18} solar masses. This is an extremely large concentration. It would have a gravitational orbital velocity at its outer surface of 7,200 km/sec, considerably larger than that of any observed mass concentration in the universe, which tend to have characteristic gravitational velocities of at most 2,000 km/sec.

Could such a large structure form? We would expect that for a body of plasma to contract gravitationally, it must be collisional. As shown in Lerner (1986) a body that is collisional with ion velocities equal to the gravationally orbital velocities, it must be less than a certain mass. If we define collsional as having a mean free path of $2\pi R$, where R is the radius of a spherical body, the maximum mass turns out to be about 1.5×10^{18} solar masses, 5.6 times smaller than the postulated large scale concentration of our model.

However, this may not be as much of a problem as it appears, since if the effective thermal velocities of ions are only 2/3 the gravitational orbital velocity, or thermal energy is about one half gravitational binding energy, the entire mass would still be collisional. This issue can only be resolved with more detailed modeling or possibly simulation.

In any case, the calculations in this paper clearly show that the Big Bang is not by any means the only theory compatible with current observations of the CBR.

References

Alfvén, H. and Klein, O.: 1962, *Arkiv fur Fysik*Vol. no. **23**, pp. 187
Becker-Szend, R.: 1990, *Phys. Rev. D.*Vol. no. **42**, pp. 2974
Bennett, C.L. *et al.*: 1994, *Astrophys. J.*, submitted
Fanti,R. and Perola, G.C.: 1977, *Radio Astronomy and Cosmology,* edited by D. L. Jauncey, D. Reidel, Dordrecht
Hancock, S., *et al.*: 1994, *Nature*Vol. no. **367**, pp. 333
Jackson, J.D.: 1963, *Classical Electrodynamics*, Wiley, New York, p.437
Kellerman, K.I. *et al.*: 1989, *Astrophys. J.*Vol. no. **98**, pp. 1195
Kim, R.,*et al.*: 1989, *Nature*Vol. no. **341**, pp. 720
Lerner, E.J.: 1986, *IEEE Trans. Plasma Sci.*Vol. no. **14**, pp. 609
Lerner, E.J.: 1988, *Laser and Particle Beams*Vol. no. **6**, pp. 457
Lerner, E.J.: 1990, *Astrophysical J.*Vol. no. **361**, pp. 63
Lerner, E.J.: 1992, *IEEE Trans. Plasma Sci.*Vol. no. **20**, pp. 935
Lerner, E.J.: 1993, *Atrophys. and Space Sci.*Vol. no. **207**, pp. 17
Mather, J.C., *et al.*: 1993, *Astrophys. J.*
Olive, K. *et al.*: 1990, *Phys. Lett B.*Vol. no. **236**, pp. 454
Peter, W. and Peratt, A. L.: 1990, *IEEE Trans. Plasma Sci.*Vol. no. **18**, pp. 49
de Rutier, H.R., *et al.*: 1990, *Astronomy and Astrophys*Vol. no. **227**, pp. 351
Saunders *et al.*: 1991, *Nature*Vol. no. **349**, pp. 32
Tully, R. B. and Fisher, J. R.: 1987, *Atlas of Nearby Galaxies*, Cambridge University Press
Walker, R.C., Benson, J.M., and Unwin, S.C.: 1987, *Astrophys. J.*Vol. no. **316**, pp. 546

A WINDOW ON THE PLASMA UNIVERSE: THE VERY LARGE ARRAY

KEVIN R. HEALY
Very Large Array Operations Center
National Radio Astronomy Observatory
Socorro, New Mexico, USA

Abstract. For nearly two decades, the aperature synthesis telescope known as the Very Large Array (VLA) has been providing a wealth of data on all types of astronomical objects. This telescope fills the need for more powerful and more sensitive instruments to gather data on a vast range of plasma phenomena in astrophysical settings. Currently the VLA is capable of observing at eight radio frequency bands (four corresponding to important spectral lines) with sensitivities on the order of $10^{-29} W m^{-2} Hz^{-1}$, or 10^{-3} Jansky (1 mJy), and noise levels ten times lower. This paper describes the VLA and its capabilities, and describes how researchers interested in plasma astrophysics can use the instrument.

Key words: Instrumentation, Radio Astronomy, Plasma Astrophysics

1 Introduction

The Very Large Array is a 27–element radio interferometric array on the Plains of San Agustin, sixty miles west of Socorro, New Mexico. Construction of the VLA began in April 1974, with the first astronomical research in May 1977. Construction was completed in 1980, on time and nearly on budget. The instrument was dedicated on October 10, 1980, with all twenty–eight antennas operational. By early 1981, regular scheduling of observations began and the current operating routine was established (NRAO Staff, 1983). While twenty–seven antennas make up the full Array, the twenty–eighth allows each antenna to be cycled out of the active Array for regular preventative maintenance.

2 Array Design

The twenty-seven antennas of the VLA are arranged in a wye, with nine antennas in each arm. The arms are known as North, West and East and are seperated by 120° (see Figure 1).

There are four standard configurations of the VLA. These configurations all have the same antenna pattern, but are scaled in size by the ratios 1:3.2:10:32. The distance from the center of the Array to the nth antenna is $R \approx 13.65(n \times m)^{1.716}$ meters for the east and west arms and $R \approx 12.31(n \times m)^{1.716}$ meters for the north arm, where m = 8, 4, 2, and 1 for the A, B, C, and D configuration, respectively. Thus, the A configuration is the largest and provides the highest angular resolution, while the D configura-

Astrophysics and Space Science **227**: 83–91, 1995.

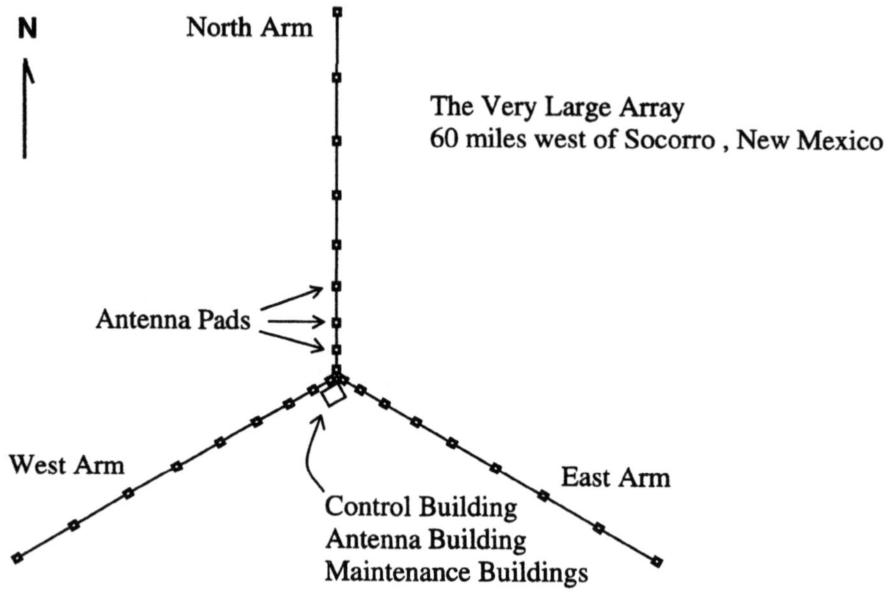

Fig. 1. The Very Large Array

tion is the most compact and has the lowest angular resolution. These four configurations have maximum baselines of 36.4, 11.4, 3.4, and 1.03 km, and minimum baselines of 0.68, 0.21, 0.073, and 0.035 km (NRAO Staff, 1983).

In addition to the standard configurations, the VLA also has three hybrid configurations. These hybrid configurations are a combination of the north arm of one configuration and the east and west arms of the next, smaller configuration. These hybrids are known as BnA, CnB, and DnC, and are used to overcome the apparent foreshortening of the north–south baselines when observing objects south of δ = -15° and north of δ = +75° (NRAO Staff, 1983).

The VLA is in each standard configuration for one "term", each lasting approximately four months with the hybrid configurations lasting about two weeks during move–time at the beginning and end of the terms. The regular configuration schedule is A $\rightarrow$ BnA $\rightarrow$ B $\rightarrow$ CnB $\rightarrow$ C $\rightarrow$ DnC $\rightarrow$ D $\rightarrow$ A. The antennas are moved between stations, or "pads", with the use of a dual-track railroad system and a 200-ton antenna transporter that travels at a modest four miles per hour. Short moves can take only a few hours, while moves over most of the length of an arm can take an entire day (NRAO Staff,

1983).

By switching between configurations in this way, the VLA can observe the sky with different angular resolutions. In A array, observing at $\lambda = 0.7$ cm, the sythesized beam–width of the Array is 0.05 seconds of arc (50 milliarcsec), while in D array, observing at $\lambda = 90$ cm, the synthesized beam–width is 200 seconds of arc. The overall field of view, however, is limited by the response of the individual antennas. The full width at half power of the antenna beam, in arcminutes, is $\theta_{PB} \approx 45/\nu_{GHz}$. Objects larger than about half of this angle cannot be observed directly by the Array. However, "mosaicing" may be used to construct a larger field (R. A. Perley, 1994).

3 Antenna Design

Each VLA antenna is a 25–meter diameter paraboloid dish on an alt–azimuth mounting. Table One summarizes the physical parameters of the VLA antennas.

TABLE I

VLA Antenna Parameters

main reflector diameter	25 m (82 feet)
antenna half–power beamwidth	~1.5 λ (λ in cm)
antenna beamwidth between first nulls	~3.6 λ
total geometric aperture	491 m^2
focal length of main reflector	9 m
total rms surface accuracy	<0.70 mm
total weight of antenna	190,455 kg (419,000 pounds)
slew rate, azimuth	40° per minute
slew rate, elevation	20° per minute
minimum elevation	8°
maximum elevation	125°
minimum zenith angle for tracking	0.5°
azimuth limits relative to track azimuth	±270°

All twenty–eight antennas are equipped with six receivers ($\lambda = 1.3$ cm, 2 cm, 3.6 cm, 6 cm, 20 cm, and 90 cm, known as K, U, X, C, L, and P band, respectively), ten of the antennas are currently equipped with a seventh receiver ($\lambda = 0.7$ cm, known as Q band), and eight of the antennas are equipped with an additional receiver ($\lambda = 400$ cm, known as 4 band). For the duration of the all–sky surveys, the 400 cm receivers have been temporarily removed.

Table Two summarizes the nominal frequency ranges of the oldest VLA bands.

TABLE II
Sensitivity ranges of VLA bands

Band	0.9 x Nominal	0.5 x Nominal	Extreme Range
90 cm	305 - 335 MHz	298 - 345 MHz	295 - 350 MHz
20 cm	1320 - 1700 MHz	1250 - 1800 MHz	1225 - 1875 MHz
6 cm	4500 - 5000 MHz	4250 - 5100 MHz	4200 - 5100 MHz
3.6 cm	8080 - 8750 MHz	7550 - 9050 MHz	6800 - 9600 MHz
2 cm	14650 - 15325MHz	14250 - 15700 MHz	13500 - 16300 MHz
1.3 cm	22000 - 24000MHz	21700 - 24500 MHz	20800 - 25800 MHz

All of the VLA antennas are designed to simultaneously collect signal at two frequencies, or "IF's". This can be done within a single band as long as the IF's are no more than about 450MHz apart, or two bands can be used. Currently the only supported, simultaneous use of two bands is limited to L and P bands, and P and 4 bands. (Others are possible, but not supported.) Multi–frequency observations with other bands require each object to be observed separately at each band by time–switching the receivers. Receiver tuning takes less than 30 seconds, so this procedure is very time–efficient. Observations using both IF's in the same band can be used to augment the widest 50 MHz bandwidth of the system, while multi–frequency observations can be used to provide spectral index or Faraday rotation information with a single observation (R. A. Perley, 1994).

After the radio radiation has been collected by the receivers a great deal of processing is done within the antenna before the signals are sent down the waveguides to the correlator. After being collected by the feed horn, the signal is split into right–hand (RCP) and left–hand (LCP) circularly polarized signals and then these are split into two nominally identical signals. These four "IF channels" are designated A (IF 1, RCP), B (IF 2, RCP), C (IF 1, LCP), and D (IF 2, LCP). Both channels of IF 1 are known as the AC "side", while both channels of IF 2 are known as the BD "side". It is these two "sides" that can be filtered to produce two passbands of the desired bandwidth and center frequency (R. A. Perley, 1994).

After the IF signals have been generated from the original signal, the IF channels are phase shifted to introduce the fine time delay required for each antenna in the Array. (Coarse delay is added by systems in the Control Building just before the signal enters the correlator.) Each IF signal is then converted to a 1025MHz signal and filtered to the desired bandwidth for the observation. The VLA can support seven standard bandwidths, $\Delta\nu = 50/2^n$ MHz, where n = 0, 1, ..., 6. An eighth bandwidth, 200 KHz, is available, but observers wishing to use it should contact VLA staff first (NRAO Staff, 1983).

After bandwidth filtering, the four IF channels are shifted in frequency again and combined with timing and monitor information for transmission down the waveguides to the Control Building.

4 Correlator Design

Once the signals have arrived at the correlator, they can be handled in a number of ways. First, the VLA correlator can process data in one of two modes: Continuum or Spectral Line. These modes are designed for observing different kinds of phenomena. Second, real–time phase corrections can be made to the data to "phase up" the Array for observation of high–time–resolution phenomena. Third, the twenty–seven antennas can be split into as many as five sub–arrays, each observing a different object at a different band.

In addition, the VLA can obtain polarization data on a source. Full polarization information can always be obtained in Continuum mode, while only circular polarization information is obtained in Spectral Line mode unless special modes are used. Polarization calibration is usually better than 0.5%, but instrumental effects add errors larger than 1% at angles greater than 10% of the primary beam width from the pointing center. Instrumental polarization effects also limit the believable fractional polarization to ~0.15%. (R. A. Perley, 1994).

In Continuum mode, the entire bandwidth of the observation is averaged together to produce a single map at a specific frequency. Novae, supernovae, and their remnants, quasars, pulsars, ionized gas clouds, and other sources of continuum emission (*i.e.* sources of thermal radiation, bremsstrahlung sources, and sychrotron sources) are usually observed in this mode. However, faint spectral–line emission is also observed in this mode to improve signal–to–noise.

In Spectral Line mode, the bandwidth of the observation is centered on the rest, or in many cases the doppler-shifted, frequency of a known spectral line (*i.e.* atomic hydrogen, ammonia, carbon monoxide, recombination lines, etc.). This bandwidth is divided into 2 to 512 "channels", where the bandwidth of each channel has a corresponding velocity–width in the moving gas. The data in each channel is then made into an image and the images are collectively called a "cube", with dimensions of right ascension, declination, and velocity (frequency). Gas–rich galaxies (*i.e.* spirals and irregulars), and atomic and molecular galactic clouds are typically observed in this mode.

However, P–band observations sometimes experience terrestrial interference and Spectral Line mode can be used to selectively remove contaminated channels.

In addition to correlator mode, signal integration time can be selected on the basis of Array configuration and observing mode. The standard integration time is 30 seconds, while the minimum varies from 1 1/3 seconds in Continuum mode to 20 seconds for 512 channel Spectral Line mode. An integration time of 0.4 seconds can also be used for solar observations (R. A. Perley, 1994).

5 VLA Data Acquisition

Those wishing to use the VLA in their research should submit an observing proposal that includes a scientific justification. Proposals are accepted year–round, but the deadline for a specific configuration is the 1st of February, June, or October of the preceding four month term. Table Three shows the approximate long–term schedule for the VLA.

TABLE III
Approximate Long–Term Schedule

	Quarter 1	Quarter 2	Quarter 3	Quarter 4
1995	D	D,A	A,B	B
1996	C	D	D,A	A,B
1997	B	C	D	D,A
1998	A,B	B	C	D
1999	D,A	A,B	B	C

Proposal applications are available by mail through Lori Appel. Applications are also available through anonymous ftp at NRAO (see section 9). In the subdirectory pub/vlacover are three files: coverl.tex (the front side of the application), cover2.tex (the back side), and logo.ps (the NRAO logo). Copy these files and follow the instructions in them. Those with World Wide Web access can also get the files from the NRAO homepage. Select "Very Large Array", and select the first item, "instructions".

It is now encouraged that proposals be submitted by e–mail. Simply concatenate the completed cover.tex files, text files, and figure files together, and e-mail them to *propsoc@nrao.edu.* This file must be printable on the NRAO system. Successful printing will be acknowledged, so problems will be reported and can be fixed if the file is received at least 5 days before the deadline. Proposals can also be mailed to Paul A. Vanden Bout, Director NRAO, Edgemont Road, Charlottesville, VA, 22903-2475 (R. A. Perley, 1994).

Proposals are reviewed by referees external to NRAO who are expert in the relevant field (*i.e.* galactic, extragalactic, etc.). Their comments are passed on to the VLA scheduling committee. Comments from both groups and a decision from the scheduling committee are passed on to the proposer. Competition between highly rated proposals is high, especially for LST ranges corresponding to certain areas of the sky, like the Galactic Center and the Virgo Cluster (R. A. Perley, 1994).

The VLA all-sky surveys currently affects scheduling. The D–Array survey is running only at night, so observing time between 13:00 to 01:00 LST during Summer 1996 will be in short supply. The B–Array survey will take up much of the time between 07:00 and 17:30 LST for late 1995 and early 1997 (R. A. Perley, 1994).

Once observing time has been granted, an "observe file" must be written. This file includes all parameters for the observation and a schedule of the sources to be observed. In addition to the target source(s), observations of flux and phase calibrators are required for post-observation calibration of the data. Observations including polarization require additional observations of a polarization calibrator.

The observe file should be e–mailed to the Data Analysts at NRAO, who will check the file for problems and forward it to the site. For straightforward observations, nothing further is needed. For observations involving special equipment (*i.e.* the high time–resolution processor), the observer must be present to set up and operate the auxiliary systems.

The software package *observe* is used to construct the observe file. SUN4, SUN5, and IBM versions are available by anonymous ftp at NRAO from the directory pub/observe.

6 VLA Data Analysis

After observations are complete, the data can be processed at the observer's home institution or at the Array Operations Center in Socorro. Travel to Socorro is strongly recommended for first–time observers, however, as computer facilities are available for use and staff are available for advice and discussion.

The *A*stronomical *I*mage *P*rocessing *S*ystem (AIPS) is the most commonly used software package for handling VLA data. Specifically written to process VLA data, AIPS not only includes software to load and copy VLA datasets, remove corrupted data, trasform the data into images and clean the images, but also includes programs for analyzing the data. Specific information on AIPS can be found in the AIPS Cookbook, available upon request from

Theresa McBride at NRAO, or in the VLA section of the NRAO homepage.

Those wishing to process their data at home can request a copy of the data. Initial data storage at the VLA is made on exabyte tape in VLA archive format, but conversion of the data to FITS format is possible. The Data Analysts can mail a tape (standard 1/2 inch, Exabyte or DAT) or provide the data on disk for ftp retrieval.

7 Conclusion

With the continuing emergence of serious difficulties in the "standard models" of astrophysics, emphasis shifts to the agreement between theory and observation. With the rise of the importance of plasma physics in the description of many astrophysical systems, the VLA is a perfect instrument to provide the observational support for laboratory, simulation, and theoretical work in plasma astrophysics. Its unprecedented flexibility and sensitivity provides a wealth of information on any radio emitting region of the universe.

8 Documentation

Further information on VLA design, observing techniques, observe file preparation, and data reduction and analysis is available in the following publications and manuals. Those marked by an "r" can be requested from Theresa McBride at NRAO by mail or e–mail. Those documents marked by a "w" are available through the VLA section of the NRAO homepage on the World Wide Web.

1. PROCEEDINGS FROM THE 1988 SYNTHESIS IMAGING WORKSHOP (w): This collection of lectures detailing synthesis imaging with a focus on the VLA was published by the Astronomical Society of the Pacific as Volume 6 of their Conference series. It is currently out of print.

2. A SHORT GUIDE FOR VLA SPECTRAL LINE OBSERVERS (r,w): This manual provides fine detail and a general overview of the use of the VLA for spectral line observations.

3. THE OBSERVE MANUAL (r,w): This manual provides instruction in the use of the *observe* program, and provides general information into the use of the VLA.

9 Contact with NRAO

The following items provide information auxiliary to that provided in the text.

- *ftp.aoc.nrao.edu* is the NRAO anonymous ftp site. Use "anonymous" as your login and use your e–mail address as the password.

- The NRAO homepage on the World Wide Web is at http://info.aoc.nrao.edu.
- Lori Appel, Theresa McBride, and the Data Analysts can be reached by mail at the NRAO Array Operations Center, P.O. Box O, Socorro, NM 87801.
- Lori Appel can be reached by e–mail at lappel@aoc.nrao.edu. She can be reached by phone at (505) 835–7310. Theresa McBride can be reached by e–mail at tmcbride@aoc.nrao.edu. She can be reached by phone at (505) 835–7245.
- The NRAO Data Analysts can be reached by e–mail at analysts@aoc.nrao.edu. They can also be reached by phone at (505) 835–7238, –7239, and –7359.

The National Radio Astronomy Observatory is operated by Associated Universities, Inc., under cooperative agreement with the National Science Foundation.

References

NRAO staff, edited by Robert Hjellming: 1983, *An Introduction to the NRAO Very Large Array*, Internal NRAO Publication

R. A. Perley: 1994, *Very Large Array Observational Status Summary, June 6, 1994 edition*, Internal NRAO Publication

INTERGALACTIC PLASMA

GROTE REBER
Bothwell
Tasmania Australia 7030

Abstract. A 144 meter antenna array located west of Ottawa, Canada has been used to survey the sky. The north galactic pole has the same intensity as the south galactic pole. We observe a 30% absorption drop at the plane of the Milky Way north of Cygnus at 45° celestial latitude.

Key words: Radio Astronomy, 144 meters, Milky Way, Galactic Haloes

1 Introduction

Quite some years ago a map of the southern sky at 144 meters wavelength was published (Reber, 1968). This is reproduced as Figure 1. The contours are in db above one microvolt at the receiver input terminals. At that time the intergalactic medium was deemed to consist of 10^4 electron/proton pairs per cubic meter. Presumably it was uniform and homogeneous throughout intergalactic space. Later considerations suggest this is too simplified a concept.

A more likely concept is that each galaxy is surrounded by a halo of plasma. Such halo has no well-defined edge. The density merely decreases inversely with distance from center of galaxy. Figure 1 was examined for telltale markings of this hypothesis. The simple theory, when applied to the earths position in the galaxy, predicts a maximum intensity from the galactic plasma at the galactic pole. However, the intensity at galactic pole is about 0.5 db, or 12.5% less than at galactic latitude 35 directly below the pole on meridian of galactic center. Below this the intensity drops rapidly to about 3% at the galactic center. This effect, at 144 meters, is caused by the absorption of the rf by ionized interstellar hydrogen at the center of our Milky Way.

2 Galactic Halo

Based on this interpretation of the measured data, Figure 2 has been constructed. This figure portrays a halo around our Milky Way centered on the center of Milky Way. The semicircle shown represents the half-density distance. This distance to the direction of the galactic pole as seen from earth is 0.5 db, or 12.5% less than the similar distance at 35 galactic latitude. If we take a value of 30,000 ly as the distance from the earth to the center of Milky

Astrophysics and Space Science **227**: 93–96, 1995.

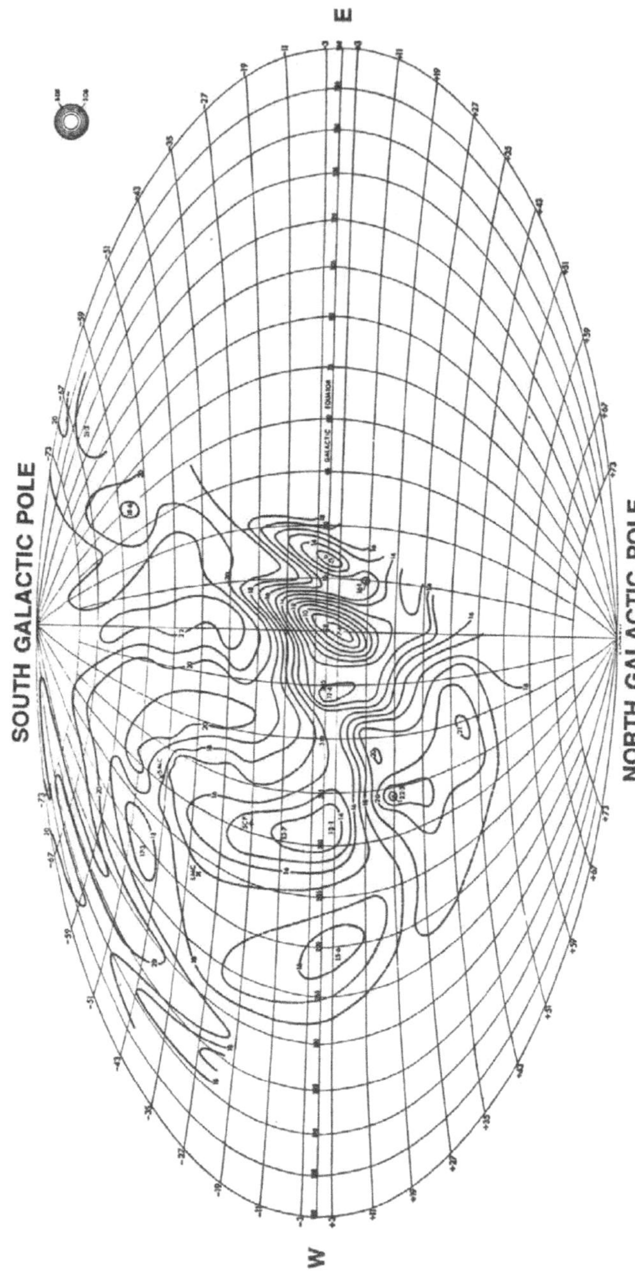

Fig. 1. Radio map of the southern sky at 144 meters

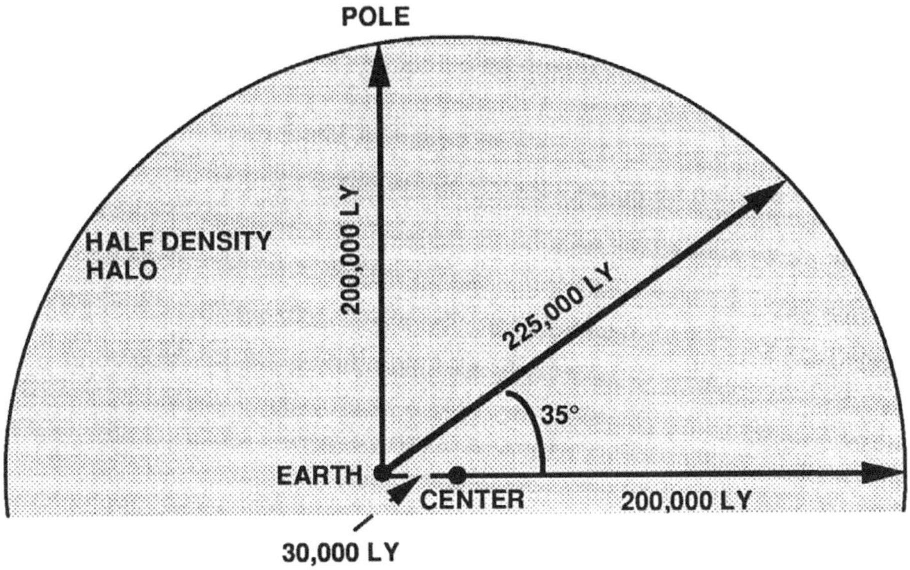

Fig. 2. Halo distribution intrepretation of Figure 1. The semi-circle represents the half-density radius

Way, the half-density radius becomes 200,000 ly. The distance to the pole is the same value, 200,000 ly. Accordingly, the distance to far edge is 225,000 ly. The ratio of latter to former is 1.125, or the 0.5 db observed. This description seems to be a more realistic interpretation of the observations in Figure 1. Accordingly there is no uniform intergalactic medium. Furthermore, it would appear that all galaxies have similar halos. We assume that these are also spread out with distance. In this picture intergalactic space is filled by an assortment of overlapping halos of varying size and density. The value 10^4 electron/proton pairs per cubic meter is probably a reasonable average approximation. Actual densities may vary above and below this, perhaps by a factor of ten at different places in space. These values will depend on proximity, or lack thereof to any given galaxy, or cluster of galaxies. While this is a simplified and incomplete picture, more data is needed to improve its accuracy.

3 144 Meter Array Measurements from Ottawa

During the winter of 1986-1987, some preliminary measures of northern sky were made using a small low-frequency antenna array located west of Ottawa, Canada. The observing frequency was again 2085 kHz. This particular frequency was chosen to reduce rf interference; it appears to be a clear channel worldwide. A few array recordings were taken at this location before an ice storm brought many of the antennas down. These recordings show that the north galactic pole has same intensity as south galactic pole. This is reassuring. There is a 30% absorption drop at plane of Milky Way north of Cygnus at 45° celestial latitude. [1]

References

Reber, G.: 1968, "Cosmic Static at 144 Meters Wavelength", *Journal Franklin Institute*Vol. no. **285**, pp. 1–12

Reber, G.: 1986, "Intergalactic Plasma", *IEEE Transactions on Plasma Science*Vol. **PS-14**, pp. 678–682

[1] A companion paper will appear in an issue of the *Journal of the Royal Astronomical Society of Canada.* This paper contains additional details about the Galaxy observations plus ionosphereic data. Optimum low-frequency viewing in the northern hemisphere seems to be near Prince Rupert in British Columbia. Observations at frequencies as low as 1 MHz should be possible during the solar activity minimum of 1997. Let's get on with it.

PLASMA AND THE UNIVERSE: LARGE SCALE DYNAMICS, FILAMENTATION, AND RADIATION

ANTHONY L. PERATT
Los Alamos National Laboratory
Los Alamos, New Mexico, USA

Abstract. One of the earliest predictions about the morphology of the universe is that it be filamentary (Alfvén, 1950). This prediction followed from the fact that volumewise, the universe is 99.999% matter in the plasma state. When the plasma is energetic, it is generally inhomogeneous with constituent parts in motion. Plasmas in relative motion are coupled by the currents they drive in each other and nonequilibrium plasma often consists of current-conducting filaments.

In the laboratory and in the Solar System, filamentary and cellular morphology is a well-known property of plasma. As the properties of the plasma state of matter is believed not to change beyond the range of our space probes, plasma at astrophysical dimensions must also be filamentary.

During the 1980s a series of unexpected observations showed filamentary structure on the Galactic, intergalactic, and supergalactic scale. By this time, the analytical intractibility of complex filamentary geometries, intense self-fields, nonlinearities, and explicit time dependence had fostered the development of fully three-dimensional, fully electromagnetic, particle-in-cell simulations of plasmas having the dimensions of galaxies or systems of galaxies. It had been realized that the importance of applying electromagnetism and plasma physics to the problem of radiogalaxy and galaxy formation derived from the fact that the universe is largely a *plasma universe*. In plasma, electromagnetic forces exceed gravitational forces by a factor of 10^{36}, and electromagnetism is $\approx 10^7$ times stronger than gravity even in neutral hydrogen regions, where the degree of ionization is a miniscule 10^{-4}.

The observational evidence for galactic-dimensioned Birkeland currents is given based on the direct comparison of the synchrotron radiation properties of simulated currents to those of extra-galactic sources including quasars and double radio galaxies.

Key words: Plasma Cosmology, Galaxies, Filamentation, Electrical Currents, Quasars, Double Radio Galaxies, Jets

1 Introduction

Among the earliest predictions about the morphology of the universe is that it be filamentary (Alfvén, 1950, 1981, 1990). This prediction follows from the fact that volumewise, the universe is 99.999% matter in the plasma state. For the most part, plasma consists of particles at high temperature, *i.e.*, an energetic state. And like all energetic plasma, the volume of plasma is inhomogeneous and consists of plasmas with differing temperatures, magnetization, degree of ionization, chemical constituency, and relative motion. It is this latter property, often the result of the first four properties, that

Astrophysics and Space Science **227**: 97–107, 1995.

produces the filamentation. Plasmas in relative motion are coupled by the currents they drive in each other.

In the laboratory and in the Solar System, filamentary and cellular morphology is a well-known property of plasma. As the properties of the plasma state of matter is believed not to change beyond the range of our space probes, plasma at astrophysical dimensions must also be filamentary.

Additionally, transition regions have been observed that delineate the 'cells' of differing plasma types (Eastman, 1990). On an astrophysical scale, these transition regions should be observable at radio wavelengths via transition radiation signatures.

The suggestion that the universe be filamentary and cellular was generally disregarded until the 1980s, when a series of unexpected observations showed filamentary structure on the Galactic, intergalactic, and supergalactic scale. By this time, the analytical intractibility of complex filamentary geometries, intense self-fields, nonlinearities, and explicit time dependence had fostered the development of fully three-dimensional, fully electromagnetic, particle-in-cell simulations of plasmas having the dimensions of galaxies or systems of galaxies. It had been realized that the importance of applying electromagnetism and plasma physics to the problem of radiogalaxy and galaxy formation derived from the fact that the universe is largely a *plasma universe.*

Any imbalance in the constituitive properties of a plasma can set it in motion [if, in fact, it has not already derived from an evolving, motional state (Bohm, 1979)]. The moving plasma, *i.e.*, charged particle flows, are currents that produce self magnetic fields, however weak. The motion of any other plasma across weak magnetic fields produces and amplifies electromotive forces, the energy of which can be transported over large distances via currents that tend to flow along magnetic lines of force. These 'field-aligned currents,' called *Birkeland currents* in planetary magnetospheres, should also exist in cosmic plasma. The dissipation of the source energy from evolving or moving plasma in localized regions can then lead to pinches and condense states. Where double layers form in the pinches, strong electric fields can accelerate the charged particles to high energies, including gamma ray energies (Alfvén, 1981). These should then display the characteristics of relativistic charged particle beams in laboratory surroundings, for example, the production of microwaves, synchrotron radiation, and non-linear behavior such as periodicities and 'flickering.'

2 Filamentation by Birkeland Currents

An electromotive force $\int \mathbf{v} \times \mathbf{B} \cdot \mathbf{dl}$ giving rise to electrical currents in conducting media is produced wherever a relative perpendicular motion of plasma and magnetic fields exist (Akasofu, 1984; Alfvén, 1986). An example of this is the (nightside) sunward-directed magnetospheric plasma that cuts the earth's dipole field lines near the equatorial plane, thereby producing a potential supply that drives currents within the auroral circuit. The discovery of these Birkeland currents in the earth's magnetosphere in 1974 (Dessler, 1984) has resulted in a drastic change in our understanding of aurora dynamics, now attributed to the filamentation of Birkeland charged-particle sheets following the earth's dipole magnetic-field lines into vortex current bundles.

The remainder of this paper is devoted to analyzing Birkeland currents in astrophysical settings (Fälthammar, 1986).

3 Field Aligned Electric Fields

Recent literature in the area of magnetospheric physics reflects considerable interest in magnetic-field-aligned electric fields Such electric fields can have important consequences in cosmic plasma (Fälthammar, 1983), including the "unfreezing" magnetic fields, the acceleration of electrons to very high energies, and filamentation of the plasma itself.

In magnetized nonhomogeneous astrophysical plasma, a number of mechanisms are present that can generate field-aligned electric fields. These include anomolous resistivity caused by wave-particle interactions, collisionless thermoelectric effects due to energy-dependent wave-particle interactions, magnetic mirror effects involving trapped particles in magnetic-field gradients, and electric double layers due to localized charge separation. While all of the above mechanisms have been studied in the laboratory and simulatied by computer, it is the last mechanism that has been found to be remarkably prolific in producing appreciable potential drops in neutral plasma. Moreover, Birkeland currents and double layers appear to be associated phenomena, and both laboratory experiments (Chan & Hershkowitz, 1982) and computer simulations (Peratt & Jones, 1986) have shown the formation of a series of double layers along current-carrying plasma filaments or beams.

When double layers (or a series of double layers) form in adjacent Birkeland current filaments, field-aligned electric fields are generated, which then serve to accelerate electrons and ions within these regions.

4 Galactic Dimensioned Birkeland Currents

Extrapolating the size and strength of magnetospheric currents to interstellar space leads to the suggestion that confined current flows in interstellar clouds assists in their formation (Alfvén, 1981).

As a natural extension of the size hierarchy in cosmic plasmas, the existence of galactic dimensioned Birkeland currents or filaments was hypothesized (Alfvén & Fälthammar, 1963; Peratt, 1986).

A galactic magnetic field of the order $B_G = 10^{-9} - 10^{-10}$T associated with a galactic dimension of $10^{20} - 10^{21}$m suggests the galactic current be of the order $I_G = 10^{17} - 10^{19}$A.

In the galactic dimensioned Birkeland current model, the width of a typical filament may be taken to be 35 kpc ($\approx 10^{21}$m), separated from neighboring filaments by a similar distance. Since current filaments in laboratory plasmas generally have a width/length ratio in the range $10^{-3} - 10^{-5}$, a typical 35 kpc wide filament may have an overall length between 35 Mpc and 3.5 Gpc with an average length of 350 Mpc. The circuit, of course, is closed over this distance (Peratt, 1990).

5 The Large Scale Structure of the Plasma Universe

Surface currents, delineating plasma regions of different magnetization, temperature, density, and chemical compositon give space a cellular structure (Alfvén & Fälthammar, 1963). As current-carrying sheet beams collect into filaments, the morphology of the surface currents is filamentary.

For the case of tenuous cosmic plasmas, the thermokinetic pressure is often negligible and hence the magnetic field is force-free. Under the influence of the electromagnetic fields the charged particles drift with the velocity

$$v = (\mathbf{E} \times \mathbf{B})/E^2 \tag{1}$$

The overall plasma flow is inwards and matter is accumulated in the filaments which, because of their qualitative field line pattern, are called "magnetic ropes". Magnetic ropes should therefore tend to coincide with material filaments that have a higher density than the surroundings. The cosmic magnetic ropes or current filaments are not observable themselves, but the associated filaments of condensed matter can be observed by the radiation they emit and absorb.

It is because of the convection and neutralization of plasma into radiatively cooled current filaments (due to synchrotron losses) that matter in the plasma universe should often display a filamentary morphology (Figure 1).

6 Confining and Interacting Forces Between Cosmic Currents

If the cosmic current is cylindrical and in a rotationless, steady-state condition, it is described by the *Carlqvist Relation*:

$$\frac{\mu_0}{8\pi} I^2(a) + \frac{1}{2} G\bar{m}^2 N^2(a) = \Delta W_{Bz} + \Delta W_k \tag{2}$$

for a current of radius $r = a$ where μ_0 is the permeability of free space, G is the gravitational constant, $\bar{m}$ is the mean particle mass, N is the number of particles per unit length, and ΔW_{Bz} and ΔW_k are the differential beam

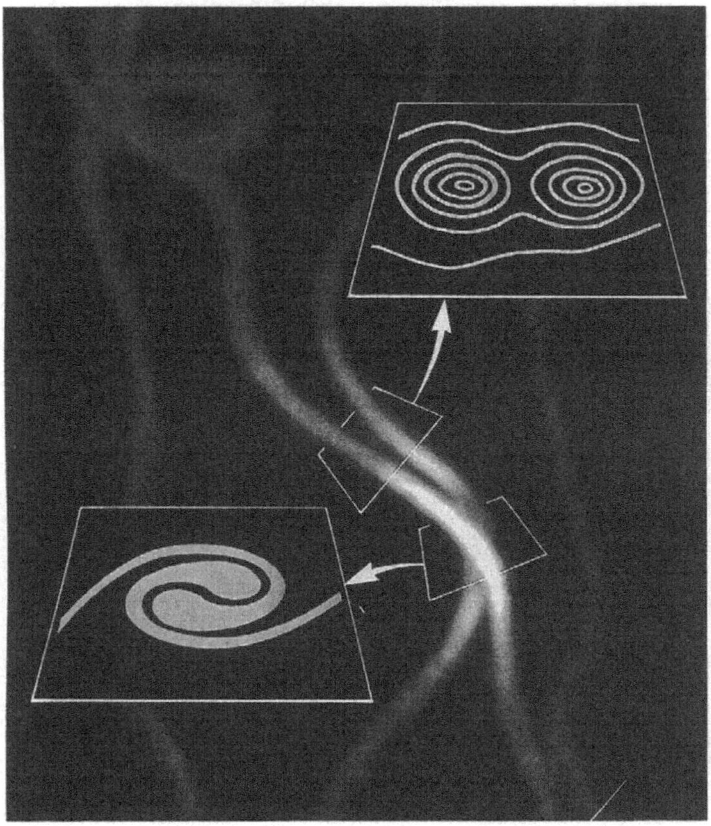

Fig. 1. The plasma universe may be eternal and infinite. Large scale field-aligned filaments may extend hundreds of megaparsecs or more in space. Where pairs of these spaghettilike structures interact, the particles gain kinetic energy and at narrow pinch regions produce the entire range of galaxy types as well as the full spectrum of electromagnetic radiation. Thus galaxies must lie along filaments, much as they are observed to do on a large scale. The bulk of the filaments are invisible from a distance, like the Birkeland currents that circle the Earth but are unobservable from its surface.

magnetic and kinetic energies, respectively (Peratt, 1992a)[1]. Thus, whether

[1] When current rotation and transient phenomena are important, the *Generalized Ben-*

or not a current or beam is gravitationally balanced, electromagnetically balanced, or force-free, depends on the magnitude of the individual terms in Eq.(2). Applications of the Carlqvist Relation are presented in this journal (Verschuur, 1995).

In contrast to the gravitational and electromagnetic forces that determine the characteristic of an individual beam, interactions between beams are always dominated by electromagnetic *Biot-Savart* forces,

$$\mathbf{F_{21}} = \int \mathbf{j_2} \times \mathbf{B_{21}} d^3r \tag{3}$$

for all space, where $\mathbf{j_2} \times \mathbf{B_{21}}$ is the Lorentz force between the field B_{21} induced by a current I_1 on the current density j_2 at current I_2.[2]

Parallel axial currents within the filaments are long-range attractive, while circular (helical) currents within the filaments (as the electrons gyrate along the axial magnetic field) are short-range repulsive. If the axial currents are able to bring the filaments close enough together so that the repulsive component of the Lorentz force becomes important, the circular currents repulse and brake, and release energy in the form of synchrotron radiation.

While a complete description of the evolution of interacting galactic currents is given elsewhere (Peratt, 1992a,b), it is useful to reproduce the evolutional sequence in this paper. Figure 2 illustrates the cross-sections of the filaments over a 10^9 yr period. The remainder of this paper is devoted to the radiation properties of the interaction as shown in the first frame of Figure 2, approximately 60 Myr.

7 Synchrotron Emission from Pinched Particle Beams

One of the most important processes that limit the energies attainable in particle accelerators is the radiative loss by electrons accelerated by the magnetic field of a betatron or synchrotron. This mechanism was first brought to the attention of astronomers by Alfvén and Herlofson (1950); a remarkable suggestion at a time when plasma, magnetic fields, and laboratory physics were thought to have little, if anything, to do with a cosmos filled with isolated "island" universes (galaxies). Synchrotron radiation is characterized by a generation of frequencies appreciably higher than the cyclotron frequency of the electrons; a continuous spectra (for a population of electrons) whose intensity decreases with frequency beyond a critical frequency (near intensity maxima); increasing beam directivity with increasing relativistic factor γ $\left(\gamma = (1-\beta)^{-1/2}\right)$; and polarized electromagnetic wave vectors.

nett Condition may be used in place of Eq.(2) (Peratt, 1992a, Chap. 2)

[2] The Biot-Savart force varies as r^{-1} and thus dominates gravitational attraction which varies as r^{-2}. 'Great Attractors', often attributed to gravitational forces between 'missing masses' display Biot-Savart, not mass attraction, characteristics.

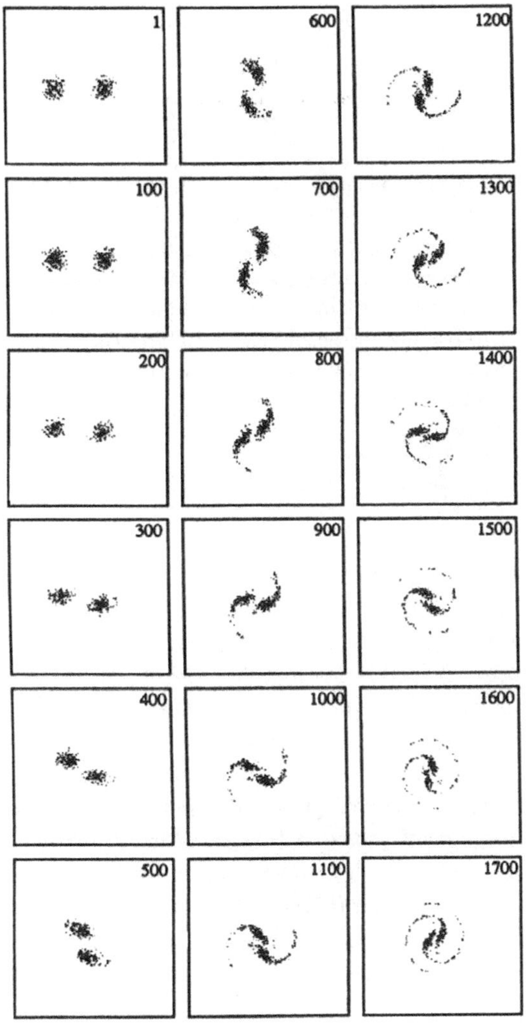

Fig. 2. Single frame stills of plasma in the simulation of two adjacent Birkeland filaments: ω_c/ω_p=3.0, $T_{e0} = T_{i0}$=32 keV, E_{z0}=62 mV/m. Total time elapsed: $\approx 10^9$ yr. The initial dimensions in frame 1 are: radius of filaments $r_{filament}$=17.5 kpc, distance between filaments $d_{filaments}$=80 kpc. The length over which E_{z0} exists in the filaments is taken to be $\approx$10 kpc.

Z-Pinches are among the most prolific radiators of synchrotron radiation known. In this regard, the Bennett-pinch, or Z-pinch, as a sychrotron source has been treated by Meierovich (1984) and Newberger (1984).

The radiation produced from the plasma configuration shown in the sec-

TABLE I
Simulation derived parameters based on the radiation properties of the double radio galaxy Cygnus A.

Parameter	*Simulation Value*
Galactic current, I_G	2.4×10^{19} A
Galactic magnetic field, B_θ	2.5×10^{-4} G
Galactic magnetic field, B_z	2.0×10^{-4} G
Plasma temperature, T	2.0 – 32.0 keV
Plasma density, n_e	1.79×10^{-3} cm^{-3}
Electric field strength, E_z	62 mV/m
Synchrotron power, P_{syn}	1.16×10^{37} W
Radiation burst duration	1.28×10^{14} s
Total energy	6.3×10^{62} J

ond frame of Figure 2 replicates both the isophotal and power spectra from double radio galaxies (Figure 3). Table I delineates the basic parameters used in the interacting galactic filament simulation.

Because the highly relativistic electrons depicted in Figure 2 flow in direction outwards from the plane of the figure, the synchrotron radiation is also beamed in this direction (Johner, 1988).

The monochromatic power of quasars and double radio galaxies span a range of about $10^{33}W - 10^{39}$W (Peratt, 1986b). For example, the "prototype" double radio galaxy Cygnus A has an estimated radio luminosity of 1.6-4.4 $\times 10^{37}$W. Together with the power calculated, the simulation isophotes are very close to those observed from this object (Peratt, 1986a). The left column of Figure 3 suggests that previously apparently unrelated double radio galaxies all belong to the same species but are simply seen at different times in their evolution.

8 Conclusions

The importance of applying electromagnetism and plasma physics to the problem of radio galaxy, galaxy and star formation derives from the fact that the universe is largely matter in its plasma state, i.e., a universe of plasma. The motion of this plasma in local regions can lead to pinches and ultimately condense states of matter. Where double layers form, strong electric fields can accelerate particles to high energies. The intensity and patterns of synchrotron radiation observed in the model simulations are in excellent agreement with those observed from double radio galaxies.

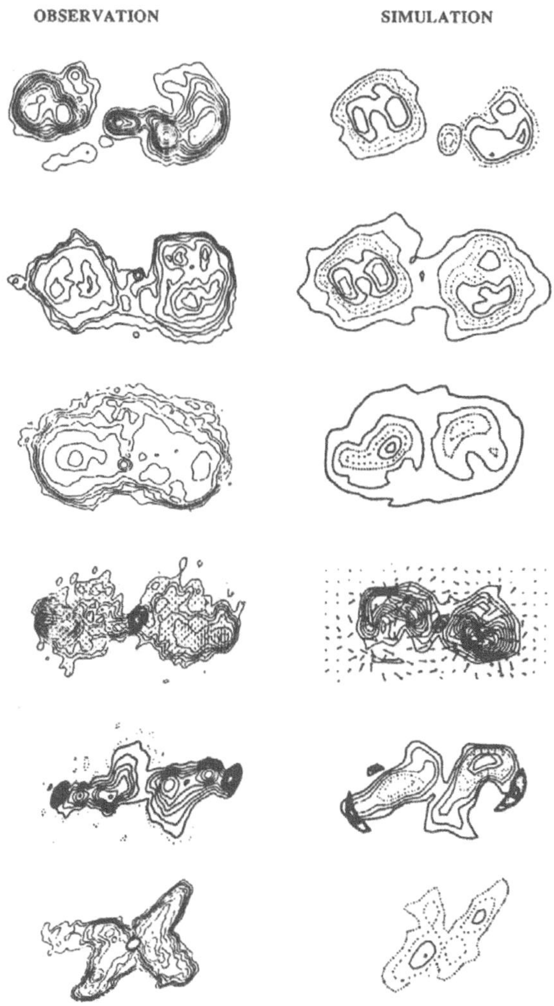

Fig. 3. (a) Synchrotron isophotes (various frequencies) of double radio galaxies, (b) Simulation analogs at time 10.4 Myr to 58.7 Myr. These times correspond to the epoch shown in frames 1 and 2 of Figure 2. Time increases from top to bottom.

This paper has summarized previous research relating to the morphology and large-scale dynamics of a plasma universe. It has also addressed the special case of the radiation seen by an observer when the observer happens to be located in the directed pattern of a synchrotron source. Many sources with this orientation can be expected in various regions of the sky from the spaghetti of radiating filaments surrounding the viewer. The background spectrum caused by an extremely large number of synchrotron radiating filaments, when the observer is not in the directed beam, has been treated by Peratt & Peter (1988).

9 Acknowledgments

This project was supported by the Institute of Electrical and Electronics Engineers, the Division of Atmospheric Sciences of the National Science Foundation, The European Space Agency, and NASA Headquarters, Space Physics Division.

References

Akasofu, S.-I.: 1984, "The magnetospheric currents: An introduction to magnetospheric currents", in *Magnetospheric Currents*, T.A. Potemra, Ed. (Geophysical Monograph No.28), Washington, DC: Amer. Geophys. Union, pp.29–48

Alfvén, H.: 1950, *Cosmical Electrodynamics* Oxford, London

Alfvén, H.: 1981, *Cosmic Plasma* D. Reidel, Dordrecht

Alfvén, H.: 1986, "Double layers and circuits in astrophysics" *IEEE Trans. on Plasma Sci.*, bf Vol. PS-14, pp. 779-793

Alfvén, H.: 1987, "Electric currents in cosmic plasma", *Rev. Geophys. and Space Sci.* **Vol. 15**, pp. 271–284

Alfvén, H.: 1990, "Cosmology in the Plasma Universe", *IEEE Trans. Plasma Sci.* **Vol. 18**, pp. 5-10

Alfvén, H. and Herlofson, N.: 1950, *Phys. Rev.* **Vol. 78**, pp. 616

Alfvén, H., Fälthammar, C.-G.: 1963, *Cosmical Electrodynamics*, Oxford University Press

Bennett, W. H.: 1934, "Magnetically self-focusing streams", *Phys. Rev.* **Vol. 45**, pp. 890–897

Bohm, D. J.: 1979, in *A Question of Physics: Conversations in Physics and Biology*, Buckley, P., Peat, F. D. (eds.), University of Toronto Press, Toronto

Chan, C., Hershkowitz, N.: 1982, "Transition from single to multiple double layers", *Phys. Fluids* **Vol. 25**, pp. 2135–2137

Dessler, A. J.: 1984, "Evolution of arguments regarding existence of field aligned currents", in *Magnetospheric Currents*, T.A. Potemra, Ed. (Geophysical Monograph No.28), Washington, DC: Amer. Geophys. Union, pp.26–32

Eastman, T.: 1990, "Transition Regions in Solar System and Astrophysical Plasmas", *IEEE Trans. Plasma Sci.* **Vol. 18**, pp. 18-25

Fälthammar, C.-G.: 1983, "Magnetic-field aligned electric fields", *ESA J.* **Vol. 7**, pp.385–401

Fälthammar, C.-G.: 1986, "Magnetosphere-ionosphere interactions - near-earth manisfestations of the plasma universe", *IEEE Trans. on Plasma Sci.* **Vol. PS-14**, pp. 616–628

Johner, J.: 1988, "Angular distribution of the total cyclotron radiation of a relativistic particle with parallel velocity", *Phys. Rev. A* **Vol. 36**, pp. 1498–1501

Meierovich, B. E.: 1984, "Electromagnetic collapse, problems of stability, emission of radiation and evolution of a dense pinch", *Phys. Reports* **Vol. 104**, pp. 259–347

Newberger, B. S., M. I. Buchwald, R. R. Karl, D. C. Moir, and Starke, T. P.: 1984, "Synchrotron radiation from bennett beams", *Bull. Am. Phys. Soc.* **Vol. 29**, p. 1435

Peratt, A. L.: 1986a, "Evolution of the plasma universe: I. double radio galaxies, quasars, and extragalactic jets", *IEEE Trans. on Plasma Sci.* **Vol. PS-14**, pp. 639–660

Peratt, A. L.: 1986b, "Evolution of the plasma universe II. the formation of systems of galaxies", *IEEE Trans. on Plasma Sci.* **Vol. PS-14**, pp. 763–778

Peratt, A. L.: 1990, "The evidence of electrical currents in cosmic plasma", *IEEE Trans. on Plasma Sci.* **Vol. 18**, pp. 26–32

Peratt, A. L.: 1992a, *Physics of the Plasma Universe*, Springer-Verlag, New York

Peratt, A. L.: 1992b, "Plasma Cosmology", *Sky and Tel.* **February**, p. 136

Peratt, A. L., Jones, M. E.: 1986, "Particle-in-cell simulations of heavy ion double layers", *IEEE Conf. Record.* Saskatoon, Canada

Peter, W., Peratt, A. L.: 1988, "Synchrotron radiation spectrum for galactic-sized plasma filaments", *Laser and Particle Beams* **Vol. 6**, Part 3, pp. 493–502

Verschuur, G. L.: 1995, "Interstellar Neutral Hydrogen Filaments at High Galactic Latitudes and the Bennett Pinch", *Astrophys. Space Sci.*

THE PILLARS OF COSMOLOGY: A SHORT HISTORY AND ASSESSMENT

JEFF KANIPE*
Austin McDonald Observatory
University of Texas
Austin, TX, USA

Abstract. Is the big bang model of the universe ascendant and unshakable, or declining and outdated? This paper cursorily explores the state of cosmology today.

Key words: Plasma Cosmology, Big Bang, Electromagnetism, Dark Matter, Cosmic Background Radiation

1 Introduction

In the early part of the twentieth century, the universe, as put together with over two centuries' worth of theories and hard–won astronomical observations, simply came undone.

Until the mid–1920s, the cosmos was conceived of as serene, static, and mostly empty. Stars wandered through space randomly, planets revolved unperturbed around the Sun, and disklike whorls, thought to be nearby stars in the making, were known only as "spiral nebulae." Today we know the spiral nebulae as galaxies–immense and remote systems of $10^9 - 10^{10}$ stars–and the ordered and serene universe has been replaced by one that appears dynamic, fluctuating, and infinitely more voluminous.

A new dominant model of the universe has emerged. In its simplest form, the most widely accepted model describes an expanding universe filled with countless galaxies rushing away from an explosive, seminal event called the big bang. Mainstream astronomers believe the big bang clarifies the picture of universal genesis by predicting what otherwise would be unpredictable and explaining what otherwise would be unexplainable.

And yet, there is a growing alliance of physicists and astronomers who feel the big bang model not only fails to describe what is observed, but fails altogether as a cosmology (Alfvén, 1981; Peratt and Green, 1983; Arp, 1987; Lerner, 1991; Peratt, 1992).

---

* Editor of *Star Date* and Contributing Editor to *Astronomy*.

Astrophysics and Space Science **227**: 109–118, 1995.

2 Three cosmological pillars

Astronomers today cite three "pillars" shoring up the big bang cosmology. The first, the expanding universe, was hefted into place in the early part of this century, when American astronomer Edwin Hubble, using Mount Wilson Observatory's 100–inch reflecting telescope, discovered that the more distant a galaxy was from the Earth, the faster it was receding. The universe appeared to be blowing up like a balloon, expanding. The implications were astonishing: For the entire universe to be expanding, a propulsive force of unimaginable magnitudean explosion of some sort must have set matter on its outward–flowing course.

The expanding universe is probably the best known of the big bang pillars. There are two others–equally formidable, say supporters; equally flawed, rejoin the critics.

The second of the three pillars is the measured total abundance of light elements in the universe, "light elements" being hydrogen, deuterium (an isotope of hydrogen), helium, and lithium. Measuring the cosmic abundance is analogous to ascertaining the ingredients of an immense cake after it has been baked.

Like a cake, the matter composing the universe today retains a substantial imprint of its initial cosmic "batter." These primeval elements are commonly sampled by assaying the oldest stars, most of which are found in globular clusters. The results: Study after study shows that the ratio of hydrogen to helium in these old stars is what the big bang theory predicts should have been forged by thermonuclear reactions 100 s after the primordial explosion. Hydrogen accounts for roughly three–quarters of the total mass and helium one–quarter, with deuterium, lithium, and the heavier elements occurring in traces.

3 The Cosmic Background Radiation

The final big bang pillar is the cosmic background radiation, thought to be the remnant radiation of the big bang itself. It was reasoned in 1948 by the American physicists Ralph Alpher, Robert Herman, and George Gamow (Gamow, 1948; Alpher and Herman, 1948) that the most monumental fireball the universe ever produced should have left a residual all–sky glow detectable 10, even 20, Gyr afterward. And, indeed, when radio telescopes are tuned to the short end of the radio spectrum–at the same wavelengths as those used in the home to pop popcorn and bake potatoesastronomers detect a "glow" that appears to emanate from all directions in space. The temperature of this ubiquitous radiation is about 2.73 K, just barely above absolute zero, and some 10^8 times fainter than the heat produced by a typical birthday candle.

One of the most striking characteristics of the cosmic background radiation is its smoothness. No matter where you point your radio telescope from the Earth's surface, this primordial glow, at large scales anyway, is uniform in every way.

Nevertheless, there is a problem. When we look out at the universe today, we note that there are of order 10^9 galaxies strung out across walls and sheets shot through with vast voids of apparently empty space. The universe may have started out smooth, but it did not end up that way. Today it resembles Swiss cheese or a froth of soap bubbles. Astrophysicists reason that whatever disturbed the smoothness of the big bang and resulted in the formation of the first galaxies must have left its imprints in the 2.73 K background radiation.

For over 20 years astronomers searched for these fluctuations in the cosmic background radiation but to no avail. Then, in April 1992, astronomers announced that the orbiting Cosmic Background Explorer (COBE) spacecraft had finally detected slight temperature fluctuations in the background radiation. Very slight: The fluctuations amounted to deviations of just 30×10^{-6} of a degree, or 1 part in 10^5 of the total signal. Still, they were enough to elicit hosannas from the scientific community. Headlines appeared in every newspaper around the country: "Astronomers See the Face of God," or words to that effect (S. Begley and D. Glick, 1992). It was, perhaps, a bit dramatic.

COBE's microwave instrument could detect only the largest–scale structures, larger than any known today. Now scientists using many other instruments are searching for smallerscale fluctuations in the cosmic microwave background that are comparable to the clusters of galaxies that we see today.

Nonetheless, COBE's unprecedented measurements of the character of the cosmic background radiation are, in the minds of most astronomers, a clear victory for the big bang. According to COBE project scientist John Mather, the results should cast off any doubt scientists may have had about the big bang. "COBE has confirmed every prediction made by the hot big bang theory about the [smooth nature of the] cosmic microwave background radiation."

4 The Cold Dark Matter Dilemma

One of the most serious shortcomings in the big bang cosmology cited by critics is that it is uncomfortably silent about how galaxies and stars materialized during the first one Gyr. And yet there must be a link. One theory that attempts to connect the big bang with galaxy formation is called the "cold dark matter" model. This hypothesis asserts that at least 90 to 99% of the early cosmos was composed of some sort of dark matter in the form of slow–moving (hence cold) hypothetical elementary particles such as axions, gravitinos, or photinos. Because these particles were sluggish, goes the the-

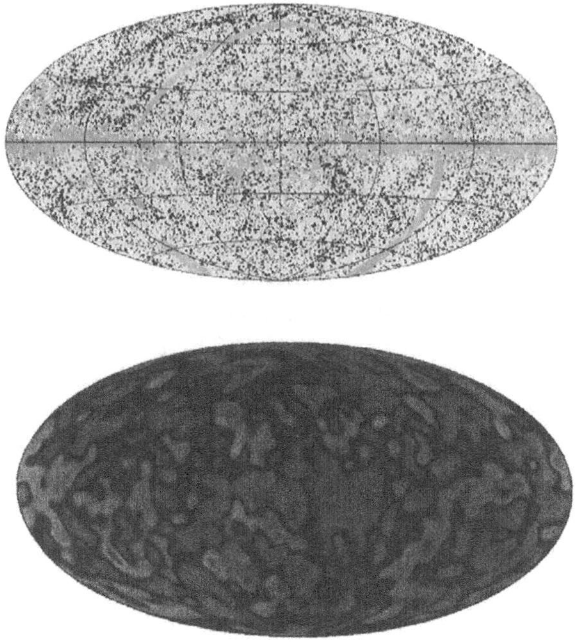

Fig 1 (Top) This infrared all–sky map, looking far beyond what the eye can see confirms to the trained observer that galaxies are clumped together in superclusters spread in some cases over hundreds of millions of light years The gray regions indicate insufficient data, with the Milky Way having obscured the data along the equator Larger symbols indicate nearer galaxies (Bottom) The big bang model predicts that the universe should be bathed in a very smooth background radiation This overall smoothness was confirmed by the Cosmic Background Explorer (COBE) orbiting observatory In addition, as shown in this all–sky map, COBE's microwave instrument detected minute fluctuations in the background radiation that may have been the precursors of the structure in the universe

ory, gravity had an easier time forming matter into clumps at small scales during the first Gyr or so. These clumps attracted more matter, until galaxy–sized masses were built up. The galaxies then collected into the elaborate clusters we see today.

Unfortunately, cold dark matter has never been detected, except perhaps indirectly by what are thought to be its gravitational effects on normal matter specifically the motions of galaxies in clusters and the rotations of spiral galaxies.

What is worse, even the most generous cold dark matter models cannot account for the astonishing dimensions of large–scale structure turning up in galaxy surveys today, structures that in some cases span 10^8 lyr, without stretching physics to the theoretical breaking point.

Replying to critics who argue that the failure to discover cold dark matter calls the big bang model into question, Princeton cosmologist James Peebles asserts that it is not so much an alternative cosmology that is needed as a good theory of galaxy formation. "Cold dark matter is a theory about the beginning of the formation of structure," he says. "The big bang is something quite different. The problem is that the press has conflated the two. They take the discussions about the beginnings of structure and confuse them with the beginning of the universe." (Flam, 1992)

Others have taken exception to this view. Peratt, for example, argues that the big bang, if it is to count as a cosmology at all, should account for the galaxies and stars we observe. "It's been interesting," says Peratt, "to watch big bang cosmologists struggle to find a mechanism for galaxy formation in spite of their arguments that this is not their problem."

5 A Quasi–Steady State Universe

Two alternate theories stand in contrast to the big bang cosmology. One is the steady state theory, which was first proposed by Cambridge University astronomers Thomas Gold, Herman Bondi, and Fred Hoyle in the late 1940s but has recently been "modernized." The second is plasma cosmology, which has come onto the scene only within the last 25 years or so.

In recent years, Hoyle, and astronomers Geoffrey Burbidge of the University of California, San Diego, and Jayant Narlikar of the Inter–University Center for Astronomy and Astrophysics in Pune, India, have labored to revise the steady state theory to reflect the latest observations. They call the new incarnation the quasi steady state cosmology (QSSC).

Their group proposes that the background radiation is actually normal radiation from stars that has been scattered by iron "whiskers," needle–shaped particles sown throughout the interstellar medium by supernovae explosions (Hoyle *et al.*, 1984).

The most contentious aspect of the original steady state model was its hypothesis of a "creation field," a region in which matter and radiation are created. Recently, Hoyle, Burbidge, and Narlikar redefined the creation field (Hoyle *et al.*, 1993). By modifying Einstein's general relativity equations, the astronomers found that creation fields can exist around high–mass, high–density regions, such as in and around the cores of active galaxies and quasars. These creation fields would produce an explosive discharge of high–energy particles and radiation that would be visible across $\approx 10^9$ lyr. Hence, they would account for the "cosmic jets" of radiation seen spewing from the centers of some galaxies and quasars, without having to invoke the existence of black holes whose magnitude is on the order of a billion solar masses.

They further assert that the universe has undergone steady expansion for about 10^{12} yr, but that superimposed on the overall expansion are relatively

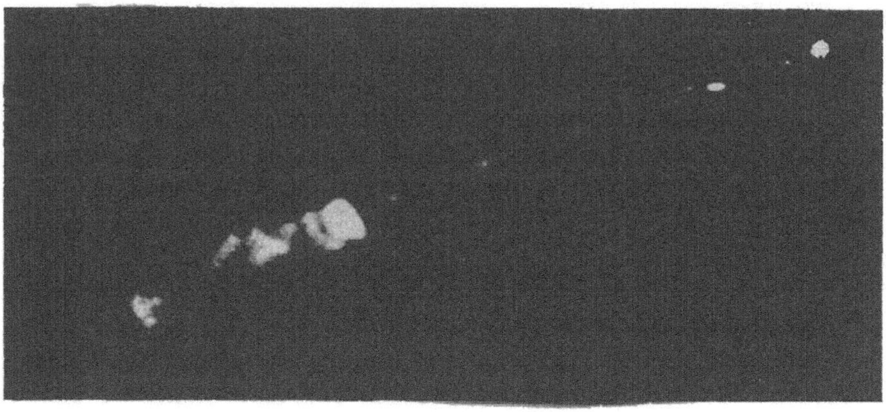

Fig. 2. Cosmic jets such as this one spewing forth from the massive galaxy in the upper right corner of the photo have been explained by the big bang model as being caused by a black hole at the center of the galaxy. In contrast, the quasi steady state cosmology (QSSC) claims that such jets are evidence of a creation field (a white hole?) where maner is being created.

short–term fluctuations, some of them 40 Gyr in duration. If the universe is as ancient as the QSSC implies, it logically follows that there must be countless burned–out stars, galaxies, and dead planets out thereat least a 10^{12} years' worth. Hoyle, Burbidge, and Narlikar suggest that this accumulation of ordinary matter is the "dark matter" the big bang theorists have been questing for.

6 The Plasma Universe

In both the quasi steady state and big bang cosmologies, gravity is the main sculptor of the universe. According to the tenets of plasma cosmology, however, electromagnetism, not gravity, makes the universe what it is. The medium in which electromagnetic forces are played out is plasma, which is believed to constitute 99% of the universe.

Plasma, the fourth state of matter, is distinct from solids, liquids, and gases. Although affected by gravity, the constituents of the plasma–charged particles–are affected more strongly by electromagnetic fields (by a factor of 10^{39} over gravity). A universal characteristic of current-conducting plasma is that it pinches along the flow because of the inward pressure of its self-magnetic field. It is at the pinch points in the cosmic flux where bodies of gravitationally significant mass, galaxies consisting of $10^9 - 10^{10}$ stars, coalesce. In these regions, gravitation is a dominant force.

On Earth, the most dramatic forms of plasma are lightning and aurorae. The Sun and the stars are gravitationally bound spheres of plasma. On larger scales, plasma gas has been detected at the center of the Milky Way and pooling around radio galaxies. The largest plasma structure was detected between two superclusters of galaxies in the late 1980s, suggesting that galaxy clusters, are embedded in a warm sea of plasma.

Plasma cosmology sprang from the pioneering work of Hannes Alfvén. Stemming from his studies in the 1950s of synchrotron radiation–emission caused by electrons spiraling at nearly the speed of light in a magnetic field (Alfvén and Herlofson, 1950b), Alfvén proposed that sheets of electric currents must crisscross the universe (Alfvén, 1950a; Alfvén and Fälthammar, 1962). Interaction with these electromagnetic fields would enable plasmas to exhibit complex structure and motion. Thus, at the grandest scales, the universe would have a cellular and filamentary structure.

When filaments and walls of superclusters of galaxies were charted in the mid–1980s, they bore a strong resemblance to the cosmic structure predicted by Alfvén. Even then, however, Alfvén's work was not taken seriously. Instead, big bang theorists invoked cold dark matter to explain large–scale structure, a concept even they are apparently questioning now.

According to this view, there is no real problem with the existence of large scale structure. This will be resolved when a satisfactory galaxy formation theory comes along. It is argued that where plasma cosmology fails is in explaining the existence of the microwave background. Plasma cosmologists disagree. The cosmic background radiation, they say, results when local fields and currents scatter microwave radiation from pervasive plasma sources–like radio galaxies and quasars just as a dense fog scatters a car's headlights.

As for the "wrinkles" detected in the cosmic background by COBE, the ones considered to be the first gravitational gathering of matter into galaxies after the big bang, these fluctuations do not explain how so much large–scale structure could have formed so rapidly in the early universe. According to plasma physicists, the discovery does not even prove that the cosmic background radiation is, in fact, the echo of the big bang (Lerner, 1995).

The second pillar, cosmic nucleosynthesis, is also being challenged. An accumulation of observations, beginning in 1991 by independent researchers, indicate the abundance of helium in the early universe before stars formed is lower than the lowest value allowed by the standard big bang model (Lerner,

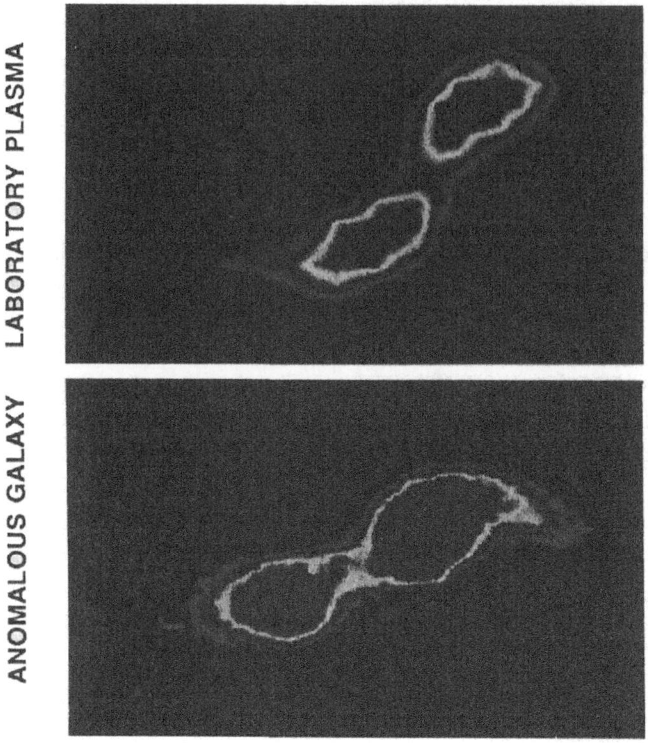

Fig. 3. Plasma cosmologists assert that the universe, which is known to consist primarily of plasma, is sculpted largely by electromagnetic forces, not gravitational forces. Recent evidence is contained in this pair of similar photos, one of a laboratory plasma and one of a distant galaxy. Surely similar processes must have been at work to produce such similar images, claim the plasma cosmologists.

1995).

7 Cosmic expansion or Wolf shift?

Finally, what other possible explanation can there be for the first pillar, the apparent recession of the galaxies? As mentioned earlier, the spectral lines of most galaxies display a redshift, which is commonly interpreted as meaning the galaxies are receding from each other with the flow of cosmic expansion. It has long been assumed that only two effects could account for spectral shifts: Either the light source is moving away from (red shift) or toward (blue shift) the observer, or light from the source is passing through the gravitational field of an intervening object, such as a galaxy or black hole.

In recent years, however, a theory has emerged that seems to show how red and blue spectral shifts can arise independently of either motion or gravity. The theory, proposed in 1989 by the physicist Emil Wolf of the University

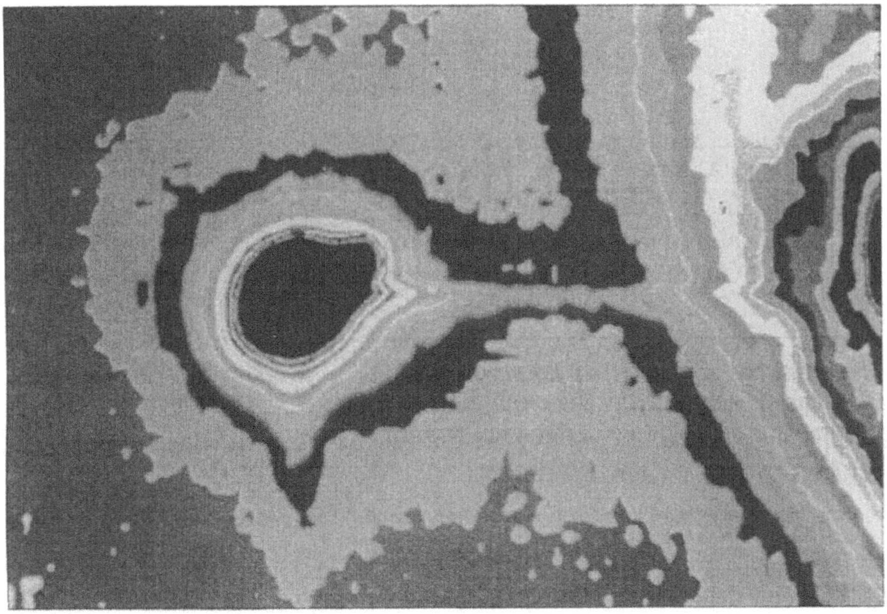

Fig. 4. One major pillar of the big bang cosmology is the expanding universe, based on redshift evidence that the farther away an object is from the earth the faster it is receding from the earth. This image of a smaller object, a quasar, linked with a larger object, a galaxy, challenges the expanding universe pillar because the redshift of the quasar suggests that it is much farther away than the galaxy.

of Rochester, is endorsed by most plasma cosmologists.

Known as the "Wolf shift," the theory states that spectral shifts can be generated by a scattering process. When the light from a galaxy passes through a suitably fluctuating plasma cloud and bounces off atoms that are moving in random directions within the cloud, the frequency of the light can be reduced so that the wavelength is lengthened toward the red end of the spectrum. The actual shift of the frequency depends on the observer's orientation with respect to the sources.

The Wolf shift, which is now supported by many laboratory experiments, may help explain the phenomenon of discordant redshifts, in which two objects that appear to be connected have different redshifts (James and Wolf, 1990).

If the Wolf shift is the cause of these discordant redshifts, the assumption that all quasars lie at remote distances may be incorrect. That would, in turn, call into question the entire redshift–distance correlation, the basis for the expanding universe. Mainstream astronomers, however, strongly doubt that discordant redshifts are anything more than perception effects or statistical inaccuracies.

8 Conclusions

Looking at all of the supposed challenges to the pillars of their cosmology, big bang theorists remain confident that there is not enough evidence to justify abandoning the big bang model.

"The big bang theory is the best we have," says Steve Maran, an astronomer at the NASA Goddard Space Flight Center and press officer for the American Astronomical Society.

Eric Lerner, author of *The Big Bang Never Happened* (Lerner, 1991), says it may not be a matter of how many pillars one must knock down before the big bang framework loses its integrity and collapses. "The big bang is a metaphysical structure, and any metaphysical struture can stand indefinitely without a visible means of support."

References

Alfvén, H. and Herlofson, N.: 1950a, *Phys. Rev.* **Vol. 78**, pp. 616
Alfvén, H.: 1950b, *Cosmical Electrodynamics* Oxford, London
Alfvén, H.: 1981, *Cosmic Plasma* D. Reidel, Dordrecht
Alfvén, H., Fälthammar, C.-G.: 1962, *Cosmical Electrodynamics* Oxford, London
Arp, H.: 1981, *Quasars, Redshifts, and Controversies* Interstellar Media, Berkeley
Alpher, R., Herman, R.: 1948, "Evolution in the Universe", *Nature* **Vol. 162**, pp. 774–775
Begley, S., Glick, D.: 1992, "The Handwriting of God", *Newsweek* May 4, 1992, pp. 76–77
Flam, F.: 1992, "COBE Sows Cosmological Confusion", *Science* **Vol. 257**, pp. 28–30
Gamow, G.: 1948, "The Evolution of the Universe", *Nature* **Vol. 162**, pp. 680–682
Hoyle, F. Narlikar, J. V., Wickramasinghe, N. C.: 1984, "The Radiation of Microwaves and Infrared by Slender Graphite Needles", *Astrophys. Space Sci.* **103**, pp. 371–377
Hoyle, F. Burbidge, G., Narlikar, J. V.: 1993, *Astrophys. J.* June 20
Lerner, E.: 1991, *The Big Bang Never Happened* Times Books, Random House, New York
Lerner, E.: 1995, "Intergalactic Radio Absorption and the COBE Data", *Astrophys. Space Sci.* This issue
James, F.V., Wolf, E.: 1990, "Doppler-Like Frequency Shifts Generated by Dynamic Scattering", *Phys. Lett. A* **146**, pp. 167–171
Peratt, A. L., Green, J. C.: 1983, "On the Evolution of Interacting Magnetized Galactic Plasmas", *Astrophys. Space Sci.* **Vol. 91**, p. 19
Peratt, A. L.: 1992, "Plasma Cosmology", *Sky and Tel.* **February**, p. 136

GRAVITATION, GLOBAL FOUR-MOMENTUM CONSERVATION AND THE STRONG EQUIVALENCE PRINCIPLE

DAVID F. ROSCOE
School of Mathematics and Statistics
University of Sheffield
Sheffield S10 2TN, UK
d.roscoe@uk.ac.shef

Abstract. It is shown that global four-momentum conservation provides all the necessary structure to *derive* a metric gravity theory which conforms to the requirements of the Strong Equivalence Principle (Will, 1981), and which satisfies all empirical tests up to, and including, those derived from the binary pulsar measurements. Significant consequences arising from this theory are: concepts of curved spacetimes become strictly superfluous to the function of describing gravitational physics; gravitational processes become direct particle/particle interactions; these interactions are arbitrated by wave processes of a kind familiar in electromagnetism; gravitational waves carry energy-momentum in the direction of their propogation vector; the essential singularities at gravitational origins, which are features of both Newtonian gravitation and General Relativity, do not exist.

Key words: General Relativity, Strong Equivalence Principle, Electromagnetism, Pulsars

1 Introduction

The General Theory of Relativity (GR) is one of the best tested of modern physical theories and, in the arena of the critical observation, has proved its superiority over rival theories many times during the past seventy years; in particular, the binary pulsar observations have recently proved decisive in this context, leaving GR as almost the only credible theory available. See Will (1981) for a comprehensive discussion. As a consequence of these circumstances, it is easy to understand why, within the narrow context of purely gravitational phenomena, GR is considered to be securely founded. The only obvious problem facing the theory is the apparent difficulty of reconciling the gravitational force (as understood within the context of GR) with the three other forces of nature, as these are understood within their respective theoretical contexts; as a measure of this difficulty, we can reflect that, although a considerable effort has been expended in the attempt to effect such a reconciliation - particularly over the past fifteen or twenty years - it is still not clear that it is possible *even in principle.* In view of this uncertainty, it would seem prudent to admit the need for the investment of a modest effort in the continued questioning of the theoretical basis of gravitation theory, and this paper is intended to represent such an effort.

An analysis of the various theories which have failed one test, or another,

Astrophysics and Space Science **227**: 119–132, 1995.

and a comparison of these with the one theory, GR, to have past all tests to date, has led to a general concensus concerning what constitutes the *essence* of a 'good' theory of gravitation; this *essence* is the Strong Equivalence Principle (SEP), Will (1981), which is satisfied uniquely by GR amongst extant theories, and which states

- *The Weak Equivalence Principle (WEP) is valid for self-gravitating bodies, as well as for test bodies;*

- *The outcome of any local test experiment is independent of the velocity of the freely falling test apparatus;*

- *The outcome of any local test experiment is independent of where and when in the Universe the experiment is performed,*

and it is even conjectured that the SEP *implies* GR (Will (1981), p83). However, the basic result of this paper is to show how the physics of four-momentum conservation provides all the structure that is required to arrive at a description of gravitation which is distinct from GR, but which satisfies the SEP, and which reduces gravitational processes to particle/particle interactions of a familiar kind.

Specifically, the idea that physics is only locally Lorentz invariant is a theory-dependent idea which is not supported by any purely empirically informed arguments. In the absence of such arguments, it is perfectly legitimate to adopt the position that, since physics is locally Lorentz invariant, then there is justification in persisting with the assumption that physics is globally Lorentz invariant for as long as this assumption leads to a realistic description of observed physics. We are then led to a Universe of global four-momentum conserving interactions in which it is found that the gravitational interaction can be considered arbitrated by retarded scalar potentials, where the passage of one such potential through any test particle causes the test particle to register a disturbed infinitessimal proper time which, effectively, means that the retarded potential has transferred momentum to the test particle. Consequently, it acts exactly as a 'gravitational wave' might be expected to act, if such a thing existed. More particularly, the effect of the retarded potential field on the infinitessimal proper time of any test particle can be made identical (at $O(1/R^3)$) to the corresponding effect described by the Eddington form of the Schwarzschild metric of GR so that all the standard classical tests of a gravitation theory are satisfied.

The presented formalism shares - uniquely amongst all competing theories - with GR the property of conforming to the SEP; according to common conjecture (eg Will (1981)), this latter property is necessary and sufficient for a theory to predict that absence of monopole and dipole gravitational radiation which is required of a theory for it to be consistent with the binary pulsar observations. In any case, since the presented theory inherently conserves global four-momentum, then the absence of monopole gravitational radiation is guaranteed by mass-energy conservation, whilst the absence of dipole gravitational radiation is guaranteed by momentum/angular momentum conservation (eg Thorne (1980)); consequently, consistency with the binary pulsar observations is guaranteed anyway.

2 A Lorentzian Model Universe: A

The presented theory is developed by arguing for the observer who takes the philosophical position that space has no meaning beyond an operational definition of distance between material objects, time has no meaning beyond an operational definition of material process and 'Lorentzian spacetime' is a convenient metaphor which serves to encapsulate the supposed connections between operationally defined distance and time in the observed world of material and process. An observer of such a operationalist persuasion would argue that since physics appears to be locally Lorentz invariant then, in the absence of empirically informed arguments to the contrary, there is justification in persisting with the assumption that physics is, in some useful sense, globally Lorentz invariant for as long as such an assumption leads to realistic descriptions of observed physics.

Suppose that a detector (the observer plus measuring instrument) is attached to a lump of material which is extremely remote from all other Universal material. Noting that there exists an operational definition of the concept 'a local state of rest with respect to the average distribution of Universal material' (for example, one can require that the cosmic background is not blueshifted in any direction), this detector can be brought into a 'local state of rest with respect to the Universe', and this state can be considered as semi-permanent. Using this local state of rest of a material system as a reference state, it is then possible to define a class of frames *in the locality of the reference state* every member of which, for all practical purposes, can be considered to be in uniform arbitrary motion with respect to the reference state for some minimum period of time, T^* say. Suppose that this whole class, including the reference state, is labelled $\mathcal{I}_{T^*}$. ¿From the point of view of the detector, the surrounding Universe is a swirling mass of fiercely interacting material within which the detector itself, or any detector attached to any member of $\mathcal{I}_{T^*}$, has a purely passive role.

Now imagine the detector observing any cosmological object in this surrounding interactive Universe: in principle, according to this detector, such an object will generally have three motional degrees of freedom but, in practice, the great distance between the detector and the object means there is no direct method of observing such motions. The only indication of any kind of process occurring between detector and observed objects are the measured redshifts of the objects concerned, and the observation of Hubble that, out to limited distances, measured object redshifts and independently estimated object distances are positively correlated; however, the only way of relating these redshift measurements with any kind of radial evolution is by inference through some theory. This absence of any direct measure of radial evolution, combined with the practical impossibility of measuring angular displacements of cosmological objects - and hence of inferring their transverse spatial displacements over accessible periods of time - means the detector is free to hypothesise a cosmology which constrains the spacetime displacements of distant objects, defined in any of the $\mathcal{I}_{T^*}$ frames, to transform in a Lorentz-invariant way between these frames. In other words, for an infinitessimal interval along the worldline of any cosmological object, $d\mathbf{x}$ say, then $ds^2 \equiv \gamma_{ij}dx^i dx^j$ can, as a working hypothesis, be assumed invariant for Lorentz transformations between members of $\mathcal{I}_{T^*}$. It follows that any detector in $\mathcal{I}_{T^*}$ can, using the assumed cosmology, assign a four-vector and subsequently a four-momentum, to any object in the observable Universe; it then becomes possible to define momentum-conservation laws, valid on a global scale, between members of $\mathcal{I}_{T^*}$. This, of course, is in direct contrast to the situation within conventional theories, like General Relativity, where global conservation laws are, *in principle* undefinable.

Now suppose that the detector is brought into proximity with some large gravitating mass, and consider what has changed. The absolute information content of any cosmological signals received by the detector will be independent of the presence or absence of any nearby gravitating mass so that, as a consequence, if it was possible to define global conservation laws in the absence of the large gravitating mass, then it is still possible in the presence of this mass. However, the detector will now be interacting strongly with this mass and, in particular, clock rates on the detector will have changed. This means that the relative clock rates between the detector and all external observed objects will similarly change in a way which depends purely on the *local conditions* of the detector; consequently, for example, the interval $d\mathbf{x}$ which is, in principle, associated with any given cosmological object in the absence of the large gravitating mass will become the interval $d\mathbf{y}$ in the presence of this mass. Therefore, if global conservation laws are to be defined, as before, then the invariant $ds^2 \equiv \gamma_{ij}dx^i dx^j$ - which gives the relation between the clock rates of any cosmological object and that of the inertial detectors in $\mathcal{I}_{T^*}$ - must be replaced by the new invariant $d\tau^2 \equiv g_{ij}dy^i dy^j$, where g_{ab} is locally defined, and takes account of clock-rate change at the detector arising from changed local conditions.

The problem then reduces to that of determining the form of this g_{ab}. It is shown over the next two sections how the foregoing modelling assumption of a Universe in which four-momentum is globally conserved leads to a natural determination of this tensor.

3 A Lorentzian Model Universe: B

The foregoing section shows how it is possible to conceive of a model Universe in which four-momentum is globally conserved through all processes; given such a Universe, a natural extension of conventional particle-physics experience then leads to the conception of a model Universe which exists as a sequence of four-momentum conserving interaction events, ..., $\mathbf{e}_r$, $\mathbf{e}_{r+1}$, $\mathbf{e}_{r+2}$..., etc. Suppose a particular one of these events, $\mathbf{e}_r$ say, involves a total of N particles through the event so that, if there are n particles into $\mathbf{e}_r$, having respective four-velocities $\mathbf{U}^i, i = 1..n$, then there will be $N - n$ particles out of $\mathbf{e}_r$, having respective four-velocities $\mathbf{U}^i, i = n + 1..N$. If the particles have rest masses given by $m_1..m_N$ respectively then, since four-momentum is to be conserved, we have

$$\sum_{i=1}^{n} m_i \mathbf{U}^i - \sum_{i=n+1}^{N} m_i \mathbf{U}^i = 0. \tag{1}$$

Now suppose that the first sum of (1), which represents the collection of all universal particles which enter into $\mathbf{e}_r$, is arbitrarily partitioned into two parts, $m_1\mathbf{U}^1 + ... + m_k\mathbf{U}^k$ and $m_{k+1}\mathbf{U}^{k+1} + ... + m_n\mathbf{U}^n$ respectively; given these partitions, then the first part can be written *uniquely* as $M_1\mathbf{V}^1$, where $\mathbf{V}^1$ has the structure of an ordinary four-velocity vector, whilst the second part can be written *uniquely* as $M_2\mathbf{V}^2$ where $\mathbf{V}^2$ similarly has the structure of an ordinary four-velocity vector. It follows that M_i, $\mathbf{V}^i$, $i = 1, 2$, can be considered as the notional rest-masses and four-velocities respectively of two 'conglomerated particles' which go into $\mathbf{e}_r$; the second sum of (1) can be treated in a similar way to give two 'conglomerated particles', having notional rest-masses and four-velocities M_i, $\mathbf{V}^i$, $i = 3, 4$ respectively, coming out of $\mathbf{e}_r$. Given this arbitrary four-fold partitioning, then the global four-momentum conservation statement (1) can be rewritten uniquely in the form

$$M_1\mathbf{V}^1 + M_2\mathbf{V}^2 - M_3\mathbf{V}^3 - M_4\mathbf{V}^4 = 0, \tag{2}$$

where it is important to recognize: (i) that the partition corresponding to M_1, for example, could be the collection of particles which form a particular star, or a galaxy, or any self-gravitating structure whatever, short of the whole Universe; (ii) statement (2) is a true statement about the Universe at *any* instant, given only that we are willing to accept Lorentzian physics as a globally invariant physics in the sense defined above; (iii) the *total* of the theoretical content of (2) and therefore, as we shall see, of all that follows, is this latter assumption that physics is globally Lorentz invariant in the sense defined. This assumption is based on the 'orthodox truth' that physics is locally Lorentz invariant, and the fact that there are no *empirically* informed arguments to suppose that this local truth is not also a global truth.

4 The Lorentzian Reference Scalar

In the following, it is shown how (2) provides the basis for a geometric representation of gravitational action within the Lorentzian model Universe leading, finally, to the identification of this action as a special kind of inertial process.

According to (2), the trajectories, represented by the four-velocity vectors $\mathbf{V}^1, \mathbf{V}^2, \mathbf{V}^3, \mathbf{V}^4$ of the four partitions involved in the global four-momentum conserving interaction event, $\mathbf{e}_r$, must lie in some Lorentzian invariant 3-dimensional subspace, $\mathcal{S}_r^3$ say, of the four-velocity space. In this way, the sequence of interactions, $\mathbf{e}_r, \mathbf{e}_{r+1}, \mathbf{e}_{r+2}, ...$, will generate a sequence of subspaces $\mathcal{S}_r^3, \mathcal{S}_{r+1}^3, \mathcal{S}_{r+2}^3, ...$

Any particular subspace, $\mathcal{S}_p^3$ say, can, in turn, be considered to define a tangent hyper-plane to a level-surface of any one of a whole class of scalar functions, $\mathcal{F}_p$ say, defined on the coordinate space, $\mathbf{x}$. In this notation, it follows that any scalar function, U say, which has the distinct subspaces, $\mathcal{S}_p^3$ and $\mathcal{S}_q^3$, as tangent hyper-planes to one of its level-surfaces must lie in the intersection of the function classes $\mathcal{F}_p$ and $\mathcal{F}_q$; that is, $U \in \mathcal{F}_p \bigcap \mathcal{F}_q$. In general, therefore, any function $U \in \mathcal{F} \equiv \mathcal{F}_r \bigcap \mathcal{F}_{r+1} \bigcap \mathcal{F}_{r+2} \bigcap ...$, must have every one of the subspaces, $\mathcal{S}_r^3, \mathcal{S}_{r+1}^3, \mathcal{S}_{r+2}^3, ...$, as a tangent hyper-plane to one of its level-surfaces. In other words, since the subspaces $\mathcal{S}_r^3, \mathcal{S}_{r+1}^3, \mathcal{S}_{r+2}^3, ...$, are representative of the interactions $\mathbf{e}_r, \mathbf{e}_{r+1}, \mathbf{e}_{r+2}, ...$, then the interactive structure of the Lorentzian model Universe can be modelled in terms of any $U \in \mathcal{F}$. In the following, it will be shown how the level-surfaces of any such scalar function provide a natural means for referencing the motion, under gravitational action, of any test particle chosen arbitrarily from any of the four partitions; for this reason, any scalar $U \in \mathcal{F}$ will be referred to as a 'reference scalar'.

5 Gravitation And The Reference Scalar

From the point of view of local physics, gravitational interactions could be defined as that class of interactions through which:

- *four-momentum is conserved;*
- *test particle trajectories are independent of the specific mass values of the test particles concerned.*

The first of these conditions is the basic statement of relativistic mechanics and is already satisfied in the Lorentzian model Universe by virtue of (2). The second of these conditions is a statement of the WEP which simply reflects that characteristic gravitational phenomonology which is discernible to a local observer; in the following, it will be shown how the reference scalar can be used to define a class of trajectories which satisfies this latter condition.

Consider the trajectory, $\mathbf{x}$, of a single particle chosen arbitrarily from one of the four partitions of all the particles; in general, such a trajectory will not be confined to any particular level-surface of U, but will cut across such surfaces with the consequence that it can, under a certain general condition, be described in terms of geometrical change in the level-surfaces of U. The general condition required is that any change, $d\mathbf{x}$, in the position, $\mathbf{x}_0$, of the particle can be put into a 1:1 relationship with the corresponding change, $d\mathbf{n}$, in the normal vector, $\mathbf{n}_0$, to the level-surface of U at $\mathbf{x}_0$. If this circumstance can be assumed to hold, it will follow that the change, $d\mathbf{x}$, in the position of the particle can be represented directly by the corresponding change, $d\mathbf{n}$, in the level-surface normal vector between the positions $\mathbf{x}_0$ and $\mathbf{x}_0 + d\mathbf{x}$. Since, of course, normal-vectors here are defined on a surface which must be considered as curved in some unknown way, then the phrase 'change in this normal-vector' implies, in particular, 'covariant change in this normal-vector' and, in general, the idea that all operations performed in this analysis are performed in some unspecified geometry.

To see how this can be done, first note that, since $n_a \equiv \nabla_a U,\ a = 1, 2, 3, 4$, defines normal vectors to the level-surfaces of the reference scalar, U, then

$$dn_a = \nabla_j(\nabla_a U)dx^j \equiv g_{ja}dx^j \quad \text{where} \quad g_{ab} \equiv \nabla_a \nabla_b U \tag{3}$$

gives the covariant change in n_a along an arbitrary infinitessimal arc $d\mathbf{x}$ (it is assumed at this stage that the geometrical connections which are necessarily required to give this latter expression an unambiguous meaning will be defined in due course). It is clear from (3) that dn_a is in a 1:1 relation with dx^a if g_{ab} is non-singular and uniquely defined everywhere; both of these requirements place conditions on the structure of U. Given these conditions, then dx^a is provided with a unique representation in terms of dn_a, and dn_a is provided with a unique representation in terms of dx^a. It follows from these considerations that the scalar product

$$dn_i dx^i \equiv g_{ij}dx^i dx^j \equiv ds^2,$$

which actually gives a scalar measure of the invariant resolution of this change along $d\mathbf{x}$, can be interpreted as a scalar measure of $d\mathbf{x}$ defined relative to locally specified geometrical properties of level-surfaces of U.

Now suppose p and q are two arbitrarily chosen point coordinates, and suppose the latter expression is integrated between these points to give the scalar invariant

$$I(p, q) = \int_p^q \sqrt{dn_i dx^i} \equiv \int_p^q \sqrt{g_{ij}dx^i dx^j} \equiv \int_p^q \mathcal{L}d\tau,$$

where $\mathcal{L} \equiv (g_{ij}\dot{x}^i\dot{x}^j)^{1/2}$ for arbitrary parameter τ, and further suppose that $I(p,q)$ is minimized with respect to choice of the trajectory connecting p and q; this minimizing trajectory is then geodesic in the Riemannian space which has g_{ab} as its metric tensor. Since the existence of this metric space - and hence of geodesics within it - depends only on the *general fact* of momentum-conservation and is explicitly independent of specific inertial-mass values, then this latter step makes particle trajectories *independent* of particle inertial masses, and is the modelling step defining that general subclass of trajectories which conform to the characteristic phenomonology of gravitation.

To summarize, the four-momentum conserving interactive distribution of all material in a Lorentzian model Universe can be represented as an abstract metric space within which geodesic trajectories serve as natural models for the trajectories of arbitrary particles which are gravitationally interacting with all Universal material. **In effect, this means that the collection of all material in the Universe acts as the natural frame of reference for describing gravitational action.**

6 The Connection Coefficients

In the previous section, it was assumed that the geometrical connection coefficients could be defined in some sensible way. To do this, we simply note that, in order to define conservation laws (ie to do physics) in a Riemannian space, it is necessary to be have a generalized form of Gausses' divergence theorem in the space. This is certainly possible when the connections are defined to be the metrical connections, but it is by no means clear that it is ever possible otherwise. Consequently, the connections are assumed to be metrical and so g_{ab}, given at (3), can be written explicitly as

$$g_{ab} \equiv \nabla_a \nabla_b U \equiv \frac{\partial^2 U}{\partial x^a \partial x^b} - \Gamma^k_{ab}\frac{\partial U}{\partial x^k}, \tag{4}$$

where Γ^k_{ab} are the Christoffel symbols, and given by

$$\Gamma^k_{ab} \equiv \frac{1}{2}g^{kj}\left(\frac{\partial g_{bj}}{\partial x^a} + \frac{\partial g_{ja}}{\partial x^b} - \frac{\partial g_{ab}}{\partial x^j}\right).$$

7 The SEP Is Satisfied

The WEP states that test-particle trajectories are independent of the specific mass values of the test particles concerned whilst the SEP states:

- *The WEP is valid for self-gravitating bodies, as well as for test bodies;*
- *The outcome of any local test experiment is independent of the velocity of the freely falling test apparatus;*

– *The outcome of any local test experiment is independent of where and when in the Universe the experiment is performed.*

The first of these conditions is satisfied by the presented theory because the conservation statement (2), upon which the whole analysis is predicated, is apriori a statement of four-momentum conservation between two *arbitrarily constituted* conglomerated particles going into an interaction, and two similarly arbitrarily constituted conglomerated particles coming out of the same interaction and, in general, such conglomerated particles will be self-gravitating bodies, such as stars, galaxies etc. The second and third of these conditions are automatically guaranteed in the theory simply because it is predicated upon the assumption of a global Lorentz invariance.

8 The Basic Case Of Inertial Motion

It is generally perceived that any material body which is in a state of free-fall and remote from any other substantial material body is in an (almost) perfectly inertial state even though *in principle* it must be presumed the body is interacting with the remainder of the Universe. In this case, the metric tensor, g_{ab}, of the general formalism, specified at (4), which is 'seen' by the freely falling material body can be reasonably approximated by

$$g_{ab} = \gamma_{ab} = \frac{\partial^2 U}{\partial x^a \partial x^b},$$

and the most general solution of this equation is easily shown to be given by

$$U \equiv U^{(0)} = \frac{1}{2}(x^i - x_0^i)(x^j - x_0^j)\gamma_{ij},$$

where $\mathbf{x}_0$ is an arbitrarily chosen origin. Consequently, the class of all possible trajectories of inertial particles is represented by that whole class of reference scalars characterized by $U^{(0)}(\mathbf{x})$. For convenience, any member of this class will be referred to as an 'inertial field'.

9 A Simple Gravitating System

The simple gravitating system to be considered is a perturbation of the inertial system discussed in the previous section. The reference scalar which is 'seen' by any body, mass M say, which is remote from any other substantial material body, is one of the inertial fields, $U^{(0)}(\mathbf{x})$, of the previous section. Now suppose that a material particle, mass $m << M$ say, which is a subordinate component of the 'rest of the Universe', is allowed into close proximity to M; since $m << M$, the trajectory of m will generally not be inertial, but the trajectory of M will be almost unaffected so that it can still be considered to be in an inertial state. Now imagine M did not exist - it is manifestly the case that, because of m's proximity to the original position of M, it would simply adopt the inertial state that was previously held by M. Consequently, it can be concluded that the non-inertial motion of m in the presence of M is due entirely to the proximity of this latter body which can therefore be reasonably designated as the 'gravitating source'.

The problem now is to identify the reference scalar which is 'seen' by m, and which enables a description of its motion to be given. To do this, the model Universe is redefined to consist of the particle m into and out of an interaction event together with the 'rest of the Universe' into and out of the same interaction event, where the 'rest of the Universe' now includes M. In the absence of M, the corresponding reference scalar would simply be one of the inertial fields, $U^0(\mathbf{x})$, of the previous section; however, from the point of view of m, the presence of M disturbs this inertial field in some fashion and, in the following, the modelling assumption is made that this disturbance has the structure of a retarded wave possessing the spatial symmetries of the source, M.

10. An Exact Solution For A Simple Special Case

Suppose the inertial field is perturbed by a source which is a rotating spherically symmetric extended matter distribution of mass M, and that we wish to model the trajectories of test-particles which are exterior to the rotating source, but which are confined to its equatorial plane. Obvious characteristic features of this system are

(a) reflective symmetry of the source about an 'equatorial plane';
(b) rotational symmetry in this equatorial plane about a specific point in it (which, in general, implies the existence of source angular-momentum);
(c) confinement of test particles to the equatorial plane,

and we make the modelling assumption that the gravitational effects of the actual source distribution can be reasonably modelled by any material distribution which possesses these general characteristics. The most simple such distribution is one which has rotational symmetry in the (x^1, x^2) plane of an inertial (x^1, x^2, x^3) rectangular system, and translational symmetry about the x^3-axis. Since the plane $x^3 = 0$ is a plane of relective symmetry in the system, we designate it as the equatorial plane, and consider all test particles to be confined to it. This describes a system possessing a pure cylindrical symmetry, and can be considered to describe any matter distribution which looks the same from any point on the x^3-axis and which, in general, has angular momentum about that axis.

We make the additional modelling assumption that the inertial field disturbance generated by this cylindrical matter distribution has the structure of a retarded relativistic wave which possesses the spatial symmetries of the matter distribution. Consequently, defining $R^2 = (x^1)^2 + (x^2)^2$, then this matter distribution will give rise to a reference scalar having the most general form

$$U(\mathbf{x}) = U^{(0)}(\mathbf{x}) + \frac{H(R-ct)}{R},$$

where $U^{(0)}(\mathbf{x})$ is the inertial field and H is any twice differentiable function.

The objective now is to find the metric tensor, g_{ab}, which this form of U gives rise to via equation (4), and it is at this stage that the simplifying benefits of the cylindrical model become manifested: the geometry on the surface of a cylinder is Euclidean (ie unlike the surface of a sphere, the surface of a cylinder can be 'opened out' to form an ordinary flat plane) and so it follows immediately that the connection coefficients in (4) are identically zero. Consequently,

$$g_{ab} = \frac{\partial^2 U}{\partial x^a \partial x^b} = \gamma_{ab} + \frac{\partial^2}{\partial x^a \partial x^b}\left(\frac{H}{R}\right). \tag{5}$$

Expanding this expression, forming the corresponding proper time element, and transforming to cylindrical polar coordinates ($x^1 = R\cos\phi$, $x^2 = R\sin\phi$) leads to

$$d\tau^2 = \Delta_0 + \Delta_1 + \Delta_2 \tag{6}$$

$$\Delta_0 \equiv c^2dt^2 - dR^2 - R^2d\phi^2,$$

$$\Delta_1 \equiv \frac{\ddot{H}}{R}\left(dR^2 - 2cdRdt + c^2dt^2\right),$$

$$\Delta_2 \equiv \frac{\dot{H}}{R^2}\left(-2dR^2 + R^2d\phi^2 + 2cdRdt\right) - \frac{H}{R^3}\left(-2dR^2 + R^2d\phi^2\right).$$

The Δ_0 component represents the infinitessimal proper time registered by a test particle moving in the undisturbed inertial field, whilst Δ_1 and Δ_2 describe the disturbance of this proper time which is due to the cylindrical matter distribution.

In the following section, the approximation $d\tau^2 \approx \Delta_0 + \Delta_1$ to (6) is shown to give rise to a two-dimensional form of the conventional Schwarzschild gravitational physics whilst, in section 11, it is shown that the exact form, $d\tau^2 = \Delta_0 + \Delta_1 + \Delta_2$, leads to a gravitational structure which has an event horizon at the usual Schwarzschild radius, and displays the usual Schwarzschild-type physics outside of this radius, but which *has no singularity* inside this radius.

11. Schwarzschild Physics As An Approximate Solution

Suppose that the disturbance which gives rise to (6) is simply one of a continual train of identical disturbances passing through the particle with a regular frequency measured in the rest frame of the source (the source is considered as an oscillator), then $\ddot{H}$, $\dot{H}$ and H can be replaced by their mean values between successive minima in the train of disturbances, and $d\tau^2$ of (6) can be redefined as a measure of the mean proper time registered by the test particle between these successive minima. With this understanding, and writing the mean value of $\ddot{H}$ as λ, then, to $O(1/R)$, the disturbance in the inertial field passing through the test particle causes the particle to register a mean infinitessimal proper time, given by

$$d\tau^2 \approx \Delta_0 + \Delta_1 = c^2 \left(1 + \frac{\lambda}{R}\right) dt^2 - \frac{2c\lambda}{R} dR dt - \left(1 - \frac{\lambda}{R}\right) dR^2 - R^2 d\phi^2,$$

measured over the interval for which the disturbance can be said to be passing through the particle. Now, since the integrated magnitude of the deviation of the disturbed proper time from the undisturbed proper time, measured over any finite interval, will depend on the frequency with which the disturbances arrive at the test particle then it can be deduced that the undetermined parameter λ must be a measure of the frequency, and hence rest mass, of the disturbing source. Consequently, $\lambda = \beta m$, for constant β and rest mass m, and the proper time element approximation $d\tau^2 \approx \Delta_0 + \Delta_1$ is reduced to the form of that arising in any conventional metric theory.

Finally, it is noticed that the general form of the latter proper time element approximation is identical to that of the Eddington form of the Schwarzschild proper time element of General Relativity, and that the two forms match exactly if $\beta = -2\gamma/c^2$; consequently, under the transformation,

$$cdt \rightarrow cdt + \left(1 - \frac{c^2 R}{2\gamma m}\right)^{-1} dR$$

then

$$d\tau^2 \approx \Delta_0 + \Delta_1 \equiv c^2 \left(1 - \frac{2\gamma m}{c^2 R}\right) dt^2 - R^2 d\phi^2 - \left(1 - \frac{2\gamma m}{c^2 R}\right)^{-1} dR^2,$$

and this form is identical to the Schwarzschild proper time element, as required.

Since Schwarzschild physics arises from GR via a modelling of a spherically-symmetric process, it a-priori excludes the effects of source angular-momentum in gravitational physics. It follows that the approximation $d\tau^2 \approx \Delta_0 + \Delta_1$, condidered in this section, similarly excludes source angular-momentum effects, but that the exact form, $d\tau^2 = \Delta_0 + \Delta_1 + \Delta_2$, which models the *most general* cylindrical symmetry must be inclusive of such angular-momentum effects. This exact form is considered in the following section.

12. The Absence Of Singularities In The Presence Of Source Angular-Momentum

In the previous section, λ was defined as the mean value of $\ddot{H}$ between successive minima in the train of inertial disturbances generated by the central massive source. If α and β are now defined, in a similar way, as the corresponding mean values of $\dot{H}$ and H, and the definitions

$$X = 1 + \frac{\lambda}{R}, \quad Y = 1 - \frac{\alpha}{R^2} + \frac{\beta}{R^3}, \quad Z = \frac{\lambda}{R} - \frac{\alpha}{R^2},$$

are made, then the full line element, given at (6), can be written as

$$d\tau^2 = Xc^2dt^2 - (4 - X - 2Y)dR^2 - YR^2d\phi^2 - 2ZcdRdt,$$

and this leads fairly easily to the radial velocity equation

$$\frac{dR}{cdt} = \frac{\pm XW}{Yk_0 \pm ZW}, \quad \text{for } W \equiv \sqrt{\frac{Y(-k_0^2Y + XY + k_1^2X/R^2)}{X(X+2Y-4) - Z^2}} \tag{7}$$

The choice of signature '$\pm$' determines if radial motion is towards, or away from, the gravitating source. Note that $\alpha = \beta = 0$ yields the corresponding GR equations.

In the previous section, it was shown that $\lambda = -2\gamma m/c^2$ and so the Schwarzschild radius is at $R_s = -\lambda$. Consequently, from the definition of X, it can be seen that $X = 0$ in (7) on the Schwarzschild boundary, and an analysis of the equation in the region of this boundary shows that test particles can cross it when inward bound (signature choice is '$-$'), but cannot cross it when outward bound (signature choice is '$+$'). In this sense, of course, the presented theory is in direct accord with the GR model.

The real differences between the presented theory and GR emerge when the radial equation (7) is analysed in the limit of $R \to 0$. Specifically, it is found that

$$\frac{dR}{cdt} \approx \pm\left(\frac{\lambda}{\alpha}\right)R \text{ so that } \frac{d^2R}{c^2dt^2} = \left(\frac{\lambda}{\alpha}\right)^2 R.$$

Since there is certainly an *inward* acceleration for $R \geq R_s$, this latter equation implies that gravitational attraction 'turns off' at some $R = R^*$ for $0 < R^* < R_s$, and becomes gravitational repulsion in the region $0 < R < R^*$. The origin, $R = 0$, is a point of unstable equilibrium, and so can be considered as the top of a 'potential hill'. Consequently, the essential singularity which exists in the Schwarzschild model at $R = 0$ disappears when the effects of source angular-momentum are included. This distinction can be expected to have profound consequences on our understanding of the physics of gravitational collapse.

13. Conclusions

It has been shown that 'metric gravity', as a general concept within the context of relativistic physics, is implicitly contained within the concept of global four-momentum conservation, and is completely independent of the curved spacetime concepts which characterize most other modern theories.

The derived theory cannot be distinguished from General Relativity for all local tests and the essential singularities at gravitational origins, which are features of both GR and Newtonian theory, do not exist here when the effects of source angular-momentum are included. Furthermore, unlike all other competing theories, except for General Relativity, the presented theory satisfies the Strong Equivalence Principle which, it is conjectured (eg Will (1981)), is a necessary and sufficient condition for a theory to pass the test of the binary pulsar observations. The truth of this latter conjecture receives support in the present case because, since the theory is globally Lorentz invariant and is predicated of the global statements (1) and (2), mass-energy and momentum/angular momentum are automatically conserved through gravitational interactions; consequently, the absence of monopole and dipole components to gravitational radiation is guaranteed (eg Thorne (1980)), and these absences are the precise conditions imposed by the binary pulsar (eg Will (1981)). It follows that the theory is indistinguishable from General Relativity at the level of the binary pulsar also.

Finally, because the model Universe in which this theory is developed is globally Lorentz invariant, then gravitational processes are finally reduced to particle/wave/particle interactions of a kind familiar to the rest of physics. The wider implication of this is simply that the conceptual barriers which have previously prevented the assimilation of gravitational physics into quantum physics do not exist here.

References

Thorne, K.S., *Gravitational Wave Research: Current Status and Future Prospects*, Rev. Mod. Phys.**Vol 52**, pp. 285-297 (1980).

Will, C.M., *Theory and Experiment in Gravitational Physics*, CUP (1981).

A MODEL COSMOLOGY BASED ON GRAVITY–ELECTROMAGNETISM UNIFICATION

JOHN E. BRANDENBURG
RSI Incorporated
635 Slaters Lane
Suite G101
Alexandria, VA, 22314 USA

Abstract. In this article the GEM (Brandenburg, 1992; Brandenburg, 1988) theory is applied to the problem of the cosmos in which most of the matter is hydrogen, spacetime is flat, and a Cosmic Background Radiation CBR field exists. Using the two postulates of the GEM theory: 1. That gravity fields are equivalent to an array of $\mathbf{E} \times \mathbf{B}$ drifts or a spacially varying Poynting field, such that spacetime is determined by EM fields so that the stress tensor of ultrastrong fields is self-canceling; 2. That EM and gravity fields and protons and electrons are unified at the Planck scale of lengths and energies and split apart to form distinct fields and separate particles at the "Mesoscale" of normal particle rest energies and classical radii. A new derivation is made of the formula for G found previously: $G = e^2/(m_p m_e)\alpha \exp(-2R^{1/2}) = 6.668 \times 10^{-8}$ dynes cm^2 g^{-2}where m_p and m_e are the proton and electron masses respectively, $R = m_p/m_e$ and α is the fine structure constant, shows that quantum processes may occur which make the vacuum unstable to appearance of hydrogen thus allowing matter creation and a steady state universe to occur. The value for the Hubble Time calculated from this model is $T_o = (3/((2\alpha)(\alpha R^{1/2})^4))^{1/3}(r_e/c)(e^2/Gm_pm_e) = 19$ Gyr where $r_e = e^2/\left(m_e c^2\right)$ and follows the form first hypothesized by Dirac(1937). The CBR is traced to this process of matter creation and its temperature is calculated as being $T_{CBR} = ((3/4)G{m_e}^2c/(\sigma^2\sigma_o))^{1/4} = 2.66$K where σ is the Thomson cross section of the electron and σ_o is the Stefan-Boltzman constant.

Key words: Plasma Cosmology, Gravitation, Electromagnetics

1 Introduction

The Gravity Electro-Magnetism (GEM) (Brandenburg, 1992; Brandenburg, 1988) theory is based on the founding assumption of the Plasma Universe (Alfvén, 1990; Peratt, 1992) description of Cosmology: that electrodynamics plays at least an equal role with gravity in shaping the universe both large and small. The GEM theory goes from this to say that gravity is in fact a manifestation of electrodynamics so that the Plasma Universe description is in fact complete, in that the electrodynamics of large numbers of free particles is the key concept for understanding the universe on all scales.

Unity of scales, from the Cosmic of sizes and energies, through the Mesoscale of classical particle radii and rest masses, to the Planck scale (Kar'orolyhazy *et al.*, 1986), the scale where quantum uncertainty allows the formation of a "virtual plasma" (Wells *et al.*, 1989) of blackholes and which is the most

Astrophysics and Space Science **227**: 133–144, 1995.

violent portion of the vacuum turbulence, is at the heart of the GEM theory, for its alternative meaning is the Latin phrase *Grandis Et Medianis*, meaning the unification of the great, meaning both Cosmic and the Planck scale, with the medium or Mesoscale. So the GEM theory attempts to define and understand the relationship of the very smallest parts of the universe known to humanity, to the very largest. Given this goal it is reasonable to try to understand cosmology through the GEM theory.

The GEM theory is based on two postulates:

1. That gravity fields are equivalent to an array of $\mathbf{E}\times\mathbf{B}$ drifts or a spacially varying Poynting field, such that spacetime is determined by EM fields so that the stress tensor of ultrastrong fields is self-canceling.

2. That EM and gravity fields and protons and electrons are unified at the Planck scale of lengths and energies and split apart to form distinct fields and separate particles at the "Mesoscale" of normal particle rest energies and classical radii.

Following Sakharov, Zeldovich, and Puthoff (Sakharov, 1968; Putfoff, 1989), the GEM theory defines Gravity as a form of ultra strong Electro-Magnetic radiation pressure arising from the vacuum turbulence predicted by Quantum Mechanics and says that the separate appearance of Gravity and Electro-Magnetic forces is mirrored in the separate appearance of the electron (leptons) and protons (baryons). In a separate discussion of the GEM theory (Brandenburg, 1992) the focus was on the transition from the Planck scale to the Mesoscale. In this discussion the GEM theory will be applied to include the Cosmic scales.

We will now use the GEM theory to construct a cosmology for a Plasma Universe. The starting point for this is a derivation for the value of G.

2 A Derivation of G

2.1 A Gedanken Experiment

In a Gedanken experiment we can place a single atom of hydrogen in a small enclosure and shrink the enclosure. As the enclosure shrinks quantum uncertainty will cause the proton and electron to gain energy so the atom will ionize. If we continue the shrinking the electron and proton will become relativistic and grow strongly in mass. Finally we will shrink the enclosure to the Planck length $r_p = (G\hbar/c^3)^{1/2}, \hbar = h/2\pi$ at this point the enclosure will become a black hole and both the electron and proton will assume the Planck mass $M_p = (G/\hbar c)^{1/2}$. The black hole will be unstable however and evaporate into a shower of particles and antiparticles that will be torn out of the vacuum just outside the black hole event horizon. The electron is a lepton and the proton is a baryon and yet the shower of particles created by the black hole evaporation is of particles and antiparticles so the original lepton number of one and baryon number of one have disappeared and been replaced by zeros. Thus our Gedanken apparatus for shrinking an atom of hydrogen apparently destroys lepton and baryon number.

2.2 A DERIVATION OF G

According to the second GEM postulate protons and electrons (the stable massive lepton and the stable isolated baryon) will merge at the near the Planck length so the Gedanken result is not unexpected. However, the GEM theory says this should happen smoothly. We can write a model for the electron and proton as a particle doublet, transformations of a single particle which we will call the "union" particle. We adopt the standard formula for the doublet from particle theory (Raby *et al.*, 1988)

$$m = m1 \sin\theta + m2 \cos\theta \tag{1}$$

However since we believe only in a union particle $[m1] = [m2] = m_o$. Since the angle θ determines lepton and baryon separation it must have a different character than the usual transformation angle whose rotations can be accomplished at accessible energies, since the lepton and baryon numbers are conserved at these energies. Therefore we choose the angle to be imaginary so that θ =iW. This means that $m2 = im1$ and we have

$$m = im_o \sin iW + m_o \cos iW \tag{2}$$

thus

$$m = m_o \exp(W) \tag{3}$$

We then have a proton as the W = +Q "up" particle and the electron is the down particle $W = Q$ where $m_o = (m_p m_e)^{1/2}$ and $W = \ln((m_p m_e)^{1/2})$. We now must have W be a function of the radius of curvature r_c such that it changes very little until it becomes close to the Planck length where W disappears. The simplest way for this to occur is

$$W = \ln(\ln(r_c/r_p)) \tag{4}$$

This means that when r_c gets to 2.7 times the Planck length that the proton and electron masses merge to become the union mass m_o, whereas when, $r_c = r_o$ the Mesoscale range of size, $r_o = e^2/m_o c^2$ of classical particle radii. This also means

$$(m_p/m_e)^{1/2} = \ln(r_o/r_p). \tag{5}$$

This formula can be then written

$$r_o/r_p = \exp((m_p/m_e)^{1/2}) \tag{6}$$

which simplifies to become

$$G = e^2/(m_p m_e)\alpha \exp(-2(m_p/m_e)^{1/2}) \tag{7}$$

where $\alpha = e^2/\hbar c$, which yields $G = 6.668 \times 10^{-8}$ dynes cm^2 g^{-2}, which is within 700 ppm of the measured value $G = 6.673 \times 10^{-8}$ dynes cm^2 g^{-2}(Cohen and Taylor, 1989). The difference between the derived formula for G and its actual value is of order $1/R$ (554ppm) where $R = m_p/m_e$, indicating that the formula for G may be the leading term in a series with terms in increasing powers of $1/R$. It is worth noting that in a previous paper (Brandenburg, 1992) we derived a formula for α

$$\alpha^{-1} = R^{2/3} - R^{1/3} - 2/3 + 8/(9R) - 32/(27R^2)... \tag{8}$$

where again powers of $1/R$ entered. Therefore the GEM theory postulates that the splitting of masses of the proton and electron is related to the splitting of the strengths of Gravitation and EM forces on these two particles and this leads to a formula accurate to leading order in the inverse mass ratio of the particles.

There are two important results from this derivation 1. the electron and proton split apart smoothly from the Planck length indicating that the virtual plasma of Planck masses can undergo decay into a proton and electron, transitioning from a state in which they behave like antiparticles with canceling quantum numbers into a state where they cannot annihilate each other and hence must form hydrogen. 2. Gravity and electromagnetism can be understood as being in a physical relationship.

2.3 A Physical Interpretation of Gravity

In obtaining a formula for the Gravitation constant we obtain insight into the relationship of gravity to EM forces. This occurs because the Newton's Constant can be interpreted as if it were a quantum mechanical operator and the portions of the formula can then have physical interpretation, as parts of an operation. To calculate the force between two masses then we have

$$F = -GM_1M_2/D^2 \tag{9}$$

where M_1 and M_2 are two masses and D is a distance between their centers.

Under GEM this can be written as

$$F = -e^2/{m_o}^2\alpha \exp(-2(m_p/m_e)^{1/2})M_1M_2/D^2 \tag{10}$$

or rather

$$F = (iM_1/m_o)(iM_2/m_o)P_{GEM}e^2/D^2 \tag{11}$$

where

$$P_{GEM} = \alpha \exp(-2(m_p/m_e)^{1/2}) = Gm_pm_e/e^2 \tag{12}$$

This says that G acts on the two masses and their separation to yield a force by first measuring each mass in units of m_o, the intrinsic mass of electrons and protons when they merge at the Planck length and form unions. When the measurement is accomplished, an imaginary electron charge quantum is assigned per union mass, so the resulting EM interaction is completely symmetric but attractive. The EM interaction is then weakened by a dimensionless factor P_{GEM} which can be interpreted as a probability of emission or absorption of a quantum of radiation that carries momentum and energy in gravity fields for a particle of charge e, or as a dielectric constant of the vacuum for imaginary E fields. In the classical interpretation the vacuum has an enormous linear dielectric constant, and nearly neutralizes the actual charge of the particles by polarization.

Following a quantum interpretation the quantity P_{GEM} is analogous to the quantity α in Quantum Electrodynamics, indeed it is linear in α, and thus corresponds to a probability for the emission and an absorption of a quantum carrying energy and momentum in gravity fields between elementary charges, which we will term a GEM photon, because it is best understood as a photon, but whose appearance and absorption is very unlikely compared to ordinary photons that would carry energy and momentum between charges.

The quantity α is taken to be the probability of a charged particle of charge e undergoing the process of an EM photon emission and absorption (Misner *et al.*, 1971). This can be seen by assuming that r_e/c, the light transit time of a classical electron radius, is the time for photon–charged electron interaction and that the quantum "jitter" of an electron in a Compton wavelength, $r_e/\alpha c$, represents the absorption and emission of a photon. The electron thus constantly Compton scattering off a photon and then emitting a bremsstrahlung photon, as is allowed by quantum uncertainty. The time for the absorption-emission process is then $r_e/\alpha c$, the light transit time of a Compton wavelength. Therefore, assuming the absorption and emission events to be independent, the probability for the photon emission-absorption is, during any given photon interaction time r_e/c,

$$P_{EM} = (r_e/c)(r_e/(\alpha c)) = \alpha \tag{13}$$

Since the classical radius cancels, this model should apply to all charged particles of charge magnitude e.

Therefore, the formula for G would say that gravity can be understood as a form of electrodynamics, where masses are normalized by the unified mass m_o and assigned imaginary units of electronic charge per unit mass and where the interaction of the charges is much diminished by polarization of the vacuum. This formula also indicates that the mass unit m_o is important at the Mesoscale despite the fact it does not appear as the rest mass of any stable particle.

The fact that we derived a highly accurate formula for G using a model where the electron and proton split apart smoothly from the Planck length indicates that the virtual plasma of Planck masses may be able to undergo an irreversible decay into a proton and electron, from a state in which they behave like antiparticles with canceling quantum numbers into a state where they cannot annihilate each other and hence must form hydrogen. It can be seen that this nonconservation of lepton-baryon number coupled with the possibility of quantum fluctuation of the vacuum must have cosmological implications.

The physical consequences of this formula and conceptual derivation for G on Cosmology will now be discussed.

3 Vacuum Decay and the Hoyle Steady State Cosmology

The derivation of G based on the GEM postulates assumed that the proton and electron split apart smoothly from the union particle. This suggests the possibility of an irreversible process, involving the splitting apart of the union particle pair, which function as each others antiparticles, into a proton and an electron. Momentary nonconservation of mass energy is necessary for any Cosmology that has an expanding universe, the possible models for this differ only in the degree of spacetime concentration of such events. Therefore, the splitting of a virtual union pair to form hydrogen does not seem unreasonable. Such a process would satisfy the second law of thermodynamics since at the Planck length only one specie of particle and one type of force exist, thus the vacuum is highly ordered, whereas the only way entropy, the spread of systems over all possible states, can increase is for the virtual particles of the vacuum to find a way to appear as real particles, and greatly expand the number of achievable states. This would mean a virtual particle pair could suddenly appear as two real particles out of the vacuum. If this could occur it would have to be an extremely unlikely process, since we know that in space very low number densities exist. However, we know of a process that involves nonconservation of lepton and baryon number, a unified gravity and EM or GEM process of GEM photon absorption and emission, the same process that leads to gravity between ordinary bodies, and we know that this process has a quantum probability that is very small. We also know that this process would not be energetic, since gravitational interactions between bodies at cosmic length are not energetic.

If we assume the probability P_{GEM} describes this process, we can estimate a rate of vacuum decay from the GEM theory.

The decay of the vacuum through a GEM process occurs as an extremely unlikely event during a fluctuation of the vacuum. The fluctuation begins as virtual pair of union particles of mass m_o with a Compton radius of

$$r_{co} = h/(2\pi m_o c) = r_e/(\alpha R^{1/2}) \tag{14}$$

where $r_e = e^2/(m_e c^2)$ is the classical electron radius and $R = m_p/m_e$. During a quantum fluctuation the two merged particles must undergo two GEM interactions, one for the negative union particle to turn into an electron and one for the positive union particle to turn into a proton. Since they are an electron and proton in close contact, much closer than a Bohr radius of a stable hydrogen atom, they must also emit an EM photon to form hydrogen, in either an atomic or ionized state. Thus we can write an approximate rate of density increase with time due to vacuum decay for empty space.

$$n_{,t} = 3/(4\pi {r_{co}}^3)(c/r_{co}){P_{GEM}}^2\alpha \tag{15}$$

where for every union Compton volume, during every union quantum fluctuation time, there is a P_{GEM}^2 probability of a double GEM photon emission and absorption event. In an another interpretation, P_{GEM} can represent the probability of a quantum mechanical "tunneling" of a union particle from the Planck scale to the Mesoscale. Since this event requires the appearance of two charged particles it must involve the absorption and emission of EM radiation, therefore it is multiplied by the probability α of an EM photon emission and absorption.

The rate of hydrogen appearance from the vacuum is thus approximately

$$n_{,t} = 3/(4\pi {r_e}^3)(c/r_e)(\alpha R^{1/2})^4(Gm_p m_e/e^2)^2\alpha = 1.57 \times 10^{-23}\,\text{cm}^{-3}\text{sec}^{-1} \tag{16}$$

A fundamental question arises at this point as to why the plus and minus unions form a proton electron pair upon undergoing GEM photon interactions, instead of an antiproton-positron pair. In this context, the question is related to the question as to why matter, rather than antimatter, dominates the universe [Although there is no physical evidence that antimatter does not exist in quantities equal to matter (Alfvén, 1990)]. The answer cannot be given completely within the GEM theory at this time. However, a partial answer can be given: the plus and minus union pair must communicate with the rest of the universe through the GEM photon field in order for the vacuum decay to occur, and through this GEM photon interaction, the universe must communicate to the pair that they must become matter rather than antimatter. Possibly this is communication is via the spin polarization of the GEM photons which, like neutrinos and antineutrinos, may come in only one polarization or spin state relative to their motion. This would make the vacuum decay similar to the decay of an isolated neutron or antineutron whose process of origin is unknown and whose identity as a neutron or antineutron cannot be determined until it decays. The neutron decays into a proton and an electron and emits an antineutrino whose spin relative to its motion is in only one configuration, in contrast to EM photons, whose spin can be either plus or minus relative to their motion. The emission of the antineutrino communicates to the rest of the universe at the speed of light that a hydrogen has formed from the decay of a neutron,rather than the corresponding antimatter process. In a similar way the GEM photons may communicate to the union pair that the universe is predominately matter, and force them to emulate this condition. This is probably fundamentally related to the fact that GEM photons produce only attractive forces between masses but EM photons produce both attractive and repulsive forces. That is, EM photons can be in two states related to two signs of charge, but GEM photons come with only one state related to charge. The GEM photons may thus force one relationship of charge to mass state, a massive positive charge and a light negative charge. If the GEM photons can be shown to have such a requirement as neutrinos in this regard, then simple laboratory tests of GEM may result.

The appearance of hydrogen from the vacuum that GEM predicts will be extremely difficult to observe directly. The estimated rate of hydrogen production per unit volume is very low due to the extremely low probability of a double GEM event during a quantum fluctuation time. It amounts to roughly one hydrogen atom appearance per cubic meter every 6 billion years. This is too low to observe in any terrestrial vacuum chamber, even if the vacuum could be made perfect. The only place to observe such a process is in the vast vacuum regions of the cosmos itself, by looking for its indirect effects.

Hoyle and Bondi (Hoyle, 1948) proposed that the cosmos was in a steady state so that new matter was added to maintain the same density and hence the same expansion rate. No Big Bang occurs in the Hoyle cosmology. Hoyle's theory was not generally accepted because it could not explain a source for the new matter in the cosmos and it could not explain the CBR. However, the GEM theory can explain both the new supply of matter and the CBR.

We can find the rate of mass density growth necessary for a Hoyle Steady state universe with an expansion rate $V = D/T_o$, and assuming a uniform steady state density number density n, from first assuming particle number conservation after appearance:

$$(nD^3)_{,t} = n_{,t}D^3 + 3nD^2V = 0 \tag{17}$$

we then obtain the rate of decrease in number density because of expansion:

$$n_{,t} = -3n/T_o \tag{18}$$

In order to be a flat cosmos we must have a steady state number density

$$n = 3/(8\pi G m_p T_o^2) \tag{19}$$

where the Hubble Time, T_o, is no longer the age of the universe but a parameter of the cosmic expansion. This then gives the number density production rate of hydrogen required to maintain density in a Hoyle steady state universe:

$$n_{,t} = 9/(8\pi G m_p {T_o}^3) \tag{20}$$

As can be seen it places a requirement on the number density appearance rate in terms of the Hubble Time.

We can test the vacuum decay rate derived from the GEM theory, under the assumption that it leads to a Hoyle steady state cosmos.

$$3/(4\pi {r_e}^3)(c/r_e)(\alpha R^{1/2})^4(G m_p m_e/e^2)^2\alpha = 9/(8\pi G m_p {T_o}^3) \tag{21}$$

Simplifying and solving for the Hubble Time we obtain

$$T_o = (3/((2\alpha)(\alpha R^{1/2})^4))^{1/3}(r_e/c)(e^2/G m_p m_e) = 5.9 \times 10^{17}\,\text{seconds or}$$

$$T_o = 19\,\text{billion years.} \tag{22}$$

This is quite consistent with estimates of 18 billion years, plus or minus 2 billion years, for the Hubble time (Sandage 1972). The steady state density of hydrogen is predicted as

$$n = 3.1 \times 10^{-6}\,\text{cm} \tag{23}$$

or $n = 3.1\,\text{m}^{-3}$ which is well within the range of present estimates of 0.1-10 per cubic meter.

This expression for the Hubble time is very close to that first noted empirically by Dirac (Dirac, 1937). The fact that the scale of the cosmos can be expressed in terms of Mesoscale and Planck scale quantities, as they now exist, suggests that whatever process allowed the appearance of matter in the cosmos, this process was consistent with present laws of physics. In addition,the fact that the Hubble time can be expressed closely in terms of quantities thought to be constants, suggests that it is either a fixed parameter of the cosmos, not an age, or else Planck and Mesoscale are all expanding together with the cosmos. In either case, the possibility of GEM vacuum decay as a process in our contemporary cosmos must be admitted. Such ideas as vacuum decay have already been proposed as part of inflationary cosmos (Linde, 1987) where the vacuum is unstable to producing particles. This leaves the CBR as the remaining objection to a Hoyle cosmology.

4 The CBR

The sudden appearance of a hydrogen atom from the vacuum can be likened to a radiative recombination in a laboratory plasma. This event is not likely, because it requires a large amount of momentum and energy to be carried off by the EM field. In other words, radiation must be produced for a recombination to occur. Under GEM this radiation accompanying the appearance of hydrogen from the vacuum is the CBR. The CBR is then the evidence not of a past Big Bang but of myriad little "pops" of hydrogen out of the virtual plasma, with an accompanying burst of radiation. If this model is correct much of the "missing mass" or "dark matter" in the cosmos is simple hydrogen in the plasma state at very low density. It is this thin hydrogen plasma that probably fills the voids present between galaxies. This would mean that hydrogen appears in an ionized state in the voids and is thus very difficult to detect because it gives no atomic absorption or emission. Thus if this vacuum decay model is correct then hydrogen in the plasma state makes up far more of the mass of the universe than presently believed. So this model may explain the origin of the CBR, but why does the CBR have such a low temperature if it represents a process originating at the Planck scale?

We can derive a value for the CBR temperature by assuming that when the electron proton pair appears and expands from the Planck scale that the electron, being lighter, is the active radiator, and that the expansion is described by the Einstein equation for a miniature flat universe

$$(8\pi/3)\ G\rho a^2 = v^2 \tag{24}$$

where ρ is the mass density and where we equate the radius a to be the electron classical radius and v^2 to be equal to the velocity induced by a radiation pressure coming from an electron acting on the Thomson scattering cross section σ of the electron

$$G{m_e}^2/r_e = 1/2 m_e v^2 = \sigma^2 P/r_e \tag{25}$$

Thus the electron expands like a tiny flat universe full of radiation pressure and then joins the rest of the universe with a local radiation density that is released. This radiation, on release, is consistent with the dynamics of a miniature universe. This gives a released radiation pressure when the electron has expanded to its classical radius:

$$P = G{m_e}^2/\sigma^2 \tag{26}$$

This can be also understood as the scattering of the electron off a preexisting radiation field pressure P. When this radiation pressure is assumed to be due to a Planckian radiation field we can recover its temperature as

$$T_{CBR} = ((3/4)cG{m_e}^2/(\sigma^2\sigma_o))^{1/4} = 2.66\,\mathrm{K} \tag{27}$$

where σ_o is the Stefan Boltzman constant and the measured value is $T_{CBR} =$ 2.73 Kelvin (Schwarzschild 1990) so the predicted value is within 3 percent.

Thus the CBR can be explained as the "radiation of appearance" of hydrogen from the virtual plasma of the vacuum in a Hoyle steady state universe, a sort of "recombination" radiation from a virtual plasma. Since this calculation is a classical one, higher order quantum effects may account for the error.

5 Conclusions

Perhaps new matter comes into being from the decay of the vacuum itself. The vacuum contains the virtual plasma at the Planck length. Under GEM this plasma is actually highly compressed hydrogen, where electrons and protons are merged and act as each others antiparticles. Occasionally a proton and electron in the virtual plasma depart from their condition of being merged and are suddenly slightly different. They can no longer act as each others antiparticles and expand violently to form an atom or puff of hydrogen plasma. Thus the virtual plasma suffers a rare "recombination" sort of event and out of the sea of invisible particles two particles, an electron and a proton, suddenly pop out, to become visible. Somethingness from nothingness.

This process of vacuum decay means perhaps the cosmos did not begin with a bang, but rather with a whisper, and its global structure springs from the structures of the proton and electron and finally the vacuum itself. This relationship between the Cosmic and the subatomic was first noted by Dirac (Dirac, 1937) who found that G and T_o seemed related by constants

$$e^2/G{m_e}^2 \approx T_o c/r_e \tag{28}$$

and assumed that T_o was a time since the Big Bang. From this Dirac suggested that G must change with time. However, it has been compellingly shown by Hellings et. al. (1982), using Viking Mars lander distance data, that G is a true constant on the timescales of T_o. Therefore the Dirac relation appears to be a relationship of constants describing a Hoyle-Bondi cosmology.

As was pointed out by Arp (1991) the only difference between a Big Bang Cosmology and a Steady State Cosmology is the degree of concentration of nonconservation of mass. If the hypothesis that the whole universe can spring into being in one event is to be admitted, why not admit the possibility that it is continually generated? If the universe can appear in a zero energy state, with gravitational energy canceling mass energy, as is suggested by Hawking (1988), then the same condition can occur as a diffuse process. The fact that several cosmic quantities can be expressed closely in terms of quantities that describe the present universe, suggests that the structure and evolution of the cosmos depends on physics that is operating continuously up until the present rather than a singular event outside of present physics.

In any case, it appears the GEM unification theory, a theory flowing from the work of Sakharov, Zeldovich, and Putoff, allows a model of the cosmos much like that proposed by Hoyle, Bondi, and Dirac, in which the Cosmic and the Planck scales, as well as the Mesoscale, are all related and unified.

6 Acknowledgments

The author wishes to acknowledge the help and encouragement of many people: my dear wife Suzie, Dr. Richard F. Post, Dr. Bruce Amlicke, and Dr. Anthony Peratt who have made the formulation and publication of this theory possible.

References

Alfvén, H.: 1990, "Cosmology in the Plasma Universe", *IEEE Trans. Plasma Sci.* **Vol. 18**, pp. 5-10

Arp, H.: 1991,"Exploding the Big Bang Hypothesis", *Physics Today* **Vol. 44**, pp. 85-86

Brandenburg, J. E.: 1988,"The GEM Unification Theory: Part III", *Bull. Am. Phys. Soc.* **Vol. 32**, p. 1049

Brandenburg, J. E.: 1992,"Unification of gravity and electromagnetism in the plasma universe", *IEEE Trans. Plasma Sci.* **Vol. 20**, pp. 944-957

Cohen, E. R. and B. N. Taylor: 1989,"The Fundamental Physical Constants", *Physics Today* **Vol. 42**, pp. 8

Dirac, P.A.M.: 1937, "The Cosmological Constants" *Nature*, **Vol. 139**, p. 323. Also in Lee Smolin, "Stochastic Mechanics, Hidden Variables and Gravity," in *Quantum Concepts in Space and Time.*

Hawking, S. W.: 1992, *A Brief History of Time* Bantam Books, New York, pp. 151

Hellings, R. W., P.J. Adams, J.D. Anderson, M. S. Keesey, E. L. Lau, E. M. Standish, V.M. Canuto and I. Goldman: 1982,"Experimental Test of the Variabilty of G Using Viking Lander Ranging Data", *Phys. Rev. Lett.* **Vol. 51**, pp. 1609-1612

Hoyle, F.: 1948,"A New Model For the Expanding Universe", *Mon. Not. R. Astronomy Soc.* **Vol. 108**, pp. 372-382

Kar'orolyhazy F., A. Frankel, and B. Luka'cs: 1989,"On the Possible Role of Gravity in the Reduction of the Wavefunction", in *Quantum Concepts in Space and Time* Edited by R. Pervioc and C. J. Isham, Clarendon Press,Oxford, pp. 112

Linde, A.: 1992,"Particle Physics and Inflationary Cosmology", *Physics Today* **Vol. 40**, pp. 61-68

Misner, C. W., Kip S.Thorne, and John Archibald Wheeler: 1971,*Gravitation* W. H. Freeman and Company, San Fransisco,, pp. 1061

Peratt, A. L.: 1992, "Plasma Cosmology", *Sky and Tel.* **February**, p. 136

Putfoff, H. E.: 1992,"Gravity as a Zero Point Fluctuation Force", *Phys. Rev. A.* **Vol. 39**, pp. 2333

Raby, S. A., R. Slansky, and G. B. West: 1988,"Particle Physics and the Standard Model", in *Particle Physics A Los Alamos Primer* Cambridge University Press, New York, pp. 34

Sakharov, A. D.: 1968,"Vacuum Quantum Fluctuations in Curved Space and the Theory of Gravitation", *Soviet Physics Doklady* **Vol. 12**, pp. 1040-1041

Sandage, A.: 1972,"Distances to Galaxies: the Hubble constant, the Freidmann time and the Edge of the World", in *Proceedings of the Symposium on the Galaxy and Distance Scale* Essex England

Schartzchild, B.: 1990,"COBE Satellite Finds no Hint of Excess In the Cosmic Microwave Spectrum", *Physics Today* **Vol. 43**, pp. 17

Wells, D. R. and Mohammad Bourouis: 1989,"Quantization Effects In the Plasma Universe", *IEEE Trans. Plasma Sci.* **Vol. 17**, pp. 270-281

ON THE PROBLEM OF BIG BANG NUCLEOSYNTHESIS

ERIC J. LERNER
Lawrenceville Plasma Physics
Lawrenceville, New Jersey, USA

Abstract. Supporters of the standard Big Bang theory point to the abundances of light elements, predicted by Big Bang Nucleosynthesis (BBN) as one of the main observational supports of the theory. However, current data no longer confirm BBN. Instead, measurements of the abundances of He^3, He^4, and D clearly contradict BBN at more than a 3σ level, eliminating a key support of the Big Bang.

Key words: Plasma Cosmology, Cosmic Microwave Background, COBE

1 Introduction

The key problem for Big Bang nucleosynthesis (BBN) predictions of the light element abundances lies in the discrepancy between the predictions for deuterium, He^3 and He^4. The predicted abundance of He^4 decreases with decreasing baryon-photon ratio, η, while the predicted abundance of D and He^3 decreases with increasing η. Since the observed upper limits on the pre-galactic abundances of all three isotopes have been declining, the situation now exists where there is no value of the density parameter which yields predictions agreeing simultaneously with observed He^3, He^4, and D abundances.

2 Observations

To begin with deuterium, Hubble Space Telescope observations by Linsky (1992), fix the current abundance of D at $(1.65 + 0.1) \times 10^{-5}$ by number relative to H. Thus the 3σ upper limit is 2×10^{-5}. Of course, much of the pre-galactic deuterium could have been destroyed in stars. However, many authors have placed strict limits on how much of the primordial D could have been destroyed. Yang *et al.* (1984) for example, show that while D is easily burned to He^3, He^3 is destroyed only where temperatures are high enough to burn H to He^4. Considerations of the amount of He^4 produced in the galaxy, combined with calculations as to the production of He^3 in stars leads to the conclusion that the current sum of abundances of D and He^3 is at least half the pre-galactic value. This leads to an upper estimate of the pregalactic D abundances of about 8×10^{-5}. Delbourg-Salvador *et al.* (1987) conclude that a destruction of more than 2/3 of the pre-galactic D by astration would lead to great variations in current D, depending on the

Astrophysics and Space Science **227**: 145–149, 1995.

exact history of a given region. Since such variations are not observed, they calculate that primordial D abundance is less than 3 times present, or, based on the Linsky observations, less than 6×10^{-5}.

Physically, both of these arguments are related to the fact that if the whole of the galactic material is processed on average once through stars, about e^{-1} of the deuterium will not have been so processed. Deuterium destruction much greater than this requires two such processings. But that rate of nuclear processing would produce about twice as much energy, He^4 and heavier elements, such as CNO as are observed. By the BBN formula, upper limits on D abundance set lower limits on the density parameter η. A deuterium D abundance of 6×10^{-5} implies $\eta > 3.4 \times 10^{-10}$ while D abundance of 8×10^{-5} implies a limit of 3.0×10^{-10}. These limits in turn imply lower limits on the predicted primordial abundance of He^4 of 23.9% by weight and 23.6% respectively. These estimates are based on an assumed neutron lifetime of 882 secs, the current 2σ lower limit. A "best" value of 888 secs would increase these limits to about 24.1% and 23.8% respectively.

Recent observations of the abundance of He^3 (Balser *et al.*, 1993) make things worse for the Big Bang. Balser observed He^3 abundances of $(1.13 + 0.11) \times 10^{-5}$ for nebula W43 and 0.68+0.10 for W49. For W43, this implies an η of 5.8 with 1 and 2σ lower limits of 5.0 and 3.8, while for W49 η is 12.5 with 1 and 2σ limits of 10.0 and 8.5. The 1σ lower limits give BBN predictions for He^4 abundance of 24.4 and 25.1% respectively.

It is difficult to get around these observations by arguing that He^3 abundances have been reduced by stellar burning. If we assume that, generously, half of all He^3 is destroyed during stellar processing and that the whole nebula has been processed once, we have to also assume that all the initial D was also processed to He^3. Given the BBN predictions for D and He^3 abundances, He^3 will only be reduced on net if D abundance is initially less than twice He^3 abundance, which occurs only for $\eta > 6$. This would have no effect on the limits from W43 and would reduce the limit from W49 only to 24.9%.

The abundance of He^4 is therefore a crucial test of BBN. Since He^4 is produced by stars, observers have long sought to focus on those galaxies with the least stellar production, which in turn is indicated by the abundance of heavier elements, C, N and O. It is assumed, based on theories of stellar evolution, that He^4 abundance should increase as a function of heavy element abundance, so that the pre-galactic abundance should be determinable by extrapolating He^4 abundance to the zero-heavy element condition. There are three ways of doing this: linear extrapolation, nonlinear extrapolation, and simply averaging the galaxies with the lowest He^4 or CNO abundances.

All of these approaches show that the observed He^4 pre-galactic abundance is less than the lower limit predicted by BBN. Melnick *et al.* (1992) using a linear correlation, obtain 21.6+0.6%, which is 4.7–5.5 standard deviations below the BBN value, depending on which BBN limit is used, 24.4%

or 24.9%. Using the six galaxies with O abundance below 5×10^{-5} (the next highest galaxy in the sample has an O abundance above 8×10^{-5}) and averaging the He^4 abundances, we obtain 22.4+.3% which again rules out BBN at the 6σ level.

Using a different sample, and a nonlinear regression of He^4 on N and O, Mathews *et al.* (1990) obtained a primordial He^4 abundance of 22.3+0.2% which puts the lowest BBN limit at least 10σ too high.

Olive *et al.* (1991), obtained a value of 23.1% with a 2σ upper limit of 23.9%. Even this higher value put the lowest BBN limit at 3σ implying only a 0.5% chance of agreement with observation. Finally, Pagel *et al.* (1992) use linear extrapolation to derive an abundance of 22.7+0.5%, or 3.6–4.4σ below the Big Bang predictions.

If we use the He^4 observations to estimate η by BBN predictions, we conclude that, at the 2σ level, He^4 abundance is at most 23.0% and $\eta < 1.8 \times 10^{-10}$. The BBN prediction for deuterium abundance would then be greater than 1.74×10^{-4}. This is 8 times the upper limit for current D abundance in the ISM and twice the highest plausible estimate for a pre-galactic abundance compatible with current observations. To achieve this 8-fold decrease in D abundance would require that the average parcel of the ISM has been processed twice through stars. As Yang *et al.* (1984) point out, this would lead to overproduction of He^4, C, N and O relative to observations. The complete processing of matter through a generation of stars with the present luminosity function of the galaxy enriches the ISM by approximately 12% He^4. Two such processings would lead to an enrichment by about 24%, giving a total abundance of 46%, far above current observations. In addition, the production of this much He^4 would yield 1.7×10^{18} ergs per gram of galactic mass. This would produce a luminosity/mass ratio 1.7 times that of the sun or about 8 times that observed.

Finally, as Delbourg-Salvador *et al.* note, such an extensive destruction of D would lead to great variations in current D abundance, which has certainly not been observed. Thus, if D observations are used to predict He^4 abundance, the lowest possible BNN observations are ruled out at beyond the 3σ level, while if He^4 observations are used to predict D abundance the gap between theory and observation is at least a factor of two, grossly more than the observational uncertainty.

The He^3 observations are even more conlcusive. The most optimistic 2σ limits on He^4, 23.9% do not reach the most generous lower limits form W43, 24.4%. We must conclude that standard BBN is excluded by the observations. This conclusion has also been reached by others, such as Riley and Irvine (1991).

3 Discussion

Since the recognition by many researchers that the He^4 and D observation together are incompatible with standard BBN, a number of efforts have been made to propose "fixes" to the theory. The two possibilities discussed are inhomogeneous BBN and a very massive tau-neutrino.

Inhomogeneous BBN, as elaborated by, for example, Mathews *et al.* (1990), hypothesizes small-scale density fluctuations in the early universe that would affect BBN. It should be emphasized that this hypothesis is entirely *ad hoc*, not emerging from any physical theory. While BBN has relatively unambiguous predictions, inhomogeneous BBN introduce 4 more free variables: the radius, density contrast, spacing and shape of the inhomogeneities; which produce an enormously larger range of possible predictions. Despite this large number of free variables, the range of possible predictions does not include points that simultaneously have $He^4 < 23.0\%$ and $D < 8 \times 10^{-5}$, the requirements of current observations. Indeed, the situation is somewhat worse than in standard BBN, since with inhomogeneous BBN, lithium production is enhanced. For conditions that produce He^4 abundance of 23.0%, D is about the same as in standard BBN, but Li^7 is predicted at around 2×10^{-9}. This is in gross contradiction with an observed upper limit of 2.3×10^{-10} for Li^7 abundance in Population II stars. Arguments that hypothesize lithium depletion to produce exactly the same depleted abundance in a range of Population II stars are too implausible on their face to merit discussion. So inhomogeneous BBN also fails to resolve the conflict between theory and observation.

A second fix is to assume a very massive tau-neutrino, which would lower the effective number of neutrinos in BBN calculations below three and thus lower He^4 abundance predictions. A massive tau-neutrino would begin to pair-annihilate during BBN, thus reducing the number of neutrinos. However, Kolb and Scherrer (1982) have shown that this effect occurs only with a tau-neutrino more massive than 15 Mev, relatively close to the observational upper limit of 35 Mev from particle physics experiments. Such a massive neutrino would have to conveniently decay rapidly to avoid a rapid gravitational collapse of the universe. Alternatively, one could hypothesize the non-existence of the tau-neutrino which has never been directly observed. This would, however, contradict existing particle physics theories.

4 Conclusion

The key problem for Big Bang nucleosynthesis (BBN) predictions of the light element abundances lies in the discrepancy between the predictions for deuterium, He^3 and He^4. The predicted abundance of He^4 decreases with decreasing baryon-photon ratio, η, while the predicted abundance of D and

He^3 decreases with increasing η. There is no value of the density parameter which yields predictions agreeing simultaneously with observed He^3, He^4, and D abundances.

Thus from any standpoint, the observational evidence is incompatible with Big Bang nucleosynthesis, assuming only that there are tau-neutrinos, as particle theory predicts.

In contrast, theories that hypothesize the creation of the "primordial" elements by thermonuclear and cosmic ray processes in early galaxies (Lerner, 1989) successfully predict the observed abundances of He^4, He^3, D and Li.

References

Balser, D.S. *et al.*: 1994, submitted to *Astrophys. J.* **Vol. no. 430**, p. 667

Delbourg-Salvador, P., Audouze, J., and Vidal-Madjar, A.: 1987, *Astrophys. J.***Vol. no. 174**, p. 365

Kolb,E.W. and Sherrer.R.J.: 1982, *Modern Phys. Rev. D*, **Vol. no. 25**, p. 1481

Lerner, E. J.: 1989, *IEEE Trans. Plasma Sci.*, **Vol. no. 17**, p. 259

Linsky, J.: 1992, *Space Telescope Science Institute Newsletter***March 1992**, p. 1

Mathews, G.J., *et al.*: 1990, *Astrophys. J.*, **Vol. no. 358**, p. 36

Melnick, J., Heydari-Malayeri, M., and Leisy, P.: 1992, *Astron. Astrophys.***Vol. no. 253**, p. 16

Olive, K.A., Steigman, G., and Walker, T.P.: 1991, *Astrophys. J.***Vol. no. 380**, p. L1

Pagel, B.E.J., *et al.*: 1992, *Mon. Not. Roy. Astron. Soc.***255**, p.325

Riley,S.P. and Irvine, J.M.: 1988, *Jour. Phys. G***Vol. no. 17**, p. 35

Yang, J. *et al.*: 1984, *Astrophysical J.***Vol. no. 281**, p. 493

OBSERVATIONAL LIMITS ON INTERGALACTIC MATTER

BOGDAN WSZOLEK
Jagiellonian University Astronomical Observatory
ul. Orla 171, PL-30-244 Krakow, Poland
bogdan@oa.uj.edu.pl

Abstract. Intergalactic gas and dust are very difficult to detect. Several methods to detect intergalactic gas have been developed. The results of these methods have provided qualitative arguments and quantitative limits on the amount of intergalactic matter in the intergalactic environment. We present here arguments as to why the density of intergalactic matter is higher than previously thought possible.

Key words: Intergalactic gas, Intergalactic Dust, Intergalactic Plasma, Quasars

1 Introduction

The knowledge of the existence and amount of intergalactic material is important for the study of large-scale structure in the universe, for the study of matter evolution in the universe, and for general cosmology. The universe seems to be populated mostly by galaxies, but this appearance may result from the fact that their high luminosities make them easily visible. In view of the possible cosmological relevance of the contribution of dark matter to the mean density of the universe, it is important to find out if or how much dark matter exists. In the present context we will not discuss the possible forms of non-baryonic matter, but concentrate on baryonic matter other than galaxies, *i.e.,* intergalactic gas and dust. Dark bodies such as black holes and "Jupiters", if they exist, will also not be considered in this paper. If the enormous space between the galaxies is filled with an intergalactic medium, different physical processes may occur because of this medium. Energy generated in the galaxies might then be absorbed and re-emitted by the intergalactic matter in ways that can be used to deduce its physical conditions.

2 Observations of Intergalactic Matter

Theoretically, we can expect to find the following in the intergalactic medium:

– hot and cold electron gas

– neutral gas (mostly hydrogen)

– ionized gas, or *plasma*

Astrophysics and Space Science **227**: 151–155, 1995.

– dust

High temperature plasma is observed in X-rays. Several tens of clusters of galaxies have been found to be powerful X-ray emitters with temperatures of the intracluster gas of 10^8K and total masses comparable to the cluster masses (10 $M_\odot$).

Neutral hydrogen is observable by means of its 21 cm line. Because of this, the upper observational limits on intergalactic neutral hydrogen is placed at 10^{29} g cm^3. Note that hydrogen can exist both in atomic and molecular form. Although atomic hydrogen is relatively easy to detect at 21 cm, molecular hydrogen is not easily detectable at all. It is very probable that intergalactic hydrogen exists mainly in this undetectable molecular form. Observed Lyman–α absorption in quasar spectra (Lyman–α forests) may originate from large hydrogen clouds distributed between distant quasars and the observer.

As far as the dust is concerned, there are two theoretical arguments for the presence of dust in intergalactic space:

1. If a pre-galactic stellar population existed, its evolution would have led to an enrichment of the primeval gas with heavy elements. At the same time, during its evolution, dust would have formed in the red supergiant phase and been ejected into space.

2. Alternatively, population III stars and the subsequent populations would have evolved in galaxies and produced dust. This dust would have been partially lost into intergalactic space.

Dust in our Galaxy exists in detectable clouds with gas to dust ratios of about 100. Observations from the infrared to the ultraviolet show a chemical compositions associated with grain sizes. In intergalactic space we might expect clouds or a general substratum of dust. If intergalactic dust originates only in galaxies, an average dust density is about 10 g cm^3. Dust grains can scatter and absorb light in intergalactic space. They are heated by absorption of photons and by collisions with gas particles and emit radiation according to their temperature at wavelengths longer than the absorbed ones. Such behavior suggests the following possibilities when searching for dust:

1. Light extinction in remote galaxies due to dust.

2. Infrared emission from dust grains.

A number of methods has been developed for the detection of dust extinction:

1. Search for correlations between surface brightness of galaxies and parameters which appear to be correlated with distance. For transparent intergalactic space, the average surface brightness should be constant.

2. Number-magnitude counts. The number of galaxies increases according to a power–law determined only by cosmological parameters when space is perfectly transparent.

3. Counting of distant clusters of galaxies in the areas occupied by nearer clusters. Intracluster dust should lead to a deficiency of distant clusters.

4. Search for correlations between the colors of extragalactic objects and their distance. Dust scatters blue light more effectively than red light, thus, distant objects should appear redder. This effect can be masked by the reddening due to the redshift. It can also be counteracted by evolution, *e.g.* when distant galaxies are intrinsically bluer.

5. Search for effects on line ratios, *e.g.* Balmer lines. Difficulties arise because one cannot distinguish between intergalactic extinction and extinction in the galaxy or quasar.

All studies performed with the above methods have given only upper limits of the mean intergalactic dust density. Most values are of the order 10^{-33} g cm^3, which is 100 times lower than the mean density of luminous matter. However, detection of condensations of intergalactic dust (dust clouds) could provide strong evidence for the existence of dense dust regions. The major cloud criteria are:

1. Strong deficiency of galaxies of all brightness classes in a well defined and connected area.

2. Consistently weaker magnitudes and redder galaxy colors in the same region.

3. Measured lower limit of the distance to the extinguishing cloud. Suspected upper limits of dust densities in intergalactic clouds can be as much as 10^{-27} g cm^3.

An advantage of the optical absorption by intergalactic dust investigations is that there are methods by which we are able to distinguish between interstellar (within our own Galaxy) and intergalactic extinction.

If intergalactic dust absorbs light, it must also emit radiation. A major difficulty here is an absence of methods for distinguishing between the radiation of intergalactic dust and interstellar one.

Detection of intergalactic dust properties based on emission effects is not yet a developed science. At present, we attempt to use the infrared IRAS data as additional evidence for the existence of intergalactic clouds. For one of the four intergalactic dust clouds candidates, the Okroy cloud, we found a distinct maximum in the 100 μm infrared emission from its center (Wszolek et al., 1989).

As yet, there are very few known intergalactic HI clouds. The best examples lie in M96 group and in Sculptor group of galaxies. The largest HI cloud in Sculptor group has shown optical extinction (Abadi and Edmunds, 1977). We have made a detailed search for infrared emission in this cloud using IRAS data. None was found. This cloud is perhaps the best candidate for such investigations because of its rather strong HI emission, its small and rather good established distance, and its close vicinity to the South Galactic Pole, where the infrared foreground given by our Galaxy is low. Similar investigations were done for HI clouds in M96 group (Schneider *et al.*, 1989). Their results were also negative. The absolute lack of 100 μm emission attributable to this clouds allows us to conclude that they must *differ substantially* from the interstellar clouds in our Galaxy. One can conclude also, that other objects of this type are difficult to find, should they constitute an important population in the universe.

3 Discussion

The role of intergalactic dust, as well as interstellar dust, is important not only in the context of contributing to the mean density of matter, but also in the context of chemical processes in intergalactic medium. Even very small amounts of intergalactic dust may be sufficient to lead to the important reaction $H+H \rightarrow H_2$. Thus, the knowledge of the ratio between atomic and molecular amounts of intergalactic hydrogen is very important.

The intergalactic medium can be much denser than generally believed. (It was only a few decades ago that interstellar space was considered a vacuum. Today it is considered to be a mixture of cold gas and dust). The density of gas rises dramatically when most of hydrogen exists as molecules. Also dust component can be more abundant then the extinction and the infrared emission measurements currently show. Extinction decreases substantially with dust grain density. Infrared emission also depends on the temperature of dust. If the temperatures of intergalactic clouds are lower than interstellar clouds (where temperatures fall below 10 K), or if they are of the order of 3 K, the dust grains may consist of frozen hydrogen. The properties of intergalactic clouds and can be quite different from interstellar clouds. Infrared emission of such cold grains should be much weaker than that observed from interstellar clouds.

References

Wszolek, B., Rudnicki, D., de Bernardis, P., Masi, S., Salvi, A.: 1989, "Far-infrared Emission from an Tntergalactic Dust Cloud?," *Astrophys. Space Sci.* **Vol. no. 152**, pp. 29-34

Abadi, H. I., Edmunds, M. G.: 1977, "Intergalactic Dust Near NGC 300," *Astron. Astrophys.* **Vol. no. 56**, pp. 319-321

Schneider, S. E., Skrutskie, M. F., Hacking, P. B., Young, J. S., Dickman, R. L., Clausssen, M. J., Salpeter, E. E., Houck, J. R., Terzian, I., Lewis, B. M., Schure, M. A.: 1989, "Multifrequency Survey of the Intergalactic Cloud in the M96 Group," *Astron J.* **Vol. no. 97**, pp. 666-673

CLUSTER-SCALE MAGNETIC FIELDS

T. VENTURI and G. GIOVANNINI *
Istituto di Radioastronomia, CNR, Bologna. Italy

Abstract. The search for non thermal radio emission from clusters of galaxies is a powerful tool to investigate the existence of magnetic fields on such large scale. Unfortunately, such observations are scarce thus far, mainly because of the very faint large scale radio emission expected in clusters of galaxies. In the present contribution we will first review the status of the radio observations of clusters of galaxies, carried out with the aim of detecting large scale radio emission.
We will then focus on the large scale radio emission detected at 327 MHz and 610 MHz in the Coma cluster of galaxies. The features of the detected radio emission suggest that a magnetic field with an intensity of the order of $\sim 10^{-7}$ Gauss must be present on a scale of about 2 Mpc (for $H_{\circ} = 100 \; km \; s^{-1} Mpc^{-1}$). The morphology of the radio emission is similar to that of the most recent X-ray images derived with ROSAT, and follows the distribution of the galaxies in the cluster. All these pieces of information will be taken into account in the discussion on the possible origin of this large scale magnetic field.

Key words: Magnetic Fields, Clusters of Galaxies, Coma Cluster

1 Introduction

One of the most important and powerful tools in the study of large scale magnetic fields is the nonthermal radio emission in clusters of galaxies, as it reveals the presence of both relativistic electrons and magnetic fields. Radio emission from the cluster as a whole, not really associated with any individual galaxy, is usually referred to as *radio halo.* In the present contribution we will first summarise some basic knowledge on large scale radio emission in clusters of galaxies, briefly reviewing those clusters which exhibit this kind of emission (Sect. 2), then we will concentrate on the best studied *halo* source, the one in the Coma cluster. The morphology of the Coma halo will be described, and its physical properties will be given (Sect. 3). We will then concentrate on the low brightness radio emission extending beyond Coma C in the direction of the cluster A1367 (Sect. 4). The overall radio emission associated with the Coma cluster will be compared to the most recent ROSAT X-ray images and possible interpretations on the origin of the relativistic electrons and of the large scale magnetic field responsible for the radio emission in the cluster will be discussed (Sects. 5 and 6). A value of the Hubble constant $H_o = 100$ km sec^{-1} Mp^{-1} will be used thoughout; for $z_{Coma} = 0.0235$ this gives a conversion factor of 0.35 kpc/arcsec.

* also Dipartimento di Astronomia, Università di Bologna, Italy

Astrophysics and Space Science **227**: 157–165, 1995.

2 Generalities on Radio Halo Sources

The first detection of a radio emission associated with a cluster as a whole, and not to individual galaxies in the cluster, was the *halo* source located in the center of the Coma cluster (A1656), and named Coma C. Very low frequency observations carried out in the fifties and sixties, and 408 MHz observations carried out in 1970 (Willson, 1970) revealed the existence of a diffuse radio emission, with an angular extension of roughly 40′, whose center was coincident with the Coma cluster optical center. The existence of such large scale radio emission was confirmed by Valentijin (1978) at 610 MHz.
There are very few other clusters of galaxies with an extended radio emission in their center, and none of them actually resembles the radio halo in the Coma cluster, both in shape and extension. In each case it is really difficult to separate the large scale radio emission associated with the cluster as a whole from the blend of low brightness diffuse emission associated with individual galaxies:
- **A2256** is very unusual, since it contains a few head-tail radio galaxies. A radio halo is present in this cluster, but it is not so well studied as the Coma cluster halo (Bridle and Fomalont, 1976). The same is true for **A2319** (Harris and Miley, 1978).
- What we see in the Perseus cluster, **A426**, looks like a diffuse emission surrounding the central galaxy NGC1275 rather than a cluster emission (Noordam and de Bruyn, 1982; Burns *et al.*, 1992).
- In **A2255** it is impossible to tell if we see a radio halo or if the diffuse radio emission is the blend of the tail of two head-tail radio galaxies.
- Finally, the radio halo in **A1367** (Gavazzi and Jaffe, 1987) is much smaller and fainter than Coma C, it is not in the center of the cluster and is located in a region of the cluster where three galaxies exhibit trails of radio emission.
Given the interest in this kind of extragalactic radio sources, searches for radio haloes were carried out by a few authors. We mention here only the survey carried out by Hanish (1982), who observed 72 clusters of galaxies of various optical types and at various redshifts. No new radio halo, beyond those mentioned above, was found and this confirmed the rarity of the phenomenon.

3 Coma C - Radio Morphology and Physical Parameters

Recent interferometric observations at various frequencies of the Coma cluster (327 MHz, Venturi *et al.*, 1990; 610 MHz, Giovannini *et al.*, 1993; 1.4 GHz, Kim *et al.*, 1990) made it possible to study Coma C in great detail, and our knowledge on this source and on its possible formation and evolution has now considerably improved.
The radio emission coming from the central region of the Coma cluster at 327

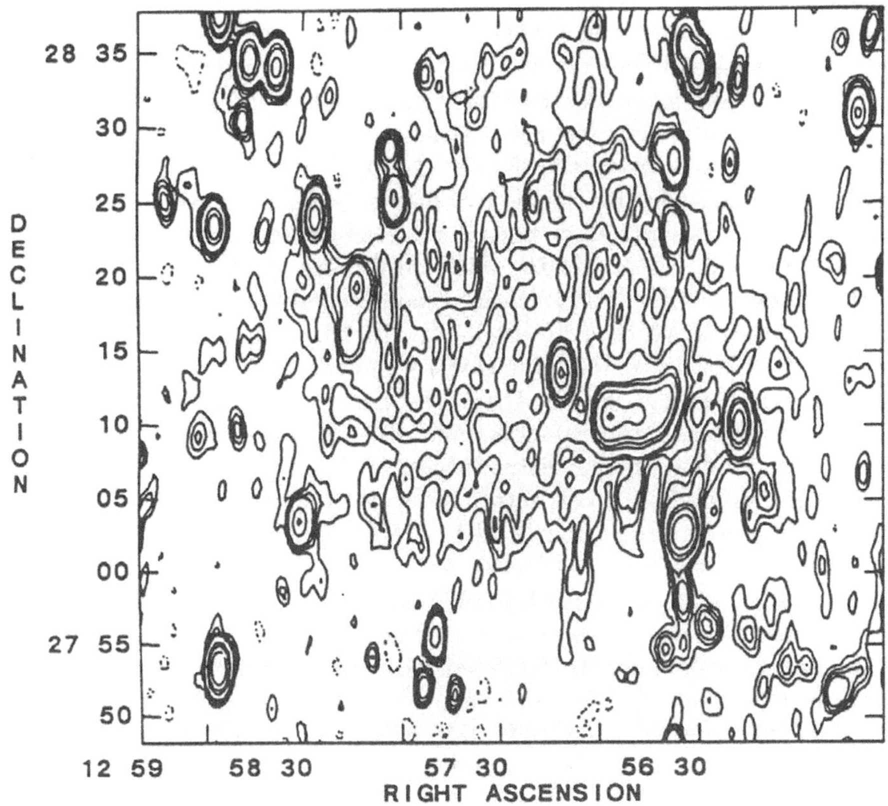

Fig. 1. Contour plots of the central region of the Coma cluster of galaxies at 327 MHz. The restoring beam of the image is $55'' \times 125''$, in p.a. $90°$.

MHz is given in Fig. 1 (see also contour plots reported in Kim *et al.*, 1989). The presence of the radio halo is outstanding. The shape of the halo is that of an ellipsoide. At this frequency the deconvolved angular dimensions are $36' \times 24'$. The center of the ellipsoide is located close to the galaxy NGC4874, one of the two giant galaxies dominating the cluster at optical wavelengths. Its brightness distribution is pretty smooth, with no peaks or other features. A comparison between this map and more recent observations at 610 MHz (Giovannini *et al.*, 1993) reveals that the emission is very similar at the two frequencies, both in extent and in the morphology. At each frequency we can recognise the radio emission coming from the head-tail radio galaxy NGC4869, in the central part of the radio halo.

A new estimate of the radio halo total spectral index, computed taking into account the new interferometric observations mentioned above, indicates that the spectrum is steep, with $\alpha \sim 1.34$ (if $S \propto \nu^{-\alpha}$) up to a frequency of 1.4 GHz, then it steepens considerably at higher frequencies (Giovannini *et al.*, 1993). The shape of the spectrum suggests that the mechanism responsible for the radio emission is non thermal synchrotron radiation.

In order to derive a more detailed information on the radio halo, we made a spectral index map, using the data at 327 MHz and those at 1.4 GHz. We used the same uv-coverage in the two data sets, and used the same gridding parameters in making the two maps, for a reliable computation of the spectral index (Giovannini *et al.*, 1993). The results are pretty interesting. The spectral index distribution is clearly separated into two different regions. There is a central plateau of about $8'$ (170 kpc in linear size) where α is constant, with a value of 0.8, then α steepens dramatically beyond this point, reaching values as high as 1.8. Given the shape of the integrated spectrum, we conclude that the spectral index must steepen considerably at frequencies higher than 1.4 GHz in the central region of the cluster.

With these values for the spectral index we derived that the age of the radiating electrons varies from 1 to 4×10^7 years going from the center to the periphery of the halo. Assuming that equipartition holds in this source, we obtained a value of the magnetic field of $\sim 0.5\ \mu G$, roughly constant all over the halo. For comparison, the intensity of the magnetic field in the most external regions of the extended radio galaxies in the Coma cluster is in the range of 2 - 4 μG (Venturi *et al.* 1989, Dallacasa *et al.*, 1989), so there is almost one order of magnitude difference in the magnetic field strength bewteen these two types of radio sources.

With an estimate for the magnetic field, the spectral index of the source and the distance of the Coma cluster, it is possible to derive the number of radiating particles in the radio halo, which turns out to be $\sim 1.6 \times 10^{60}$ (Giovannini *et al.*, 1993).

4 The Bridge and the Peripheral Extended Source 1253+275

A very important result emerging from the 327 MHz map of the Coma cluster is the extension of the low brightness radio emission beyond Coma C. As it is very clear from Fig. 2, a *bridge* of radio emission connects the outer part of Coma C to the south-western edge of the cluster, in the direction of the cluster A1367.

Even before this image was made, it was well known that the region located at about $70'$ - $90'$ towards south-west with respect to the cluster center is very peculiar. The extended radio source 1253+275 is located in this region (see Giovannini *et al.*, 1991 and references therein). Furthermore, two extended radio galaxies with distorted morphology are present in this region (Venturi *et al.*, 1989), and another extended radio galaxy, associated with the cD galaxy NGC4839 (one of the very few cD galaxies located outside the central

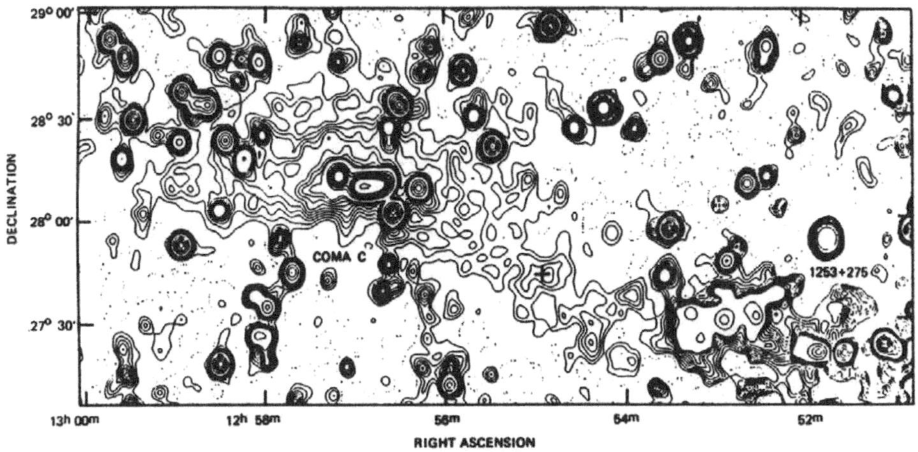

Fig. 2. Contour plots of the radio emission from the Coma cluster at 327 MHz covering an area of ~ 1.5° from the cluster center. Coma C and the extended radio source 1257+275 are labelled. The cross indicates the optical position of the cD galaxy NGC4839.

region of a rich cluster), is located midway between the Coma cluster center and the two peripheral radio galaxies.

It is now clear that a unique filament of radio emission goes from the radio halo to the most peripheral extended radio galaxy NGC4789. We can recognise three main features:

1) a *bridge* going from Coma C to the radio source 1253+275;
2) the intriguing radio source 1253+275 itself;
3) the small bridge connecting 1253+275 and the radio tail of NGC4789.

The brightness radio emission of the *bridge* is faint, but definitely real. The high resolution of the data lead us to beleive that what we see is true diffuse emission and not the blending of many discrete sources. The total extent of this long and thin feature is $\sim 50' \times 20'$, in position angle about $-140°$ and its total flux density at 327 MHz is 760 mJy (see Kim *et al.*, 1989 for further details).

A reanalysis of older data at 408 MHz in the light of the great importance of this detection (Kim *et al.*, 1990), has revealed that the *bridge* was present also in those observations, but the detection was considered unreliable. By remapping those old 408 MHz data it was possible to give an estimate of the spectral index of the bridge between 327 MHz and 408 MHz. The total spectrum turned out to be very steep in this frequency range, with $\alpha \sim 1.5$. With the standard equipartition assumptions we derived a magnetic field

strength in the range 0.2 - 0.6 μG, taking into account the uncertainties in the filling factor, in the proton/electron ratio (k factor) and in the true emitting volume. This value for the magnetic field is similar to what we found for Coma C.
The maps at 327 MHz (Kim *et al.*, 1989) and 608 MHz (Giovannini *et al.* 1991) reveal that the low brightness emission continues beyond 1253+275, in the same position angle as the main *bridge*, and it reaches the radio tail of NGC4789.
Given the very large extent of Coma C and of this global extended emission, the importance of this result is remarkable. Again, we point out that the direction of this huge radio feature coincides with the axis of the A1656 - A1367 supercluster.

The nature of the radio source 1253+275 is still matter of debate (see Giovannini *et al.*, 1991 and references therein). Its radio morphology and the lack of an optical identification make it very difficult to interpret its nature and origin. Various possibilities have been discussed thus far.
The shape of the radio spectrum is typical of a synchrotron radio source, with a total spectral index $\alpha = 1.22$ up to 5 GHz. Using the very good data at 327 MHz and 608 MHz, it was possible to make a spectral index map between these two frequencies. The map reveals that the distribution of the spectral index is unusual. The spectrum steepens going from the north-eastern ridge, where α ranges from 0.6 to 1 to the south-eastern one, where α increses up to 1.8. The physical parameters of this source are very similar to those found in Coma C and in the *bridge*, in particular the equipartition magnetic field strength is $\sim$ 0.6 μG, and the lifetime of the radiating electrons is $\geq 6.5 \times 10^7$ years.

It is interesting to point out that the numbers we derive for the physical parameters in the various emitting regions are similar, and this suggests that the plasma and magnetic field conditions in the *bridge*, 1253+275 and the mini-bridge could be very similar to those in Coma C.

5 Comparison with the X-Ray Maps

The similarity between the low frequency radio morphology of the Coma cluster and the X-ray image obtained with ROSAT observations in the 0.1 - 2.4 keV band (White *et al.*, 1993, Briel *et al.*, 1992) is remarkable. The maps in the two completely different domains of the electromagnetic spectrum coincide in the overall morphology out to a distance of more than 1° from the cluster center. The X-ray emission, produced by thermal Bremsstrahlung, is associated with the whole cluster, with few peaks of emission coincident with cluster galaxies. This means that a large amount of gas permeates this cluster, even at very large distances from the core. This was expected on

the basis of the 327 MHz maps of the two extended radio galaxies NGC4789 and NGC4827 located at the periphery of the Coma cluster (Venturi *et al.*, 1989). Our data suggested that the radio tails in both sources had to be in pressure equilibrium with the outer medium, indicating the presence of a yet undetected thermal gas.
An analysis of the optical distribution of the galaxies in the cluster (Fitchett and Webster, 1987) indicate that all cluster galaxies with associated X-ray emission are at the center of subgroups, one of which is in the process of merging with the largest central part of the cluster itself.

6 Discussion

The origin of such large scale radio emission is still matter of debate. Any model explaining the formation of radio haloes must account for the rarity of the phenomenon. Furthermore, in the case of the Coma cluster the problem is even more complicated by the presence of the *bridge* radio emission in the A1656 - A1367 direction.
The two basic questions related to the formation and lifetime of halo radio sources concern the origin of the relativistic electrons and that of the magnetic field.

A lot has been written on both subjects. A very good review on the electron reacceleration processes is given in Schlickeiser *et al.* (1987). The topic, together with possible mechanisms of cluster halo formation, is also briefly reviewed in Giovannini *et al.*, (1993). We will mention here the most recent studies made by Tribble (1993 and 1994) on the origin of the cluster magnetic fields and relativistic particles, and on their acceleration, and comment on the Coma cluster of galaxies.

It is generally agreed that the high number of radio active galaxies present in clusters of galaxies provide enough leaking electrons to fuel the radio emission on large scale. In particular, for the Coma cluster, the presence of a *head tail* radio galaxy (NGC4869) orbiting around the cluster center may be related to the existence of the radio halo (Giovannini *et al.*, 1993). The suggested correlation between tailed radio galaxies and radio haloes needs confirmation by observing other clusters with tailed sources orbiting around their centers, however this hypotesis seems very plausible. In order to compensate for radiative losses of leaking electrons, some reacceleration mechanism has to be efficient on a cluster scale. The *in situ* reacceleration mechanism seems to be the most efficient one (Schlickeiser *et al.*, 1987), even though it is not enough to amplify the primordial magnetic field at the required intensity.

The existence of large scale (cluster scale) magnetic fields, their amplification and the reacceleration of relativistic electrons were recently discussed by Tribble.

According to Tribble (1994), active galaxies in clusters should be considered the source of both intracluster magnetic field and relativistic electrons. As the radio galaxies evolve and fade away, the magnetic energy, originally frozen into the radio source, is deposited in the thermal plasma within the cluster, and spread on a much larger volume. If we take into account the total number of radio galaxies during the whole life of a cluster, and if we integrate over the luminosity and time evolution, we obtain that the magnetic energy density accumulated by the activity of all radio sources is of the order of 10^{48} J Mpc^{-1} (Tribble, 1994). Such amount is enough to provide a magnetic field intensity of the order of $\sim 1\mu G$.
This cluster magnetic field and relativistic electrons would then be accelerated by cluster merging (Tribble, 1993). It seems that the large amount of energy released (as a consequence of turbulence and shocks) when subgroups of galaxies collapse to form a cluster is enough to amplify the pre-existent magnetic field to an intensity of the order of μG, and to diffuse and reaccelerate the relativistic particles on a cluster scale.

Radio active galaxies over the whole life of a cluster and cluster merging would therefore provide an explanation for the existence of large scale magnetic fields, as well as for the relativistic electrons on such large scale. The remaining question is why radio haloes are so rare. Tribble (1993) also derives that the magnetic field strength decays slower than the electron energy, and therefore after some critical time, which he estimates in a few times 10^8 years, the radio halo fades away. Following this theoretical frame, haloes are rare because they can take place only as a consequence of merging, and they are a transient event. The electron decay time is longer than the magnetic field one, so in case a second merger event takes place to amplify the magnetic field once again, a rebirth of the radio halo type activity can be activated.
This theoretical frame seems to be supported by the observational results on the Coma cluster. As it was reported in Sect. 5, the X-ray morphology of the Coma cluster suggests that it is not dynamically relaxed, at least in the outer regions, where merging among subgroups is still going on at present. Following Tribble (1993), this means that the process responsible for the magnetic field amplification and the particle reacceleration is still active. At present the main merging event would take place between the central condensation and the subgroup of galaxies centered on the cD NGC4839 (White *et al.*, 1993), which is located in the *bridge* region (see Fig. 2). As it is pointed out in Watt *et al.* (1992), the merger of two subclusters should generate large-scale turbulence in regions of shear between the two impacting gas clouds. We think it is remarkable that we detect the *bridge* radio emission exactly in this area.

References

Bridle A.H., Fomalont E.B.: 1976, *Astron. Astrophys.*, **52**, 107

Briel U.G., Henry J.P., Bohringer H.: 1992, *Astron. Astrophys.*, **259**, L31

Burns J.O., Sulkanen M.E., Gisler G.R., Perley R.A.: 1992, *Astroph. J.*, **338**, 149

Dallacasa D., Feretti G., Giovannini G., Venturi T.: 1989, *Astron. Astrophys. Suppl. Ser.*, **79**, 391

Fitchett M., Webster, R.: 1987, *Astrophys. J.*, **317**, 653

Gavazzi G., Jaffe W.J.: 1987, *Astron. Astrophys.*, **186**, L1

Giovannini G., Feretti L., Stanghellini C.: 1991, *Astron. Astrophys.*, **252**, 528

Giovannini G., Feretti L., Venturi T., Kim K.-T., Kronberg P.P.: 1993, *Astrophys. J.*, **406**, 339

Hanish R.J.: 1982, *Astron. Astrophys.*, **111**, 97

Harris D.E., Miley G.K.: 1978, *Astron. Astrophys. Suppl.*, **34**, 117

Kim K.-T., Kronberg P.P., Giovannini G., Venturi T.: 1989, *Nature*, **341**, 720

Kim K.-T., Kronberg P.P., Dewdney P.E., Landecker T.L.: 1990, *Astrophys. J*, **355**, 29

Noordam J.E., de Bruyn A.G.: 1982, *Nature*, **229**, 597

Schlickeiser R., Sievers A., Thiemann H.: 1987, *Astron. Astrophys.*, **182**, 21

Tribble P.C.: 1993, *MNRAS*, **263**, 31

Tribble P.C.: 1994, *MNRAS*, in press

Venturi T., Feretti L., Giovannini G.: 1989, *Astron. Astrophys.*, **213**, 49

Venturi T., Giovannini G., Feretti L.: 1990, *Astron. J.*, **99**, 1381

Valentijn E.A.: 1978, *Astron. Astrophys.*, **68**, 449

Watt M.P., Ponman T.J., Bertram D., Eyles C.J., Skinner G.K., Willmore A.P.: 1992, *MNRAS*, **258**, 738

White S.D.M., Briel U.G., Henry P.: 1993, *MNRAS*, **261**, L8

Willson M.A.G.: 1970, *MNRAS*, **151**, 1

ROTATION VELOCITY AND NEUTRAL HYDROGEN DISTRIBUTION DEPENDENCY ON MAGNETIC FIELD STRENGTH IN SPIRAL GALAXIES

CHARLES M. SNELL and ANTHONY L. PERATT
Los Alamos National Laboratory
Los Alamos, New Mexico, USA

Abstract. The rotation velocity of a simulated plasma galaxy is compared to the rotation curves of Sc type spiral galaxies. Both show 'flat' rotation curves with velocities of the order of several hundred kilometers per second, modified by $\mathbf{E} \times \mathbf{B}$ instabilities. Maps of the strength and distribution of galactic magnetic fields and neutral hydrogen regions, as-well-as as predictions by particle-in-cell simulations run in the late 1970s, are compared to Effelsberg observations.

Agreement between simulation and observation is best when the simulation galaxy masses are identical to the observational masses of spiral galaxies. No dark matter is needed.

Key words: Plasma Cosmology, Galaxies, Dark Matter, Galactic Rotation, Galactic Magnetic Fields, Neutral Hydrogen

1 Introduction

Rotational velocities of spiral galaxies are obtained by measuring the doppler shift of the Hα line emitted by neutral hydrogen in the spiral arms. If the galaxy is canted toward earth, the emission-line in the arm moving away from earth is red-shifted while the line in the arm moving towards earth is blue-shifted. The first doppler shift measurements of the radial velocities indicated that these were flat, rather than following the Keplerian law. If galaxies were gravitationally bound systems, their outer mass should follow Kepler's laws of motion and be slower than the inner mass. The flat rotation curves of galaxies has been cited as the strongest physical evidence for the existance of dark matter. In this scenario a massivive halo of dark matter has been envoked to produce the flat rotation curves. However, the rotation curves are not really flat; they show appreciable structure representative of an instability mechanism within the arms. This instability precludes the existance on any external halo of matter around galaxies that, while making the rotation curves flat, would also dampen any instability growth.

In the late 1970s and early 1980s, fully three-dimenional, fully electromagnetic particle-in-cell simulations of plasmas of galactic dimensions became possible. As the plasma universe has always been considered filamentary in morphology (Alfvén, 1950), the simplest simulation in which an interaction

Astrophysics and Space Science **227**: 167–173, 1995.

current-conducting, plasma filaments. As noted very early on, filaments form spirals reminescent of astrophysical *nebulae* (Birkeland, 1912).

The justification for applying plasma physics to galaxies evolving out of cosmic plasma is the overwhelming strength of the electromagnetic field; of order 10^{36} times that of gravity and 10^7 times that of gravity in neutral hydrogen in the space environment.[1]

These early simulations showed ordered magnetic field and neutral hydrogen distributions that had no observational support at the time. It is interesting to compare their predictions against current observations of spiral galaxies.

2 The Association of Neutral Hydrogen with Galactic Magnetic Fields

Neutral hydrogen distributions are characteristic of spiral galaxies but not pre-spiral galaxy forms. Because the rotation velocities of spiral galaxies are determined by the motion of neutral hydrogen, it is desirable to know the process for neutral hydrogen accumulation in late–time galaxies.

In the plasma universe model, spiral galaxies form from the interaction of current-carrying filaments at regions where electric fields exist. The individual filaments are defined by the *Carlqvist Condition* that specifies the relationship between gravitational and electromagnetic constraining forces (Verschuur, 1995). In this model, whether or not neutral hydrogen and other neutral gases form from hydrogenic plasma depends of the efficiency of convection of plasma into the filament.

When an electric field is present in a plasma and has a component perpendicular to a magnetic field, inward convection of the charged particles occurs. Both electrons and ions drift with velocity

$$\mathbf{v} = (\mathbf{E} \times \mathbf{B})/\mathrm{B}^2$$

so that the plasma as a whole moves radially inwards. The material thus forms as magnetic ropes around magnetic flux tubes. Magnetic ropes thus contain material filaments that have a higher density than the surrounding plasma.

When a plasma is only partly ionized, the electromagnetic forces act on the non–ionized components only indirectly through the viscosity between the ionized and non–ionized constituents. For a filament, the inward radial velocity drift is

$$v_r = E_z/B_\phi$$

for the case of an axial electric field and azimuthal magnetic field (induced by the axial current I_z). Hence, at a large radial distance r, the rate of

[1] 'Neutral' hydrogen in space has a degree of ionization of the order of 10^{-4}.

accumulation of matter into a filament is

$$\frac{dM}{dt} = 2\pi r v_r \rho_m = (2\pi r)^2 \rho_m \frac{E_z}{\mu_0 I_z} \tag{1}$$

Marklund (1979) found a stationary state when the inward convections of ions and electrons toward the axis of a filament was matched by recombination and outward diffusion of the neutralized plasma (Figure 1). The equilibrium density of the ionized component normally has a maximum at the axis. However, because of the radiated loss of energy, the filament cools and a temperature gradient is associated with the plasma. Because of this hollow cylinders, or modifications of hollow cylinders of matter, will form aout the flux tubes.

Because the radial transport depends on the ionization potential of the element, elements with the lowest ionization potentials are brought closest to axis. The most abundant elements of cosmical plasma can be divided into groups of roughly equal ionization potentials: He(24 eV); H, O, N(13 eV); C, S(11 eV); and Fe, Si, Mg(8 eV). These elements can be expected to form hollow cylinders whose radii increase with ionization potential. Helium will make up the most widely distributed outer layer; hydrogen, oxygen, and nitrogen should form the middle layers, while iron, silicon, and magnesium will make up the inner layers. Interlap between the layers can be expected and, for the case of galaxies, the metal–to–hydrogen ratio should be maxium near the center and decrease outwardly. Both the convection process and luminosity increase with the field E_z.

For the case of a fully ionized hydrogenic plasma, the ions drift inwards until they reach a radius where the temperature is well below the ionization potential and the rate of recombination of the hydrogen plasma is considerable. Becouse of this "ion pump" action, hydrogenic plasma will be evacuated from the surroundings and neutral hydrogen will be most heavily deposited in regions of strong magnetic field.

When this process was discovered in the laboratory, there existed some debate as to whether galaxies possessed magnetic fields at all. Today, it is known that large-scale magnetic fields do exist in spiral galaxies. For example, the Effelsberg radio telescope has collected polarization data from about a dozen spiral galaxies at 6 to 49 cm wavelengths (Beck, 1986, 1990). Rotation measures show two different large-scale structures of the interstellar fields: Axisymmetric-spiral and bisymmetric-spiral patterns (Krause et al, 1989).

The orientation of the field lines is mostly along the optical spiral arms. However, the uniform field is often strongest outside the optical spiral arms. For example, in the case of galaxy IC 342, two filamentary structures were found in the map of polarized intensity (Peratt, 1992, Chap. 3). Their degree of polarization of $\approx$ 30 percent indicates a high degree of uniformity of the

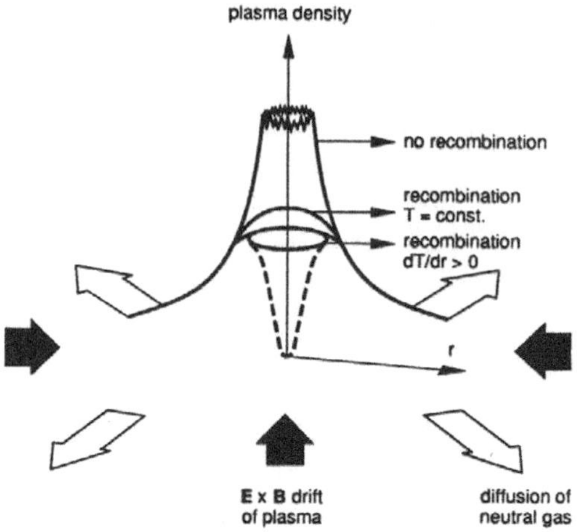

Fig. 1. Plasma density profile as a function of radius shown qualitatively for three cases: No recombination, recombination with temperature T = constant, and recombination with a lower central temperature.

magnetic field on the scale of the resolution ($\approx$ 700 pc). These filaments extend over a length of $\approx$ 30 kpc and hence are the most prominent magnetic-field features detected in normal spiral galaxies so far.

A detailed analysis of the rotation measure distribution in a spiral arm southwest of the center of the Andromeda galaxy M31 (Beck et al, 1990) shows that the magnetic field and a huge HI cloud complex are anchored together. The magnetic field then inflates out of the plane outside the cloud. The tendency for the magnetic field to follow the HI distribution has been noted in several recent observations. Circumstantial evidence has accumulated which suggests that there is a close connection between rings of CO and Hα seen rotating in some galaxies and the magnetic fields in the nuclear regions. This is particularly apparent in observations of spiral galaxies viewed edge-on. This scenario has also been invoked for our Galaxy (Wielebinski, 1989).

Neutral hydrogen is detected from galaxies via the van de Hulst radio-emission line at 21.11 cm (1.420 GHz). High-resolution observation of neutral hydrogen in irregular and spiral galaxies usually reveal extended HI distributions. Contour maps of the HI typically show a relative lack of HI in the cores of spiral galaxies but high HI content in the surrounding region, usually in the shape of a horseshoe. This region is not uniform but may have two or more peaks in neutral hydrogen content. Figure 2 (right–side) shows an

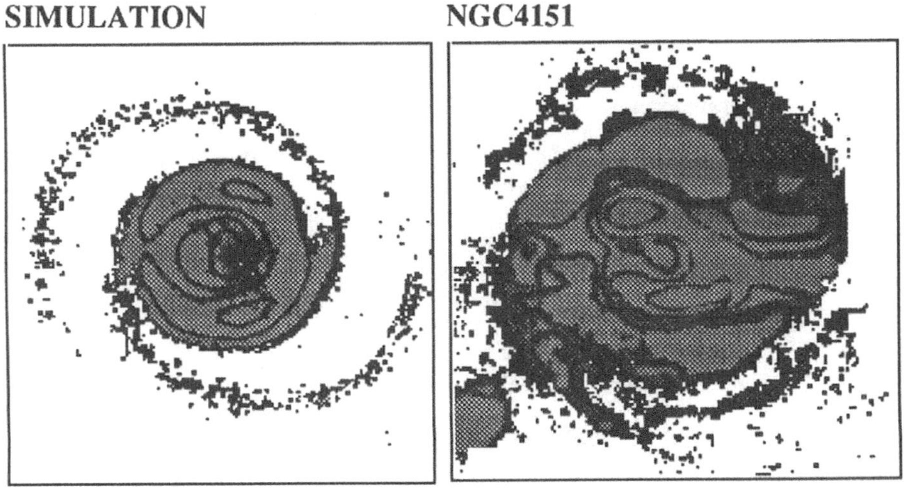

Fig. 2. (right) HI distribution superimposed on an optical photograph of NGC 4151. (left) Simulation magnetic energy density superimposed on simulation galaxy. In both cases a 'horse–shoe' shaped cusp, opening towards a spiral arm, surrounds a magnetic field/HI minima core. Within the cusp, two magnetic field/HI peaks are observed.

example of the HI distribution in a spiral galaxy.

3 Simulation Results

Figure 2 (left–side) shows the plasma spiral formed in this simulation overlayed on its magnetic field line (squared) isobars. The diameter of the spiral is about 50 kpc with a mass of 10^{41} kg, i.e., a size and mass of that observed from spiral galaxies. A direct comparison to observations is made by superimposing the HI distribution in NGC 4151 on its optical photograph. The observation shows two peaks in neutral hydrogen surrounding a void. The void is orientated towards one of the arms. The simulation allows the two peaks to be traced back to their origin. Both are found to be the remnants of the originally extended components, i.e., cross-sections of the original Birkeland filaments. The hydrogen deficient center is the remnant of an elliptical galaxy formed midway between the filaments, in the magnetic null.

Since E_z is out of the plane of the page, the column electrons spiral downward in counter-clockwise rotation while the column ions spiral upward in clockwise rotation. A polarization induced charge separation also occurs in each arm, which, as it thins out, produces a radial electric field across the arm. Because of this field, the arm is susceptable to the diocotron instability

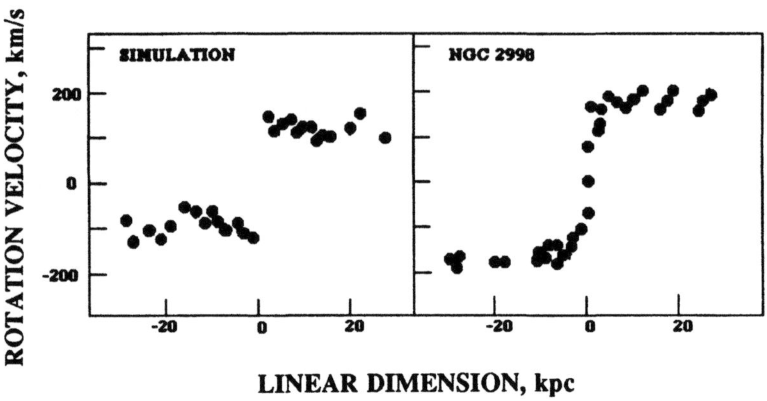

Fig. 3. Spiral galaxy rotational velocity curves. Note the well–defined structure on the 'flat' portions of the curves.

(Peratt, 1992). This instability appears as a wave motion in each arm and is barely discernable in the single frame photographs in Figure 1 of Peratt (1995) at late times. However, the instability is readily apparent in the simulation spiral rotational velocity curve (Figure 3).

The velocity consists essentially of a linearly increasing component due to a central body undergoing rigid rotation, with two flat components on either side of $r = 0$ due to the trailing arms. The diocotron instability modulates the flat components at the strong-magnetic-field, low-density instability wavelength $\lambda \approx 2.5\Delta r$, where Δr is the width of an arm.

4 Conclusions

In the late 1970s, plasma simulations of galaxies suggested that highly-ordered magnetic fields existed in galaxies, stretching for tens of thousand of light years. The strengths of the magnetic fields appearing in the simulations also suggested that appreciable amounts of nearly neutral hydrogen, known as HI regions, should collect around the field lines.

Simulation plots of the magnetic fields compared nicely with maps of observed HI regions, both showing a horse-shoe shaped distribution of gas with a bi-polar characteristic.

At about the same time, radio astronomers at the Max-Planck-Institute for Radio Astronomy in Bonn, started to measure ordered magnetic fields

At about the same time, radio astronomers at the Max-Planck-Institute for Radio Astronomy in Bonn, started to measure ordered magnetic fields in galaxies. This work had by 1988 shown unequivocally that large scale magnetic fields do exist in galaxies and do trace the distribution of neutral hydrogen.

With respect to the rotation velocities of spiral galaxies measured from the rotation of the neutral hydrogen: If galaxies were gravitationally bound systems, their outer mass should follow Kepler's laws of motion and be slower than the inner mass. The flat rotation curves of galaxies has been cited as the strongest physical evidence for the existance of dark matter. In this scenario a massivive halo of dark matter has been envoked to produce the flat rotation curves. However, the rotation curves are not really flat; they show appreciable structure representative of an instability mechanism within the arms. This instability precludes the existance of any external halo of matter around galaxies that, while making the rotation curves flat, would also dampen any instability growth.

The best agreement between the particle–in–cell maps, both magnetic field and neutral hydrogen, to the radio telescope data, and the replication of the optical features of spiral galaxies by the simulation, occurs when the observable galaxy mass is used. No dark matter is needed to explain the detailed features of a galaxy if electromagnetic forces are present.

5 Acknowledgments

This project was supported by the Institute of Electrical and Electronics Engineers, the Division of Atmospheric Sciences of the National Science Foundation, and NASA Headquarters, Space Physics Division.

References

Alfvén, H.: 1950, *Cosmical Electrodynamics* Oxford, London

Beck, R.: 1986, "Interstellar Magnetic Fields", *IEEE Trans. Plasma Sci.* **Vol. 14**, p. 470

Beck, R.: 1990, "Magnetic Fields in Spiral Galaxies", *IEEE Trans. Plasma Sci.* **Vol. 18**, p. 33

Birkeland, Kr.: 1912, *The Norwegian Aurora Polaris Expedition 1902–1903, Volumes I and II*, H. Ascheough & Co., Christiania, Norway

Krause, M., Beck, R., Hummel, E.: 1989, "The Magnetic Field Structure in Two Nearby Galaxies", *Astron. Astrophys.* **Vol. 217**, pp. 4–17

Marklund, G. T.: 1979, "Plasma Convection in Force-Free Magnetic Fields as a Mechanism for Chemical Separation in Cosmical Plasmas", *Nature* **Vol. 277**, p. 370

Peratt, A. L.: 1992a, *Physics of the Plasma Universe*, Springer-Verlag, New York

Peratt, A. L.: 1995, "Plasma and the Universe: Large Scale Dynamics, Filamentation, and Radiation", *Astrophys. Space Sci.*

Verschuur, G. L.: 1995, "Interstellar Neutral Hydrogen Filaments at High Galactic Latitudes and the Bennett Pinch", *Astrophys. Space Sci.*

Wielebinski, R.: 1989, "Magnetic Fields in the Galaxy", Max-Planck-Institut für Radioastronomie preprint series no. 347

HOW CAN SPIRALS PERSIST?

CYNTHIA KOLB WHITNEY
Visiting Industry Professor
Tufts University Electro-Optics Technology Center
Mailing address: 141 Rhinecliff St., Arlington, MA 02174, U.S.A.

Abstract. The image of a spiral galaxy is one of the most tantalizing images in nature. It demands that we ask: why do so many galaxies present this morphology? We currently have two main schools of thought concerning galaxy morphology, one based on gravitational effects and the other based on electromagnetic effects. The older gravitational models can explain how spirals might form, but they also predict that the spirals would quickly disintegrate. And the observed pervasiveness of spirals seems to imply not only their formation, but also their persistence over time. The newer plasma cosmology model is an improvement in that it explains how spirals might form and persist so long as plasma persists. But the formation of charge-neutral stars seems to return the scenario to the gravitational domain, and to subsequent dissolution. Clearly we need an additional idea to account for pervasive and persistent spiral galaxy structure. The present paper attempts to uncover a previously unrecognized gravitational mechanism that can serve as a viable candidate for sustaining persistent spiral galaxy structure.

Key words: Spiral Galaxies, Gravitation, Electromagnetics

1 Introduction

We live in a spiral galaxy, and from our vantage point we can see many other spiral galaxies as well. The empirical evidence seems to indicate that nature very much favors this particular form of galaxy morphology. The visual images are dramatic, and compel us to address the question: Why should this be so?

The state of the art in computing capability has advanced enough to allow us to propose and test realistic physical theories about galaxy dynamics. We can see how a large-scale system evolves under electromagnetic and gravitational forces, and how spiral structure can occur. The most recent theories are purely electromagnetic: the interaction between two large-scale plasma current loops can produce spiral structure (Peratt, 1986). Older theories are purely gravitational: the tides that arise in a collision between two amorphous looking galaxies can produce spiral structure (Toomre, 1977).

But we may not yet have in place the whole story, inasmuch as the mechanisms so far discussed are inherently transient: Plasmas eventually collapse into charge-neutral stars, leaving only gravitational forces. Colliding galaxies eventually complete their interactions, leaving only isolated systems. Once we have an isolated system governed only by gravity, spiral patterns should rapidly deteriorate because the orbit frequency of material in the spiral arms is not independent of the orbit radius. A simple Keplerian analysis predicts radius dependence for both orbit frequency and tangential velocity.

Astrophysics and Space Science **227**: 175–186, 1995.

Doppler measurements show tangential velocity more nearly insensitive to radius (Rubin, 1983), which implies a different radius dependence for orbit frequency, but a dependence nonetheless. So in any case, spiral structure should tend to unwind.

The observed prevalence of spiral galaxies seems to require further explanation. Do the proposed transient mechanisms that create spirals keep recurring with great frequency? Or is there another mechanism not yet considered that tends to maintain spiral structure? The first possibility cannot be discounted *a priori*, but the present paper concentrates on the second one. The additional mechanism is presumed not to be electromagnetic, because it must continue to act long after plasma collapses into charge-neutral stars. Accordingly, it is presumed to be gravitational.

The next Section introduces a candidate model for subsequent development into an explanation for galaxy morphology. The proposed gravitational mechanism has an electromagnetic analog which has to be examined in detail before the gravitational implications can be articulated. The several Sections following contain some new observations about standard electromagnetic field theory.

The gravitational model developed in the subsequent Section yields some quantitative predictions about spiral galaxy structure. These predictions are compared with observations to help assess whether the candidate model is indeed a potentially viable one.

2 A Candidate Galaxy Model

The candidate model here proposed treats observable stars as test particles in a dominant background gravitational field. This perturbation approach makes all of the calculations relatively simple, and many of them feasible analytically, or at least without benefit of extensive computing power. Naturally, success with such preliminary exercises creates a motivation for continuing with a second-generation version of the model. Given suitable computing resources, one could dispense with some of the idealizations employed here.

The proposed dominant field is created by a super massive system hidden at galaxy center. The proposed central system consists of two bodies orbiting each other. Within the context of Einstein's general relativity theory (GRT), these two bodies could be black holes. There is, however, no need to require that they be black holes. So readers averse to that particular aspect of GRT can nevertheless proceed. The important point is: the masses of the visible stars that form the galactic spirals are insignificantly small by comparison to the masses in the proposed two-body driver system.

The development of the model begins with problem of the gravitational field created by a single moving mass. In some important ways, this problem is analogous to the problem of predicting the electric and magnetic fields created by a single moving charge. Gravity is assumed to propagate at some finite speed just as light does, and the gravitational potentials are assumed to have the same functional dependence on coordinates as do the electromagnetic potentials (Einstein, 1955).

The currently accepted theory for the electromagnetic potentials created by a moving source charge was developed at the turn of the twentieth century, by Liénard (1898) and Wiechert (1900). One salient feature of the theory concerns the lines of force for the electric field **E** created by a uniformly moving charge. The E field lines are predicted to be purely radial, pointing always to the present position of the charge (see, for example, Jackson 1975). The only effect of the motion is to concentrate the E field lines in directions near normal to the motion.

Two bodies orbiting are, of course, subject to more complicated forces. Each body experiences not only the electric field, but also a magnetic field **B**. In addition, neither body is in uniform motion, so there are acceleration-dependent as well as velocity-determined contributions to the fields. Nevertheless, the attraction between two bodies is still predicted to align with the present positions of the bodies.

As a basis for predicting some local gravitation effects, the Liénard-Wiechert field-line behavior has some merit. Based on the observed behavior of bodies in the solar system, present positions of bodies do seem to determine force directions (Van Flandern, 1993). From this behavior, LaPlace concluded that the propagation speed for gravity must be very large compared to the speed of light. Newton later idealized the gravitational propagation speed to infinity. But Einstein's GRT requires that the propagation speed be the same as light speed, c. The Liénard-Wiechert analog can help resolve this conflict. It explains the present-directedness of gravitational forces, while at the same time allowing finite gravitational propagation speed; in particular, speed c.

But as a basis for developing a background field to support spiral galaxy structure, the Liénard-Wiechert field-line behavior is not promising at all. Galaxy structure arising from present-directed forces would be straight spokes rather than spiral arms (Whitney, 1988bs). Clearly, if the electromagnetic analog is to take us anywhere in regard to spiral galaxy structure, we must look into it more deeply.

3 The Standard Electromagnetic Model

Consider a source charge Q traveling with velocity $\mathbf{v}$, or normalized $\boldsymbol{\beta} = \frac{\mathbf{v}}{c}$. Let $\mathbf{R}$ be the vector from the source to the observer and $\mathbf{n}$ be a unit vector along $\mathbf{R}$, and define $\kappa = 1 - \mathbf{n} \cdot \boldsymbol{\beta}$. Let subscript "ret" denote evaluation with source coordinates taken at retarded source time; i.e. $t_{\text{ret}} = t - \frac{R_{\text{ret}}}{c}$.

The Liénard-Wiechert scalar and vector potentials are

$$\Phi = \frac{Q}{(\kappa R)_{\text{ret}}} \qquad \text{and} \qquad \mathbf{A} = \Phi \boldsymbol{\beta}_{\text{ret}} \tag{1a}$$

The electric and magnetic fields are defined as

$$\mathbf{E} = -\boldsymbol{\nabla}\Phi - \frac{1}{c}\frac{\partial \mathbf{A}}{\partial t} \qquad \text{and} \qquad \mathbf{B} = \boldsymbol{\nabla} \times \mathbf{A} \tag{1b}$$

and evaluated by application of the differentiation chain rule using

$$\boldsymbol{\nabla} t' = -\frac{1}{c}\left(\frac{\mathbf{n}}{\kappa}\right)_{\text{ret}} \qquad \text{and} \qquad \frac{\partial t'}{\partial t} = \frac{1}{\kappa_{\text{ret}}} \tag{1c}$$

Assuming only uniform motion, we have velocity-determined fields

$$\mathbf{E}_{\text{vel}} = Q\left(\frac{\mathbf{n} - \boldsymbol{\beta}}{\gamma^2 \kappa^3 R^2}\right)_{\text{ret}} \qquad \text{and} \qquad \mathbf{B}_{\text{vel}} = \mathbf{n}_{\text{ret}} \times \mathbf{E}_{\text{vel}} \tag{2a}$$

With t' standing for source time, the acceleration-dependent fields are

$$\mathbf{E}_{\text{acc}} = \frac{Q}{c}\left(\mathbf{n} \times \frac{(\mathbf{n} - \boldsymbol{\beta}) \times \frac{d\boldsymbol{\beta}}{dt'}}{\kappa^3 R}\right)_{\text{ret}} \qquad \text{and} \qquad \mathbf{B}_{\text{acc}} = \mathbf{n}_{\text{ret}} \times \mathbf{E}_{\text{acc}} \tag{2b}$$

The Liénard-Wiechert formulations present many problems that have been detailed elsewhere (Whitney, 1989rc; Whitney, 1988nc; Whitney, 1988lw; Whitney, 1988nc; Whitney, 1991). Yet another critique can be based on symmetry properties. For the case of uniform motion, the Liénard-Wiechert formulations are claimed to be in one-to-one correspondence with the predictions of SRT (which they predate by several years). To investigate this claim, we need the parameter $\gamma = \frac{1}{\sqrt{1-\beta^2}}$. Lorentz transformation requires that potentials satisfy

$$(\Phi, \mathbf{A}) = \Phi_0 \gamma (1, \boldsymbol{\beta}) \tag{3a}$$

where subscript 0 means source at rest. Lorentz transformation further requires

$$E_v = E_{0v} \qquad \text{and} \qquad E_p = \gamma E_{0p} \qquad \text{and} \qquad \mathbf{B} = \boldsymbol{\beta} \times \mathbf{E} \tag{3b}$$

where subscript v means component along $\mathbf{v}$ and subscript p means component perpendicular to $\mathbf{v}$. Thus according to SRT, the magnitudes of all potentials and fields must be insensitive to reversal of source $\boldsymbol{\beta}$.

Now consider the Liénard-Wiechert potentials 1a and fields 2a. The parameter κ is *not* insensitive to $\boldsymbol{\beta}$ reversal, so apparently none of the Liénard-Wiechert potentials and fields are insensitive to $\boldsymbol{\beta}$ reversal either.

But suppose the potentials *were* insensitive to $\boldsymbol{\beta}$ reversal. By chain-rule expansion the E field would be

$$\mathbf{E}_{\text{vel}} = -\boldsymbol{\nabla}\Phi|_{t'} - \frac{\partial\Phi}{\partial t'}\boldsymbol{\nabla}t' - \frac{\boldsymbol{\beta}}{c}\frac{\partial\Phi}{\partial t'}\frac{\partial t'}{\partial t} \tag{3c}$$

The middle term, and *only* the middle term, would be sensitive to $\boldsymbol{\beta}$ in first order. Thus $\mathbf{E}_{\text{vel}}$ would *not* be properly insensitive to $\boldsymbol{\beta}$ reversal. Evidently then, the Liénard-Wiechert expressions for potentials and fields simply cannot conform with the symmetry requirements of SRT.

4 Remedies for the Electromagnetic Formulation

A rework of the Liénard-Wiechert problem should make no unnecessary assumptions early in the derivation, but rather leave as much as possible to be determined later by the requirements of Lorentz transformation. In this spirit, we set

$$(\Phi, \mathbf{A}) = Q\left(\frac{\gamma(1,\boldsymbol{\beta})}{R_0}\right)_{\text{ret}} \tag{4a}$$

where R_0 is left to be determined later. The Liénard-Wiechert formulations 1a would require $R_0 \equiv \gamma\kappa R$, but we here retain only a weaker statement; namely, that the components of vector $\mathbf{R}_0$ are linear functions of the components of vector $\mathbf{R}$, with coefficients that depend on $\boldsymbol{\beta}$:

$$\mathbf{R}_0 = \mathbf{R}_0(\mathbf{R}, \boldsymbol{\beta}) \tag{4b}$$

In formulating the fields, we recall the fundamental idea of a field theory; namely that only the potentials *themselves* matter, not the sources from which they have arisen. So we constrain derivatives of potentials to preclude variation of their respective sources. The Liénard-Wiechert expressions 2a, 2b arise according to 1b with no constraints at all. But here we put

$$\mathbf{E} = -\boldsymbol{\nabla}\Phi|_{\rho} - \frac{1}{c}\frac{\partial\mathbf{A}}{\partial t}\bigg|_{\mathbf{j}} \quad \text{and} \quad \mathbf{B} = \boldsymbol{\nabla}\times\mathbf{A}|_{\mathbf{j}} \tag{5a}$$

The constraints affect the $\nabla t'$ and $\frac{\partial t'}{\partial t}$ previously given by 1c. Now we have

$$\nabla t'|_{\rho} = 0 \quad \text{and} \quad \nabla t'|_{\mathbf{j}} = -\frac{\mathbf{n}_{\text{ret}}}{c} \quad \text{and} \quad \frac{\partial t'}{\partial t}\bigg|_{\mathbf{j}} = 1 \tag{5b}$$

Assuming uniform motion, we then have

$$\mathbf{E}_{\text{vel}} = -\boldsymbol{\nabla}\Phi|_{t'} - \frac{\boldsymbol{\beta}}{c}\frac{\partial\Phi}{\partial t'} \tag{6a}$$

$$\mathbf{B}_{\mathrm{vel}} = \left(\boldsymbol{\nabla}\Phi |\, t' - \frac{\partial \Phi}{\partial t'} \frac{\mathbf{n}_{\mathrm{ret}}}{c} \right) \times \boldsymbol{\beta} \tag{6b}$$

With R_0 still undetermined but nevertheless substituted into Φ, we have

$$\boldsymbol{\nabla}\Phi|_{t'} = -\frac{\partial \Phi}{\partial R_{0\mathrm{ret}}} \boldsymbol{\nabla} R_{0\mathrm{ret}}|_{t'} = -Q \left(\frac{\gamma}{R_0^2} \right)_{\mathrm{ret}} \boldsymbol{\nabla} R_{0\mathrm{ret}}|_{t'} \tag{7a}$$

$$\frac{\partial \Phi}{\partial t'} = -\frac{\partial \Phi}{\partial R_{0\mathrm{ret}}} \frac{\partial R_{0\mathrm{ret}}}{\partial t'} = -Q \left(\frac{\gamma}{R_0^2} \right)_{\mathrm{ret}} \frac{\partial R_{0\mathrm{ret}}}{\partial t'} \tag{7b}$$

This gives us

$$\mathbf{E}_{\mathrm{vel}} = Q \left(\frac{\gamma}{R_0^2} \left(\boldsymbol{\nabla} R_0|_{t'} + \frac{\boldsymbol{\beta}}{c} \frac{\partial R_0}{\partial t'} \right) \right)_{\mathrm{ret}} \tag{8a}$$

$$\mathbf{B}_{\mathrm{vel}} = -Q \left(\frac{\gamma}{R_0^2} \left(\boldsymbol{\nabla} R_0|_{t'} - \frac{\mathbf{n}_{\mathrm{ret}}}{c} \frac{\partial R_0}{\partial t'} \right) \right)_{\mathrm{ret}} \times \boldsymbol{\beta} \tag{8b}$$

It is now possible to find $\mathbf{R}_0$ to fit the Lorentz transformations 3b. One can verify by direct substitution that

$$R_{0p} = R_p \qquad \text{and} \qquad R_{0v} = \gamma R_v \tag{9a}$$

With R_0 so defined,

$$\boldsymbol{\nabla} R_0|_{t'} = \left(\mathbf{n}_p + \gamma^2 \mathbf{n}_v \right) \qquad \text{and} \qquad \frac{1}{c} \frac{\partial R_0}{\partial t'} = -\boldsymbol{\beta} \gamma^2 \frac{R_{0v}}{R_0} \tag{9b}$$

It then follows that

$$\mathbf{E}_{\mathrm{vel}} = Q \left(\frac{\gamma \mathbf{n}}{R_0^2} \right)_{\mathrm{ret}} \tag{10a}$$

$$\mathbf{B}_{\mathrm{vel}} = \boldsymbol{\beta} \times \mathbf{E}_{\mathrm{vel}} \tag{10b}$$

which is manifestly consistent with 3b.

It is is possible to evaluate $\mathbf{E}_{\mathrm{acc}}$ and $\mathbf{B}_{\mathrm{acc}}$ also. One finds that, unlike the Liénard-Wiechert result, $\mathbf{E}_{\mathrm{acc}}$ is not necessarily normal to $\mathbf{n}_{\mathrm{ret}}$. But the corresponding $\mathbf{B}_{\mathrm{acc}} = \mathbf{n}_{\mathrm{ret}} \times \mathbf{E}_{\mathrm{acc}}$ is formally the same as the Liénard-Wiechert result, although of course it is different numerically.

5 Application to Galaxy Morphology

The most significant of the electromagnetic results reported here is 10a describing $\mathbf{E}_{\mathrm{vel}}$. It implies first of all a significant revision to field lines (Whitney, 1987). Instead of everywhere pointing to the present position of the source, field lines now point to the locally-perceived retarded position of the source.

The altered field directions then imply a previously unexpected effect in a two body system. Each body is attracted to the retarded position of the other. This means the forces in the system are not central. There is an overall torque on the system, causing field-to-matter energy transfer (Whitney, 1989fm).

With equal masses M at equal orbit radii R, the energy gain rate P_t due to torquing is twice the force magnitude F times the sine of the force rotation angle ϕ times the orbit velocity v. With v low, the force is Newtonian,

$$F \approx \frac{GM^2}{(2R)^2} \tag{11a}$$

and the rotation angle ϕ is given by

$$\sin(\phi) \approx \beta = \frac{v}{c} \tag{11b}$$

so we have

$$P_t \approx \frac{GM^2v^2}{2cR^2} \tag{11c}$$

The torquing effect opposes the inevitable matter-to-field energy transfer due to radiation. The radiation energy loss rate P_r can be estimated by the well-known quadrupole formula

$$P_r = \frac{32GI^2\omega^6}{5c^5} \tag{12a}$$

With moment of inertia

$$I = 2MR^2 \tag{12b}$$

we have

$$P_r \approx \frac{128GM^2v^6}{5c^2R^2} \tag{12c}$$

In fact, the torque effect can even compensate the radiation effect completely in some circumstances. With the Newtonian force and quadrupole radiation assumptions, the two mechanisms balance at v roughly equal to $\frac{c}{2}$. This result is so large that it really demands relativistic corrections to the assumptions that produced it. But those corrections do not substantially alter the result (Whitney, 1988ss).

The result that v is about $\frac{c}{2}$ is independent of the mass of the bodies in the two-body system. It is also formally independent of the numerical value of c; that is, all the same formulae hold even if light speed c is replaced by some other value c_g for gravitational propagation, as suggested by Van Flandern (1993). It is thus a very robust result from which to proceed further.

Accordingly, for the system driving a spiral galaxy, we consider two massive bodies orbiting at about half the signal propagation speed. Stars respond

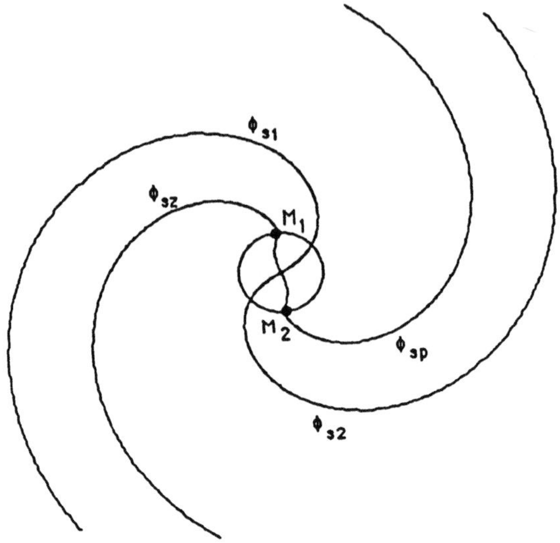

Fig. 1. Spiral structure of the background field created by a two-body system. The spiral lines are the locii of zero tangential force. The smooth spiral correlates with minimum star density, while the spiral with the central bar correlates with maximum star density.

to the two bodies as seen in their retarded positions. This situation produces tangential forces with interesting properties.

Firstly, the average taken over star angle ϕ_s is *not* zero (Whitney, 1988bs). Stars orbiting in such a background field experience time average torque, and hence secular increase in orbit energy and angular momentum. The orbits expand to form a disk-shaped system, and the system as a whole acquires large net angular momentum. These features are clearly presented by nature.

Secondly, the placement of zero crossings implies spiral structure. Figure 1 plots the approximate locii of zero crossings. The spiral structure is obvious. There is even a central bar suggesting the barred spiral structure often observed in nature (Whitney, 1988bs).

The behavior of test-particle stars in this barred-spiral background field can be described at least approximately with simple algebraic formulae derived from the model. This was done first (Whitney, 1988tp), and is here somewhat simplified and extended. We begin with the Newtonian relationship

$$R_s \omega_s^2 = F_s \propto \frac{1}{R_s^2} \tag{13a}$$

where R_s is radius to the star, ω_s is its orbit frequency, and F_s is Newtonian attractive force on the star. From 13a follows the radius dependence of orbit frequency:

$$\omega_s \propto \frac{1}{\sqrt{R_s^3}} \tag{13b}$$

and the radius dependence of star angular momentum

$$L_s \propto R_s^2 \omega_s \propto \sqrt{R_s} \tag{13c}$$

The present model gives a secular increase of angular momentum

$$\frac{dL_s}{dt} \approx F_s \phi_d R_s \tag{14a}$$

where ϕ_d is the average offset angle due to delay for signal propagation from the driving two-body system. For either body,

$$\phi_d \approx \frac{p\,(a_d)}{R_s} \tag{14b}$$

where a_d is the arc traversed by the body during the delay time, and $p\,(a_d)$ is its projection normal to the line of sight from the star to the delayed source position. The projection of the arc is absolutely limited to the diameter of the two-body system, so for large R_s, ϕ_d declines as $\frac{1}{R_s}$. This makes

$$\frac{dL_s}{dt} \propto \frac{1}{R_s^2} \tag{14c}$$

The increase angular momentum implies a corresponding increase in orbit radius. Given the power law relationship 13c, it follows that

$$\frac{dR_s}{dt} = \frac{R_s}{L_s}\frac{dL_s}{dt} \tag{15a}$$

For large R_s we have

$$\frac{dR_s}{dt} \propto \frac{1}{\sqrt{R_s^3}} \tag{15b}$$

In other words,

$$\frac{dR_s^{5/2}}{dt} = \text{constant} \qquad \text{or} \qquad \Delta R_s \propto \Delta T_s^{2/5} \tag{15c}$$

where ΔT_s is the time since the star was formed somewhere nearer the center of the galaxy and ΔR_s measures its outward progress since then. The implication is that older stars tend to accumulate toward the outer extremities of the galaxy. Since older stars may eventually explode and/or cease to radiate, this prediction is in harmony with observations indicating cold, dark matter at the outer edges of galaxies (Rubin, 1983).

Suppose a number of stars are launched in close proximity. Their spatial density ρ will evolve over time. Conservation of total matter requires

$$R_s \rho\,(R_s) \propto \left(\frac{dR_s}{dT_s}\right)^{-1} \tag{16a}$$

For large R_s,

$$\rho(R_s) \propto \sqrt{R_s} \tag{16b}$$

One may also enquire about the density variation with angle ϕ_s. This can be estimated under the assumption that orbits are very nearly circular, but with tangential velocity varying slightly in response to tangential forces. Conservation of total mass flow requires that density vary inversely to tangential velocity:

$$\rho(\phi_s) \propto \frac{1}{v_t(\phi_s)} \tag{16c}$$

The tangential velocity has an angular variation given by the integral of tangential force. The periodic nature of tangential force implies that tangential velocity must be out of phase with it. Minimum velocity, and hence maximum star density, must then occur on the barred spiral spine plotted in Fig. 1.

6 Discussion

At this point, where do we stand? We have here offered a candidate model that predicts certain otherwise perplexing features of spiral galaxy morphology. For example, we see why the galaxy forms a disk, why it displays prominent spiral arms, and even why it may exhibit a central bar structure.

Most importantly, the present model helps account for *persistence* of spiral galaxy structure. Persistence seems required by the observed prevalence.of spiral galaxy structure, but previous models seemed less than adequate in this regard. The plasma cosmology model (Peratt, 1986) provided persistence of spiral structure for as long as plasma persists, but not after plasma collapses into charge-neutral stars. The colliding galaxy model (Toomre, 1977) was fundamentally transient in nature.

The author sees the present model as the natural sequel to the plasma cosmology model. Plasma effects set up initial conditions that are bilateral. Compression and gravitation then create stars, most especially two massive stars. These form a steady-state two-body system that in turn forms a spiral shaped background field in which smaller, test-particle stars move.

But the present model is not without problems. Simplest among these is the fact that insights gleaned so far are rather sketchy. This is because only analytic type results have been extracted. It seems very desirable that future research also include computer simulations illustrating the evolution of a galaxy according to the present model. The ideal simulation would involve both electromagnetic interactions for plasma currents and gravitational interactions for stars.

The analytic results (Whitney, 1988tp) may not reconcile well with the standard "big bang" model for the universe. The putative time involved may be too short (or too long) to account for some galaxies. However, one has to recall that the "big bang" itself is not without doubters (Lerner, 1992), so discrepancies in this area can hardly be considered fatal.

Another problem lies in reconciling the spiral galaxy model with the observed behavior of the solar system. At this smaller scale, observed behavior is apparently consistent with forces alligned to present positions of sources. This could mean a speed of gravitational propagation c_g far in excess of light propagation speed c, as held by La Place, or even infinite c_g, as Newton assumed.

The phenomenon of aberration has been invoked to demonstrate a real difference between light and gravity (Van Flandern, 1993). So far as we know, aberration occurs only with light, not with gravity With light, there are distinct cases of stellar aberration, planetary aberration, and no aberration at all. The several possible phenomena are really *not* so perfectly understood as is generally assumed (Hayden, 1993). The present author sees the situation with light as presently too unresolved to support any additional conclusions about gravity.

If the gravitational propagation speed were infinite, there would be no such thing as gravitational radiation or torquing due to signal delay, and the present galaxy model would fall apart. Even if the gravitational propagation speed exceeded $2c$, the model would not work, because the driving super massive two-body system would have to orbit at speed greater than c. If the propagation speed were below $2c$, the model would be affected but not destroyed. The basic shape of the barred spiral background field for the galaxy is insensitive to gravitational propagation speed c_g, although its rotation rate $\omega = \frac{c_g}{R}$ increases with c_g, and its ability to concentrate test-particle stars into spiral arms decreases correspondingly.

References

Einstein, A.: 1955, *The Meaning of Relativity*, Fifth edition, Eq. 101, Princeton University Press, Princeton, New Jersey

Hayden, H.: 1993, "Stellar Aberration", *Galilean Electrodynamics* **Vol. 4**, pp. 89-92

Jackson, J. D., *Classical Electrodynamics*, Second Edition, Chapter 11, John Wiley & Sons, New York, New York

Lerner, E.J.: 1992, *The Big Bang Never Happened*, Vintage, Random House, New York, New York

Liénard, A.: 1898, "Champ Electrique et Magnetique", *Eclairage electrique* **Vol. 16**, pp. 5-14, 53-59, and 106-112

Peratt, A.: 1986, "Evolution of the Plasma Universe: I and II", *IEEE Trans. Plasma Sci.* **Vol. 14**, pp. 639-660 and 763-778

Rubin, V. C.: 1977, "Theories of Spiral Structure", *Science* **Vol. 220**, pp. 1339-1344 (1983).

Toomre, A.: 1977, "Theories of Spiral Structure", *Ann. Rev. Astron. Astrophys* **Vol. 15**, pp. 437-478

Whitney, C.K.:1987, "Field lines in Clasical Theory", *Hadronic Journal* **Vol. 10**, pp. 91-93

Whitney, C.K.: 1988nc, "Operator Noncommutation in Classical Field Theory", *Hadronic Journal* **Vol.11**, pp. 101-107

Whitney, C.K.: 1988lw, "On the Liénard-Wiechert Potentials", *Hadronic Journal* **Vol. 11**, pp. 257-261

Whitney, C.K.: 1988ss, "Steady State in the Gravitational Two-Body Problem", *Hadronic Journal* **Vol. 11**, pp. 153-156

Whitney, C.K.: 1988bs, "Barred Spiral Field Pattern Around a Relativistic Two- Body System", *Hadronic Journal* **Vol. 11**, pp. 157-162

Whitney, C.K.: 1988tp, "Distribution of Stars as Test Particles in a Two-Body Background Field", *Hadronic Journal* **Vol. 11**, pp. 163-167

Whitney, C.K.: 1989fm, "Field-to-Matter Energy Transfer", *Hadronic Journal* **Vol. 12**, pp. 35-41

Whitney, C.K.: 1989rc, "On The Properties of Retarded Coordinates", *Physics Essays* **Vol. 2**, pp. 246-248

Whitney, C.K.: 1991, "A Gedanken Experiment With Relativistic Fields", *Galilean Electrodynamics* **Vol. 2**, pp. 28-29

Wiechert, E.: 1900, "Electrodynamische Elementargesetze", *Archives Neerlandes* **Vol. 5**, pp. 549-573

Van Flandern, T.: 1993, *Dark Matter, Missing Planets and New Comets*, North Atlantic Books, Berkeley, California

INTERSTELLAR NEUTRAL HYDROGEN FILAMENTS AT HIGH GALACTIC LATITUDES AND THE BENNETT PINCH

GERRIT L. VERSCHUUR
Physics Department
Rhodes College
Memphis, TN, USA

Abstract. Observed properties of interstellar neutral hydrogen filaments suggest the presence of the Bennett pinch as described by the Carlqvist relationship with rotation around the filament axes included. A brief summary is first given of three ways in which a filament model for interstellar "cloud" structure was tested. Preliminary results from highresolution HI mapping of gas and dust in an apparent HI "cloud" indicate that the neutral gas and dust within and around its boundary is itself highly filamentary. An attempt to detect magnetic fields in this and similar features using the Zeeman effect technique at the 21–cm wavelength of interstellar neutral hydrogen set upper limits of a few μG. In contrast, the strength of the toroidal magnetic field expected from the examination of the Carlqvist relationship is of order 5 μG, which would be produced by a current of 1.4×10^{13} A. Zeeman effect technology is at present not able to detect toroidal magnetic fields of this order at the edge of barely resolved HI filaments. Nevertheless, currently available high–resolution HI data suggest that interstellar filament physics should take into account the role of currents and pinches for creating and stabilizing the structures.

1 Introduction

Neutral hydrogen (HI) surveys at high galactic latitudes show that the interstellar gas is filamentary; see for example Verschuur (1973, 1974a, 1974b, 1991a, 1991b) and Verschuur *et al.* (1992). The filamentary nature of the HI is also dramatically evident in the data by Colomb, Pöppel, and Heiles (1980) and the new all–sky Leiden–Dwingeloo HI survey (Hartmann, 1994). What, then, is the relationship between such fialments and magnetic fields that thread their way through interstellar space? And is it possible that the origin and stability of the filaments depends on the existence of large–scale currents as found by Carlqvist and Gahm (1992).

Verschuur (1991a) has shown that morphological waves with wavelengths of order 30° and amplitude $\sim 1/8\lambda$ exist in several HI filaments in a 540 square degrees of sky around l=230°, b=+40°. The extent of the longest filament is greater than 72°, the limit set by the boundaries of the area surveyed. The motion within the filaments, determined by analysis of velocities along their axes, suggests the presence of wave patterns with amplitude 5 to 6 km/s on an angular scale similar to that seen in the spatial wave–like structure projected on the sky. Furthermore, in this area of sky every so-called HI "cloud" (or enhanced emission feature; EEF) is found to be associated with a filament while the "clouds" are usually found where the filaments show

Astrophysics and Space Science **227**: 187–198, 1995.

changes in the orientation of their axes, as if defining kinks in the filaments. This raised the interesting possibility that a "cloud" of diffuse interstellar HI, defined by a localized enhancement of HI emission, might be a geometric illusion produced where a segment of filament twists into the line of sight. Our study began in order to test this simple model.

An example of an HI "cloud" is shown in Figure 1. Each frame plots the distribution of the antenna temperature of the HI emission at the velocity indicated at the lower right of the frame (in km/s with respect to the local standard of rest). The original maps were plotted with a resolution of 20 arcmin (Verschuur 1974a). The lower right frame shows a sketch of the derived filamentary patterns taken from Verschuur (1991b). The numbers in this frame indicate the velocity of the HI emission at various locations along Filament A whose morphology is derived by reference to the HI structure in the other frames.

The bright feature in the center of the various maps is an example of an EEF ("cloud"), this one labelled H0827+10. At the extreme velocities at which it is identified it appears to be connected to material at the top-right and lower-left of the area shown here. If filament geometry accounts for EEFs, H0827+10 may represent a direction in which the filament twists into the line–of–sight.

We have carried out three types of observations to test this hypothesis. First, if cool interstellar HI and dust are associated, we might obtain a better view of the filament geometry by examining the distribution of 100 μ emission from the dust, since the IRAS survey of 100 μ emission has a resolution of a few arcmin, considerably better than most radio telescopes used in the study of interstellar HI structure. Second, high-resolution HI mapping should show the structure more clearly. The third test involves mapping the magnetic field in the direction of the EEF using the 21–cm Zeeman effect as a probe. This experiment is sensitive to the line–of–sight component of the magnetic field and if EEFs are illusions produced where a segment of filament twists into the line–of–sight, the magnetic field should appear stronger toward in the EEF than toward its surroundings. We briefly report on the result of all three tests.

2 The Data

Figure 2 shows the 100 μ flux from cool interstellar dust in the direction of the EEF H0827+10, the bright feature at the center of this map. The presence of filaments is obvious and it is not difficult to imagine that even on this scale the brightest features are produced where filaments overlap or twist into the line–of–sight. When the central part of Figure 2 is examined more closely under different contrast, the brightest feature is seen to have a loop shape.

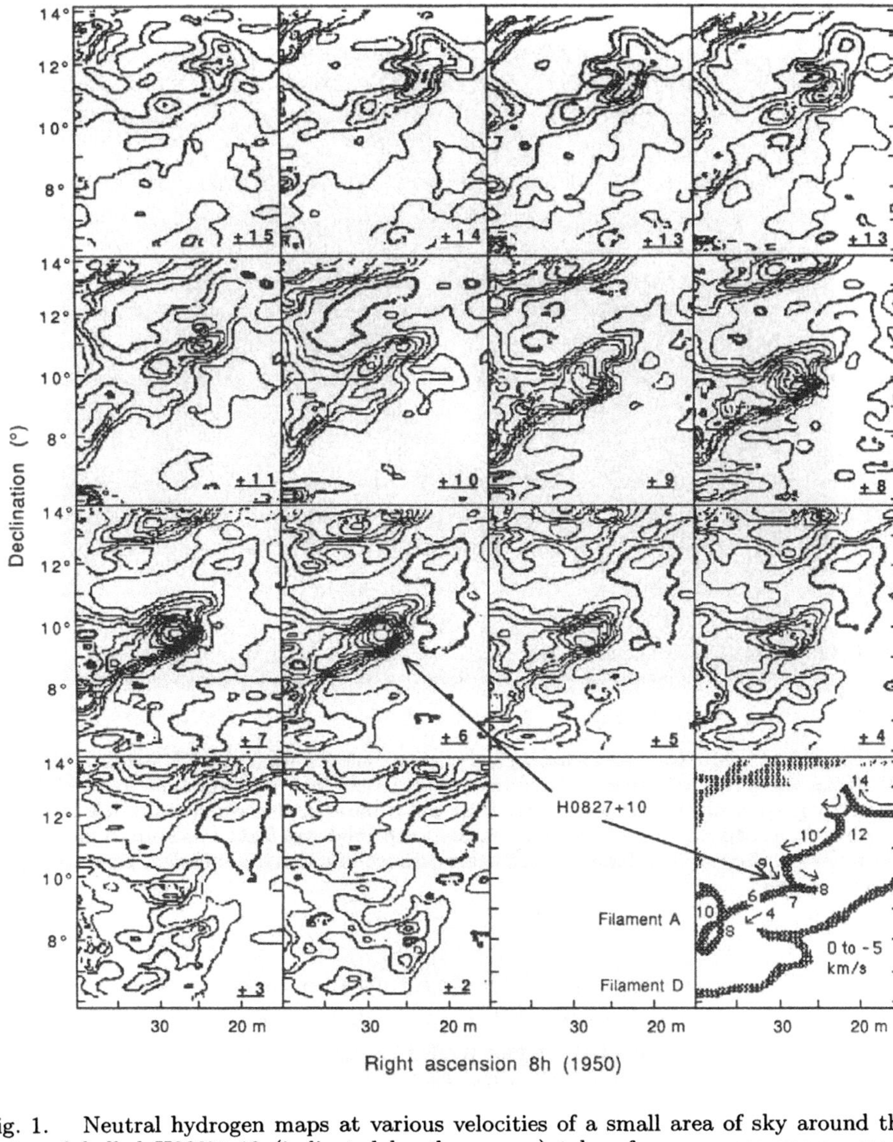

Fig. 1. Neutral hydrogen maps at various velocities of a small area of sky around the feature labelled H0827+10 (indicated by the arrows) taken from an area survey made by Verschuur (1974a). The velocity is indicated at the lower right of each frame. The lower-right frame is a schematic representation of several filaments that cross this field of view taken from Verschuur (1991b). The numbers indicate the velocity of the gas at various locations along Filament A. H0827+10 would traditionally be referred to as a "cloud."

The Arecibo radio telescope was next used to obtain high resolution (4 arcmin) maps of the HI emission from H0827+10. Our analysis technique involved the Gaussian decomposition of hundreds of profiles in order to iden-

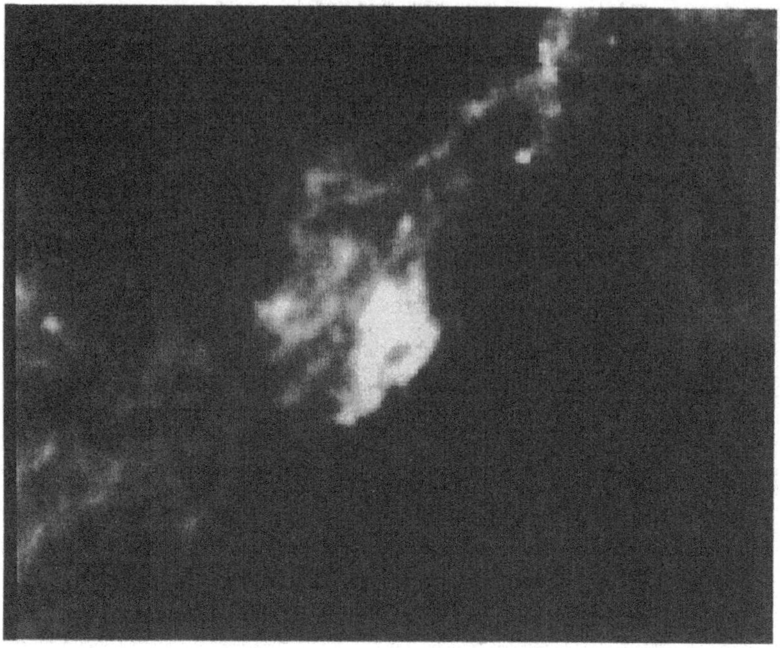

Fig. 2. The 100 HI emission from cool dust associated with the HI feature H0827+10. The image size is 4° x 5° centered on right ascension 8h 27m, declination +9° 48'. The dust shows a great deal of filamentary structure not visible in the lower resolution HI data in Figure 1. In a display using different contrast, the brightest feature is seen to be a distinct narrow loop. There is no indication of the existence of a clearly bounded, cloud-like entity in this region.

tify families of components and then making an area map of each component before comparing the results with the 100 μ data. Figure 3 shows the combined column density of cool gas found in three of the brightest Gaussian components defining the emission profiles in this region. Comparison of Figures 2 and 3 shows a great deal of similarity, certainly within limits set by the resolution and data display techniques. It is striking that the HI map appears on the verge of revealing as much detail as is seen in the 100 μ data. A similar map of the total HI column density shows far less contrast, due in part to the presence of unrelated gas along the line–of–sight, and to lesser extent to the unwanted influence of sidelobes distorting the emission profiles at low levels. The morphological similarities between the structures seen inf Figures 2 and 3 suggest that this way of analyzing the data is more powerful than the traditional method of comparing total HI column density with 100 μ brightness.

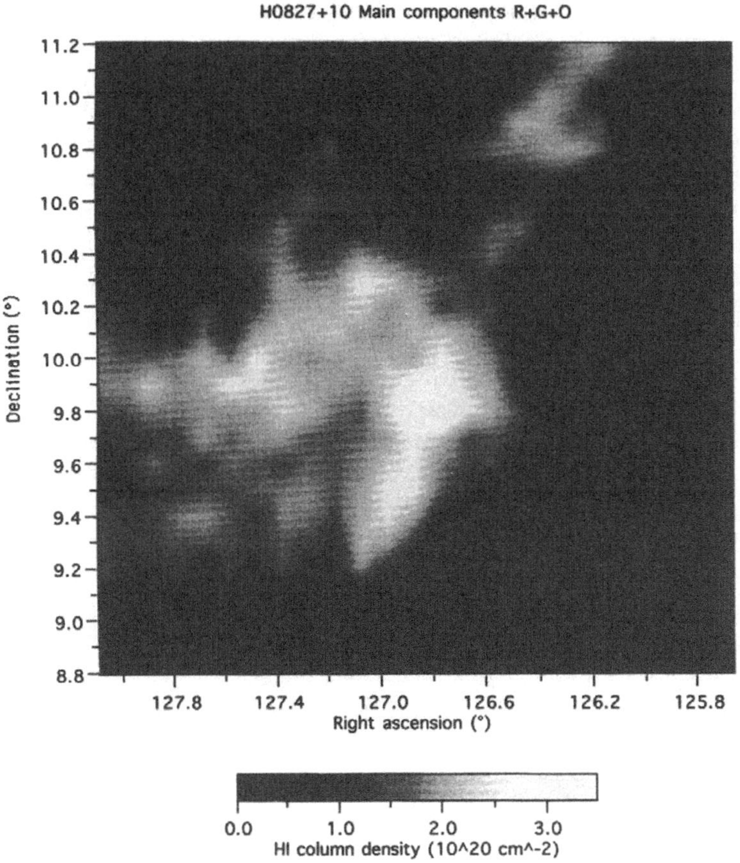

Fig. 3. Gray-scale image of the combined HI column density in units of 10^{20} cm^{-2} in three Gaussian components most clearly defining the HI feature H0827+10 based on the high-resolution survey referred to in the text. Comparison with Figure 2 shows the high degree of morphological similarity between the 100 emission and the HI structure. This HI image is on the verge of revealing as much detail as is seen in Figure 2. Two dark rectangles indicate areas of missing data at the top-left, down to declination 10.°2, and at the topcenter, down to 10.°75. Right ascension 127° is 8h 28m.

Figure 3 shows that the the HI "cloud" H0827+10 in Figure 1 contains enhanced filamentary structure and that the original "cloud" is not a physically bound, coherent entity at all, at least not in the sense that some unifying force, whether gravity or a simple magnetic field configuration, gathers the gas into a small, spherically symmetric volume of space. It seems likely that the enhancement in brightness in the lower resolution HI data is related mainly to line–of–sight geometrical effects while revealing little about the possible physics involved in creating the structure. (Bear in mind that the structure seen in these maps is not two-dimensional but has an unknown depth component).

Our third test of the twisted filament model involved observations of the Zeeman effect in HI emission profiles (Verschuur, 1995a,b). The goal was to find evidence for magnetic fields in the 5–20 μG range, which could control gas motion in the filaments (Verschuur 1991b). Various observers such as Troland and Heiles (1982) and Heiles (1989) claimed that magnetic fields this strong exist in diffuse HI. Instead of confirming their results or adding to the data base, our study revealed the presence a systematic problem that has affected all such observations. After making appropriate corrections to the data, no evidence for magnetic fields in the 5–20 μG range in diffuse HI emission is found anywhere in the sky (Verschuur, 1995a,b). The only magnetic fields in interstellar space of this order that have been observed are associated with dense HI structures ("clouds") adjacent to star forming regions, where the magnetic fields may have been amplified by shock compression (see Verschuur 1995c for a summary).

For H0827+10 the Zeeman effect observations give a value forthe line–of–sight magnetic field strength of -0.7 + 1.0 μG (one sigma error) toward the brightest feature in Fig. 3. The average magnetic field for five directions sampled in H0827+10 is -0.1 + 0.6 μG. This implies an upper limit of $\sim$2 μG. Upper limits of 1.4 μG have been set for two other EEFs, while a third gave 0.6 μG (Verschuur 1995a). Small-scale structure in the magnetic field morphology produced by field reversals within the beamwidth of the telescope would not be detected in the Zeeman effect experiment and could especially hide toroidal magnetic fields with small angular scales.

3 Discussion: Filaments as Tracers of Current Flow

Our initial study focussed on the nature of the most clearly delineated filament seen in Fig. 3, just to the left and above the map center and partly between the dark areas indicating lack of data. From the appearance of this filament, α, it is assumed that is being viewed nearly normal to its axis. Its properties were then derived by examining HI contour maps and observed profiles not shown here. The generally similarity between the dust and gas structures in Figs. 2 and 3 lends support to the notion that "clouds" in the diffuse interstellar medium, which are usually identified with brightness enhancements in HI or 100 μ area maps, are more likely to be produced by of line–of–sight filament geometry than by simple "cloud" physics. If so, this has far reaching implications for the way such data are interpreted to derive estimates of the physical parameters for interstellar HI. It was in this context that we examined whether plasma physical processes may play a role in controlling the formation and stability of interstellar filaments.

TABLE I
Summary of parameters that produce equality in the Carlqvist Relation.

Filament α in the H0287+10 region

Parameter	Value	Comments
Observed		
HI Column density	0.85×10^{20}	From HI maps at sides of filament
Angular width	$0.^\circ 2$	Data from two sides of filament
Angular rotation	1.63 km s^{-1}	From HI data
Temperature warm, neutral medium	3000 K	Verschuur and Magnani (1994)
Density of warm medium	0.3 cm^{-1}	For nonplanar directions[1]
Adjustable		
Aspect ratio	2	Takes into account geometry (los through center = 2× los through sides of filament).
Distance	100 pc	Assumed (see text)
HI gas temperature	30 K	Set equal to dust temperature
Axial magnetic field in filament	1.5μG	From pulsar rotation measures and Zeeman effect data (Verschuur 1995c).
Axial magnetic field outside filament	1.5μG	Same reference. This is the strength of the general interstellar magnetic field.
Derived		
Filament n_H	40.4 cm^{-3}	From HI data and distance
Toroidal magnetic field	5.3 μG	Produces equilibrium in filament
Required current	1.4×10^{13}A	From Eq.(1)

[1]Reynolds (1983) planar data on the warm ionized medium (8000 K, 0.1 cm^{-3}) produces virtually the same pressure.

The philosophy that evolved as the data were examined (with plasma physics concepts in mind) is that HI gas acts as a tracer of the motion and structure of plasma in interstellar space. We will next describe what emerges when the data are considered from this point of view.

Carlqvist and Gahm (1992) first demonstrated that electric currents may generate and maintain interstellar filamentary structure. Key to their discussion is the theory of a generalized Bennett pinch, which involves a cylindrically symmetric plasma filament. A helical magnetic field with a toroidal component B_φ and an axial component B_z exists in a filament with radius a inside which a current $I(a)$ flows. The linear density in the filament, $N(a)$, is defined as $\pi a^2 n$, where n is the plasma volume density, which we will take to be the same as the neutral gas volume density, n_H. The mean mass of the plasma particles is m, which includes neutrals, electrons, ions and a fraction of helium. For these calculations it is taken to be 2×10^{-24} gm. The neutral gas is assumed to be coupled to the ionized component (e.g., Kahn and Dyson, 1965).

The generalized Bennett relationship (Carlqvist, 1988) that describes the equilibrium of the filament contains five distinct terms:

$$\mu_0 I^2 (a) + 4\pi G m^2 N^2 (a) = 8\pi \left[\Delta W_k (a) + \Delta W_{Bz} (a) + W_{rot} (a)\right] \quad (1)$$

Peratt (1992) has labelled this the *Carlqvist Relation*. The two terms on the left of Eq. (1) represent the pinching forces due to the current and gravity. A current $I(a)$ in the filament generates a toroidal magnetic field B_φ related by

$$B_\varphi (a) = \frac{\mu_0 I (a)}{2\pi a} \quad (2)$$

The terms on the right represent expansive forces, except in the case where the external gas pressure overcomes internal gas pressure in the filament. This follows from the fact that the first term on the right represents the excess kinetic energy in the filament given by

$$\Delta W_k (a) = 2\pi r p_k dr - \pi a^2 p_k (a) \quad (3)$$

where

$$p_k (a) = n_W k T_W$$

is the kinetic pressure in the warm (here taken to be neutral) gas just outside the filament, and

$$p_k = n_H k T_H$$

is the gas pressure inside the filament, whose density and temperature are derived from the HI data. It was found that under certain circumstances, depending mainly on the filament distance chosen (usually an unknown parameter), the external gas pressure can be of the order of, or greater than, the internal gas pressure. Bear in mind that the filament gas density can be derived from the HI observations if the distance to the filament is known. In the analysis described below, it was noted that there is always a critical value for the (unknown) filament distance beyond which the external gas pressure is calculated to be greater than the internal pressure, so that the relative importance of the Bennett pinch in stabilizing the filament appears to be reduced. It may be no more than a coincidence that the critical distance is usually of the order judged to be reasonable for the filaments, and may in fact hint at a new way to obtain this distance.

A second point concerning how to estimate the value of $\Delta W_k (a)$ in Eq.(3) needs to be stressed. The HI in the region in question is clearly morphologically similar to associated cold dust (cirrus). The temperature of dust in atomic filaments is around 30 K (Boulanger and Perault, 1988; Verschuur *et al.*, 1992) yet the observed HI emission linewidths are of order 4 km s^{-1}, which corresponds to ~340 K. This disparity between expected gas temperature and observed linewidth has often been noted in the study of interstellar HI and turbulent motions have usually been blamed for the difference. Here we suggest that the HI linewidth is broadened by the presence of rotation around the filament axis, which relates to the last term in Eq. (1).

The excess rotational energy due to angular velocity ω is given by

$$W_{rot}(a) = \int_0^a 2\pi r \frac{mn\omega^2 r^2}{2} dr \tag{4}$$

Interstellar HI filaments are barely resolved in most current observations (e.g., Verschuur *et al.*, 1992) so that rotation about their axes could play an important role in defining emission profile linewidths. The amount of rotation in Filament α was determined by comparing profiles observed toward the filament edges; their central velocities are separated by 3.26 km s^{-1}. The emission linewidths (~4 km s^{-1}) are of this order and it may be possible that line broadening may be produced by rotation around the filament axes of ~2 km s^{-1}. For use in Eq. (1), $\omega = v/a$ was derived from the profile data, which imply $v = 1.63$ km s^{-1} for this filament.

The excess magnetic energy term in Eq. (1) is given by

$$\Delta W_{Bz}(a) = \int_0^a 2\pi r \frac{B_z^2}{2\mu_0} dr - \pi a^2 \frac{B_z^2(a)}{2\mu_0} \tag{5}$$

where $B_z(a)$ is the axial magnetic field just outside the filament.

We next attempt to derive values for the current that would generate a magnetic field to confine the filament. Our analysis using the Carlqvist relationship is slightly different from that offered by Carlqvist and Gahm (1992). They took the filament cold gas temperature to be given by the HI emission linewidth and had no information on filament rotation.

As a first order approach to the data, the above equations were applied using the physical parameters given in Table 1. Eq. (1) was entered into a spreadsheet to allow the various parameters to be adjusted so that their effect on the outcome of the calculations could be readily explored. The finally chosen parameters are indicated in the table and these values are reasonable in the light of present knowledge about conditions in interstellar space (references in table). In particular, the distance of 100 pc is of the order expected for relatively local gas at intermediate galactic latitudes. These parameters listed in Table 1 combine to produce an equality in Eq. (1). The current is estimated to be 1.4×10^{13} A which generates a toroidal magnetic field of 5.3 μG. These values are similar to those found by Carlqvist and Gahm (1992) for an unrelated set of filaments.

TABLE II
Stability of Other Filaments.

	Region A[1]	Eridanus[2]	
References for data	Verschuur (1974a, 1991a)	Verschuur *et al.* (1992)	
Filament diameter	$\sim 0.^{\circ}4$	$\sim 0.^{\circ}6$	
Rolling motion	~5.0 km s^{-1} $degree^{-1}$	~6 km s^{-1} $degree^{-1}$	
Rotation velocity	~2.5 km s^{-1}	~1.8 km s^{-1}	
HI column density, N_H	$1.5 \times 10^{20} cm^{-2}$	$3.0 \times 10^{20} cm^{-2}$	
Distance	100 pc	150 pc	400 pc
Implied toroidal magnetic field strength	7.5 μG	6.0 μG	3.4 μG
Required current	3.9×10^{13}A	7.1×10^{13}A	11.0×10^{13}A

[1] Relates to a segment of a highly elongated filament in data presented by Verschuur (1974a), see text.
[2] From Verschuur *et al.* (1992) who discuss the distance estimates, see text.

Substitution into Eq. (1) shows that gravity plays no role in controlling the filament. Most striking is the importance of rotation in determining the magnetic field required to constrain the filament. Varying the rotation velocity from 1–4 km/s changes the required constraining toroidal magnetic field strength from 3.3–12.8 μG, which in turn requires a current of 0.9–3.4 $\times 10^{13}$ A.

If we insist on an additional constraint, that external gas pressure *not* dominate internal gas pressure, then the distance of the filament has to be <140 pc. Otherwise the toroidal magnetic field and external gas pressure will combine to stabilize Filament α.

Changing the input parameters (Table 1) revealed that a 20 to 30% unbalance in Eq. (1) to make either external or internal forces dominate is produced by roughly the same percentage change in any of the parameters, except that the equality is least sensitive to the strength of the axial field outside the filament and the temperature and density of the surrounding warm plasma.

Once the relevance of introducing rotation into the Carlqvist relationship was recognized, other data sets, in which this author had previously become aware of possible rotation, were re–examined. For example, notes made during the analysis phase of the work reported by Verschuur (1991a,b) referred to "rolling motion," which implies rotation about a filament axis. Table 2 summarizes an attempt to estimate the parameters characteristic for the regions mapped by Verschuur (1974a). The values of the toroidal magnetic field strength and current along this filament axes, found by applying Eq.

(1), are given. The current is again of the order found by Carlqvist and Gahm (1992).

Another case of rolling motion in filaments was noticed during the study of the structure in Eridanus (Verschuur *et al.*, 1992). This was not mentioned at the time because it did not seem relevant. However, re-examination of the data shows that Filament A exhibits velocity gradients of $\sim$6 km s^{-1} per degree normal to the filament axis. A similar effect is seen in Filaments B and C. Considering only Filament A, a rotation velocity of 1.8 km s^{-1} for a diameter of 0.°6 is estimated. The parameters relevant to Filament A are shown in Table 2. The derived quantities have been calculated for two possible filament distances discussed by Verschuur *et al.*, (1992). It is found that for the choice of parameters shown in Table 2, the internal gas pressure exceeds that outside the filament if the distance is <210 pc, which means that that if current flow is important in stabilizing this filament it must be at the closer of the two distances considered. Otherwise the pressure term acts to confine the filament in the absence of current flow.

Examination of the data summarized in Tables 1 and 2 suggest that the dominant motion tending to disrupt the filaments may be rotation about their axes. Such rotation results from plasma flowing along the toroidal magnetic fields lines generated by the current which produces the Bennett pinch to stabilize the filament.

4 Conclusions

Our analysis suggests confirmation of the model proposed by Carlqvist (1988) and Carlqvist and Gahm (1992), that the physics of the generalized Bennett pinch applies to gaseous interstellar filaments. This has far-reaching implications for the study of the properties of interstellar matter. The role of large-scale currents may be far more important in defining interstellar structure than has generally been recognized within the astronomical community. Diffuse interstellar HI is clearly highly filamentary. In low-resolution surveys, the pervasiveness of filaments may have been overlooked by labelling localized enhancements in HI brightness as "clouds" defined as nearly spherically symmetric, bound entities. It is suggested that unless such features can be proven to have distinct three-dimensional boundaries under the pervasive influence of gravity or magnetism to maintain their morphology, the cloud concept may only apply to unusual structures in star forming regions where gravitationally-bound giant molecular clouds do appear to exist. Exploration of the model invoking current flow in interstellar space suggests that the strength of the toroidal magnetic field surrounding Filament α in the H0827+10 structure is 5 μG, relatively independent of any parameter other than distance, given a diameter and a column density of gas through the filament. The field strength is insensitive to the value of axial magnetic field strengths inside and outside the filaments, which have been assumed equal. The derived toroidal magnetic field strengths lie in the range where it was hoped that 21–cm Zeeman effect observations would be able to detect them. Currently two factors conspire to make such measurements very difficult. Toroidal magnetic fields in barely resolved filaments tend to cancel in the beam and for large angular diameter structures existing radio telescopes suffer from systematic effects that make detection of magnetic fields toward the edge of filaments especially challenging (Verschuur, 1995a,b). Works proceeds on examining available data to test the plasma model of filament creation and stabilization and plans are underway to look more carefully at specific HI features to relate structure to velocity in order to identify the existence of toroidal motion, the signature of the Bennett pinch.

5 Acknowledgements

I am profoundly grateful for the encouragement given me by Tony Peratt and the valuable insights of an anonymous referee who enabled me to penetrate the seemingly obscure realm of interstellar plasma physics. I am also grateful to the George Larson, Editor of Air and Space Smithsonian Magazine, whose generous support enabled me to attend this meeting. Some of this work was undertaken with assistance from NSF grant, AST 9022002.

References

Boulanger, F., Perault, M.: 1988, *Astrophys. J.* **Vol. 330**, pp. 964
Carlqvist, P.: 1988, *Astrophys. Space Sci.* **Vol. 144**, pp. 73
Carlqvist, P., Gahm, G. F.: 1992, *IEEE Trans. Plasma Sci.* **Vol. 20**, pp. 867
Colomb, F. R., Poppel, W. G. L., Heiles, C.: 1980, *Astron. and Astrophys.* **Vol. 40**, pp. 47
Hartmann, L.: 1994, Ph.D. Thesis, Leiden University *(to be published by Cambridge Univ. Press, 1995)*
Heiles, C.: 1989, *Astrophys. J.* **Vol. 336**, pp. 808
Kahn, F.D., Dyson, J .E.: 1965, *ARA&A* **Vol.3**, pp. 47
Peratt, A. L.: 1992, *Physics of the Plasma Universe*, Springer-Verlag, p.59
Reynolds, R. J.: 1983, *Astrophys. J.* **Vol. 268**, pp. 698
Troland, T. H., Heiles, C.: 1982, *Astrophys. J.* **Vol. 260**, pp. L19
Verschuur, G.: 1973, *Astrophys. J.* **Vol. 78**, pp. 573
Verschuur, G.: 1974a, *Astrophys. J. Supplement* **Vol. 27**, pp. 65
Verschuur, G.: 1974b, *Astrophys. J. Supplement* **Vol. 27**, pp. 283
Verschuur, G.: 1991a, *Astrophys. Space Sci.* **Vol. 185**, pp. 137
Verschuur, G.: 1991b, *Astrophys. Space Sci.* **Vol. 185**, pp. 305
Verschuur, G.: 1994a,b, *Astrophys. J.* (submitted)
Verschuur, G.: 1995c, *Astrophys. J.* (in preparation)
Verschuur, G. L., Rickard, L. J, Verter, F., Pound, M., Leisawitz, D.: 1992, *Astrophys. J* **Vol. 390**, pp. 514
Verschuur, G. L., Magnani, L.: 1994, *Astronomical. J* **Vol. 107**, pp. 287

PROTOSTAR FORMATION UNDER TWO CURRENT CARRYING GAS FILAMENTS COLLISION

BIDZINA CHARGEISHVILI
Abastumani Astrophysical Observatory of Georgian Academy of Sciences
Tbilisi 380000 Georgian Republic
and
Laboratory for Plasma Astrophysics and Fusion Science, Department of Electronics and Information Faculty of Engineering
Toyama University, Toyama, 930 JAPAN

and

JUN-ICHI SAKAI
Laboratory for Plasma Astrophysics and Fusion Science, Department of Electronics and Information Faculty of Engineering
Toyama University, Toyama, 930 JAPAN

Abstract. A model of protostar formation under two current carrying gas filaments collision is presented. The model implies MHD approach involving self-gravity and radiative cooling effects. We suppose that through the current carrying gas filament collision a magnetic field reconnection takes place. Using an appropriate self-consistent presentation for time and special dependences of physical quantities in MHD equations, we derive the full set of equations that describes time evolution of the physical quantities just after an occurrence of magnetic field reconnection. Numerical simulations reveal that the process consists of three main phases of evolution. The first is an appearance of preceding peaks in time profiles of density and temperature following by the next phase of depression of both temperature and density and the final fast condensation phase with either cooling or heating of matter depending on initial parameters of problem. Effects of initial conditions like as magnetic field strength, current strength, initial gravity energy, cooling time and a geometry of collision are investigated. Main conclusion is that protostar formation takes place within the time interval less than one free fall time and it is preceded by the appearance of dense and hot matter with lifetime much less than free fall time. The final temperature of the protostar depends on the physical conditions and mainly on the ratio between free fall time and cooling time in the colliding current carrying gas filaments.

Key words: Filaments, MHD, Star Formation, Simulations

1 Introduction

Stars are the fundamental objects of astronomy and thus their formation constitutes one of the basic problems of astrophysics. The main question confronting any principal discussion of star formation is the origin of stellar masses. Normal stars are self-graviting spheroids of gas whose central temperatures, are high enough to sustain fusion reactions, and with masses at least marginally capable of burning hydrogen. Answering the question why should the Universe ubiquitously produce objects that have just the right mass range to stably transform the primary product of the big bang–hydrogen–into heavier elements. Shu et. al (1987) have divided the various

Astrophysics and Space Science **227**: 199–215, 1995.

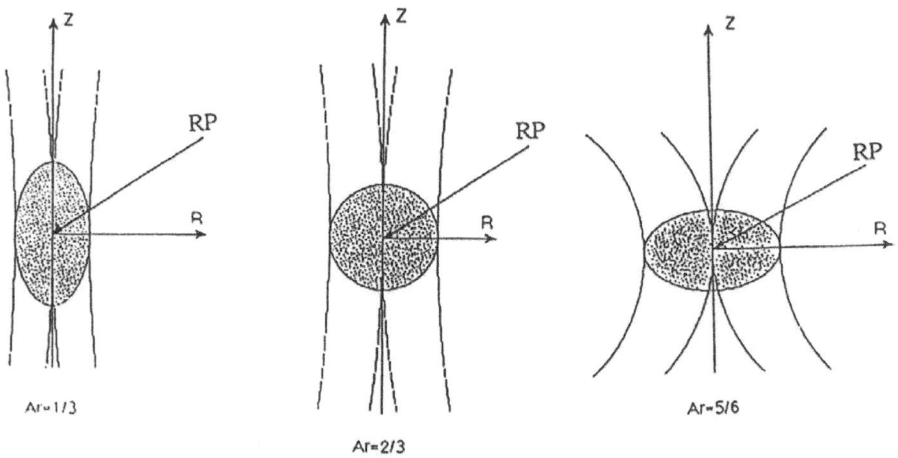

Fig. 1. Schematic configuration of loop-loop collision and coordinate system with it's origin coinciding with the RP (point where the reconnection happened) and Z axis is directed along the line of symmetry of colliding loops for three cases when A_r=1/3, 2/3, 5/6.

lines of approaches on two. The first is to suppose that the result–objects with masses sufficient for thermonuclear fusion–occurs somewhat by chance through some process of hierarchical fragmentation. The second is to suppose that star formation is an accretion process, whereby protostellar masses are built up from initially small values, and that this accumulation is eventually halted by some process, possibly a stellar wind, which is itself somehow triggered by the onset of thermonuclear fusion. Evaluation of the merits and disadvantages of these two schools of thought is out of scope of our paper. We only mention that it is doubtless that the star formation regions are situated in atomic and molecular clouds that are embodied in a lower density, intercloud gas. A variety of research has been directed toward understanding the dynamics of such the gas concentrations in both cases with considering of magnetic forces and without it. The clouds are inferred to be supported against gravity since their estimated lifetimes are generally 3-10 free-fall times. The mean density of the observed cloud is $10^{4\ 5}\mathrm{cm}^{-3}$ and the field strength is 120 mG, sufficient to magnetically support this cloud. The fields in the shells derive their energy from the large reservoir, available in the kinetic energy of the expansion. By stretching a magnetic field over very large scales by the expansion of a large gas shell in the interstellar medium, the magnetic field becomes strong enough to play a dominant role. Thus it is energetically possible that clouds formed in this way may initially be slightly magnetically dominated. The cloud can contract dynamically only when its magnetic flux is smaller than a critical value which is proportional to its mass. Nakano and Umebayashi (1986) considered the problems of magnetic fields in star formation with regard to the contraction of the cloud. They investigated the mechanism how the cloud with the supercritical flux begins dynamic contraction after it has lost some of the flux.

Molecular clouds contain cores whose properties have been extensively studied because these dense regions appear to be the actual sites of star formation. Young stellar objects are found within or very near to dense molecular cores. The cores themselves are strung out along larger filamentary structures which may extend to 10 pc or more in length. Carlberg and Pudritz (1990) have demonstrated that a strongly magnetized cloud traversed with a spectrum of MHD waves leads to cloud support and fragmentation.

Magnetic fields are important for interstellar cloud dynamics. They give cohesion to the cloud fluid because clouds on the same magnetic flux tube exchange momentum over long distances and clouds on different flux tubes exchange momentum when the tubes intersect linking up small clouds into large complexes (Elmegreen, 1988).

In the present paper we develop a model of protostar mass condensation based on the process of current carrying magnetized filamentary clouds collision. In section 2 we present basic MHD equations and derive full set of equations governing a behavior of the plasma in the current carrying cloud filaments collision region. In section 3 we present results of numerical simulations of the set of equations and discussion of the results are presented in section 4.

2 Basic Equations

We now present a schematic picture (Figure 1) of a two current filament interaction which can describe the magnetic field configuration when magnetic reconnection takes place. The reconnection region is considered as a cylinder with the axis coinciding with the RP (point where the reconnection happened) and directed along the line of symmetry of colliding filaments. We use a cylindrical coordinates (r, φ, z). Axis Z is directed along the line of symmetry of colliding filamentary structures and an origin of a reference frame coincides with the RP. The MHD equations we use are:

The continuity equation

$$\frac{\partial \rho}{\partial t} + \nabla \cdot (\rho \mathbf{V}) = 0 \tag{1}$$

The energy equation

$$\frac{\partial P}{\partial t} + (\mathbf{V}\cdot\nabla) P + \gamma P (\nabla\cdot\mathbf{V}) = (\gamma - 1)\frac{\mathbf{J}^2}{\sigma} - (\gamma - 1)\chi\rho^2 T^\alpha \quad (2)$$

The induction equation

$$\begin{array}{c}\frac{\partial \mathbf{B}}{\partial t} = \nabla\times(\mathbf{V}\times\mathbf{B}) + \frac{c^2}{4\pi\sigma}\Delta\mathbf{B} \\ \nabla\cdot\mathbf{B} = 0\end{array} \quad (3)$$

and the equation of motion

$$\rho\left(\frac{\partial \mathbf{V}}{\partial t} + (\mathbf{V}\cdot\nabla)\mathbf{V}\right) = -\nabla P + \frac{1}{4\pi}(\nabla\times\mathbf{B})\times\mathbf{B} - \rho\nabla\Phi \quad (4)$$

where ρ and P are the plasma density and pressure, respectively. $\mathbf{V}$ is the plasma flow velocity, $\mathbf{B}$ is the magnetic field, γ is the adiabatic ratio and σ is the electric conductivity, Φ is the gravitational potential and it obeys the poison equation

$$\nabla^2\Phi = 4\pi\rho G \quad (5)$$

where G is the gravitational constant.

The first term of the right-hand side of the energy equation (2) shows the Ohmic heating, and the current is given by

$$\mathbf{J} = \frac{c}{4\pi}\nabla\times\mathbf{B} \quad (6)$$

The second term of the right-hand side of the Equation (2) is radiative cooling term and χ and α are the coefficients that depend on a temperature range. We consider a time evolution of physical state just after the sudden change that has already occurred due to magnetic field reconnection and assume that the process is not steady but dynamical. For simplicity we also assume an azimuthal symmetry and so no variable depends on φ coordinate. For the components of magnetic field and velocity, we use the following self-consistent presentation:

For the velocity components,

$$V_r(t) = \frac{\dot{a}(t)}{a(t)}r,\ V_\varphi(t) = v_\varphi(t)\frac{r}{\lambda},\ V_z(t) = \frac{\dot{b}(t)}{b(t)}z \quad (7)$$

For the magnetic field components,

$$B_r(t) = b_r(t)\frac{r}{\lambda},\ B_\varphi(t) = b_\varphi(t)\frac{r}{\lambda},\ B_z(t) = b_z(t)\frac{z}{\lambda} \quad (8)$$

where the dot means the time-derivation. The time-dependent values, $a(t)$, $v_\varphi(t)$, $v_{z0}(t)$, $b_r(t)$, $b_\varphi(t)$ and $b_z(t)$ will be self-consistently determined from the MHD equations (1-6). The quantity λ is a characteristic scale length of the interaction region. The equations (7) and (8) present the lowest order terms in expansion of the values close to the RP. Using the expressions (7) and (8) in the Equations (1) and (3), one can get that the plasma density and the magnetic field components obey the following expressions:

$$\rho = \frac{1}{a^2 b} \tag{9}$$

$$b_r = \frac{B_{r/\varphi}}{a^2 b}, \; b_\varphi = \frac{1}{a^2 b}, \; b_z = -2\frac{B_{r/\varphi}}{a^2 b} \tag{10}$$

where density is normalized on ρ_0 which is a constant and we fix it with a value of density when $t = 0$. The quantities b_r, b_φ, and b_z are normalized by $b_{\varphi 0}$ and $B_{r/\varphi} = b_{r0}/b_{\varphi 0}$ (b_{r0} and $b_{\varphi 0}$ are constants which are also fixed at $t = 0$).

Using an expression of the plasma pressure as follows,

$$P(t) = P_0(t) - \frac{p_r(t)r^2 + p_z(t)z^2}{2\lambda^2} \tag{11}$$

and the expression for the ideal gas as,

$$P = nkT \tag{12}$$

the Taylor expansion gives the following expression

$$T^\alpha = \left(\frac{a^2 b}{\rho_0}\frac{m}{k}\right)^\alpha P_0^\alpha \left[1 - \frac{\alpha}{P_0}\left(\frac{p_r r^2 + p_z z^2}{2\lambda^2}\right)\right] \tag{13}$$

From Equation (2) we find that the time-dependent coefficients in the Equation (11) can be given by the following expressions:

$$\dot{p}_r = -\gamma p_r \left[2(\gamma+1))\frac{\dot{a}}{a} + \gamma\frac{\dot{b}}{b}\right] - (\gamma-1) R_{ac}\frac{\alpha P_0^{\alpha-1} p_r}{(a^2 b)^{2-\alpha}} \tag{14}$$

$$\dot{p}_z = -\gamma p_z \left[2(\gamma+1))\frac{\dot{a}}{a} + \gamma\frac{\dot{b}}{b}\right] - (\gamma-1) R_{ac}\frac{\alpha P_0^{\alpha-1} p_z}{(a^2 b)^{2-\alpha}} \tag{15}$$

$$\dot{P}_0 = -\gamma P_0 \left[2\frac{\dot{a}}{a} + \frac{\dot{b}}{b}\right] + \frac{4(\gamma-1)}{\beta S}\frac{1}{a^4 b^2} - (\gamma-1) R_{ac}\frac{\alpha P_0^\alpha}{(a^2 b)^{2-\alpha}} \tag{16}$$

In Equations (14-16), the pressures p_z, p_r and P_0 are normalized by the p_{00}, which is a constant and is a value of the P_0 when $t = 0$. The time is normalized by $\tau_A = \lambda/v_{A\varphi}$, with $v_{A\varphi} = b_{\varphi 0}/\left(4\pi\rho_0\right)^{-1/2}$. The plasma β ratio is $\beta = c_s^2/v_{A\varphi}^2$, where $c_s = (p_{00}/\rho_0)^{1/2}$ is the sound speed. The

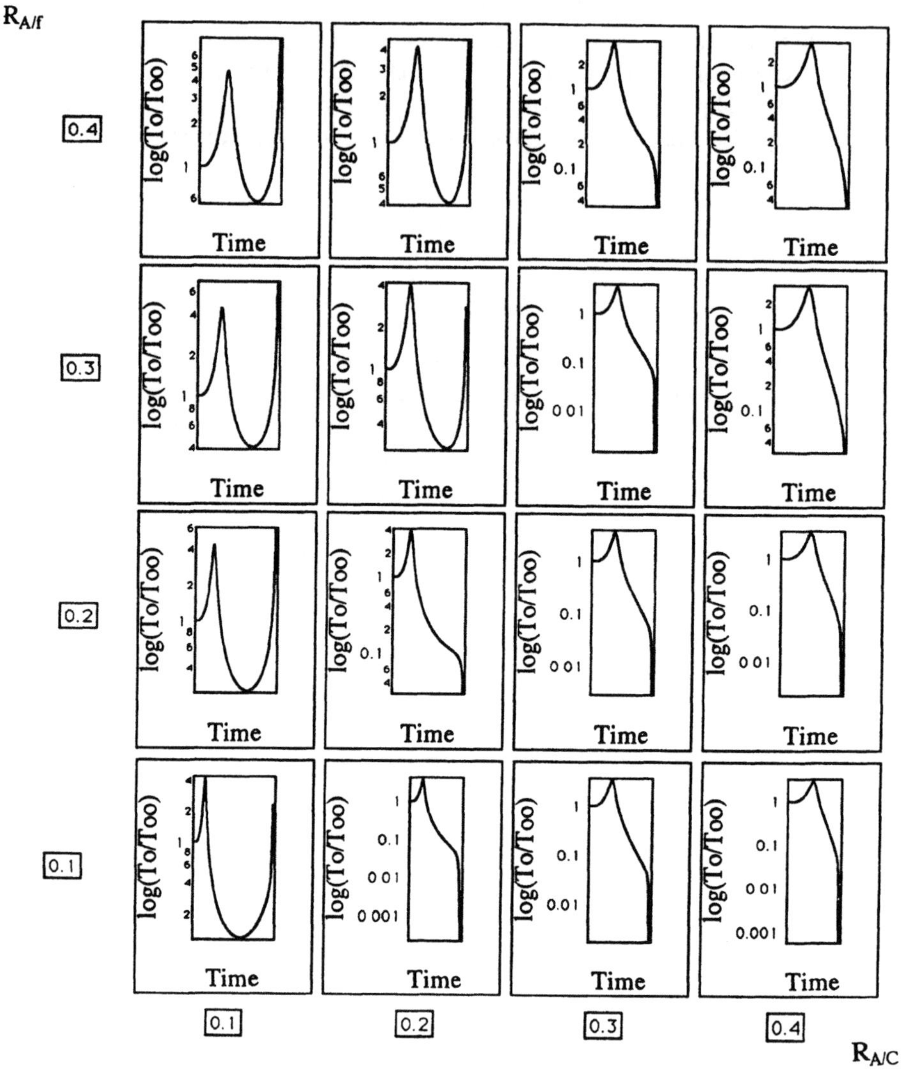

Fig. 2. Dependence of the time Histories of the temperature in the center of interaction region on $R_{A/f}$ (vertical axis) and $R_{A/C}$ (horizontal axis).

magnetic Reynolds number is $S = \tau_B/\tau_A$, and the magnetic diffusion time is $\tau_B = 4\pi\sigma\lambda^2/c^2$. The parameter $R_{A/C} = \tau_A/\tau_c$ is the ratio between the Alfvén transit time and the cooling time $\tau_c = \left[\chi c_s^{\alpha-1} (m/k)^{\alpha} \rho_0\right]^{-1}$.

Finally,from the equation of motion, Equation (4), using the expression

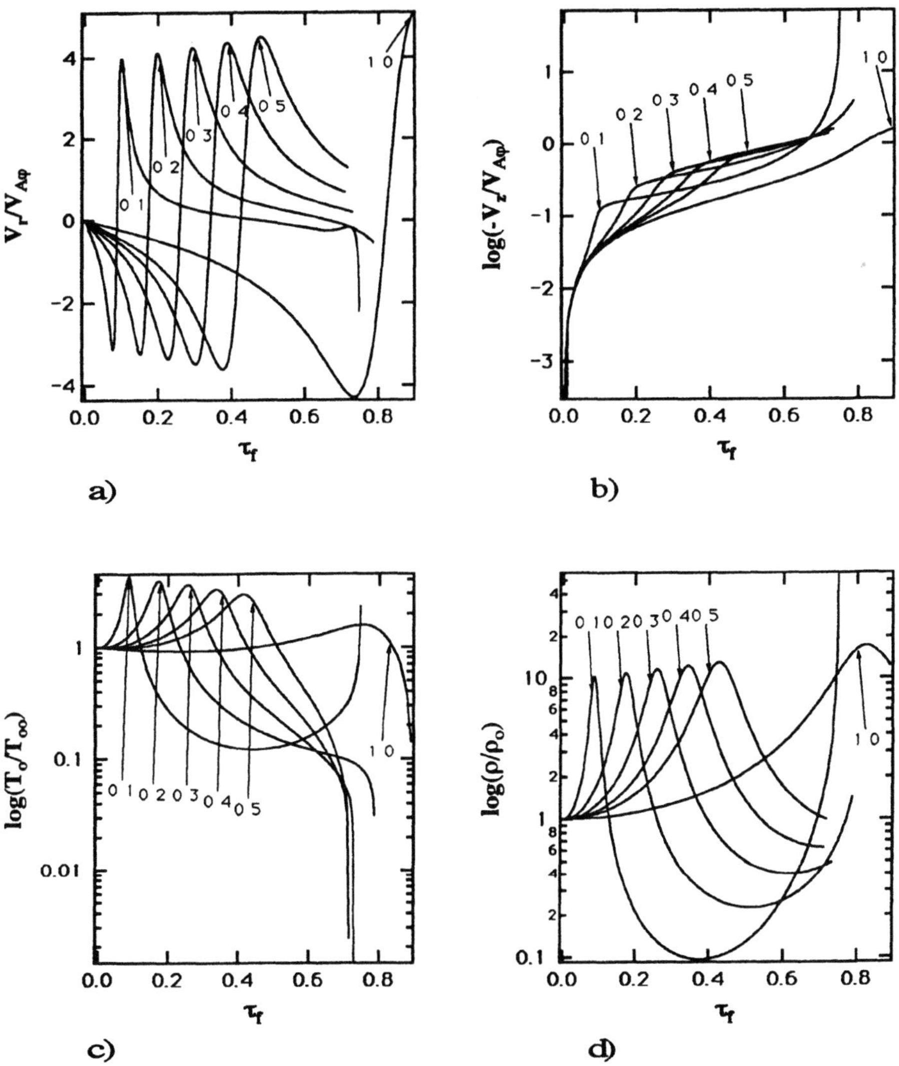

Fig. 3. Time histories of amplitudes of a) radial velocity, b) longitudinal velocity, c) temperature in the center and d) density for the initial conditions as: $T_{00} = 10^7$K, $B_{r/\phi}$=0.5, A=2/3, β=0.01, $R_{A,f/A/C}$=1.0 and $R_{A/C}$=0.1, 0.2, 0.3, 0.4, 0.5, 1.0. Numbers with arrows indicate values of $R_{A/C}$.

for the gravitational potential as follows,

$$\Phi = \pi\rho G\left(A_r r^2 + A_z z^2\right) \tag{17}$$

$$A_z = 2(1 - A_r) \tag{18}$$

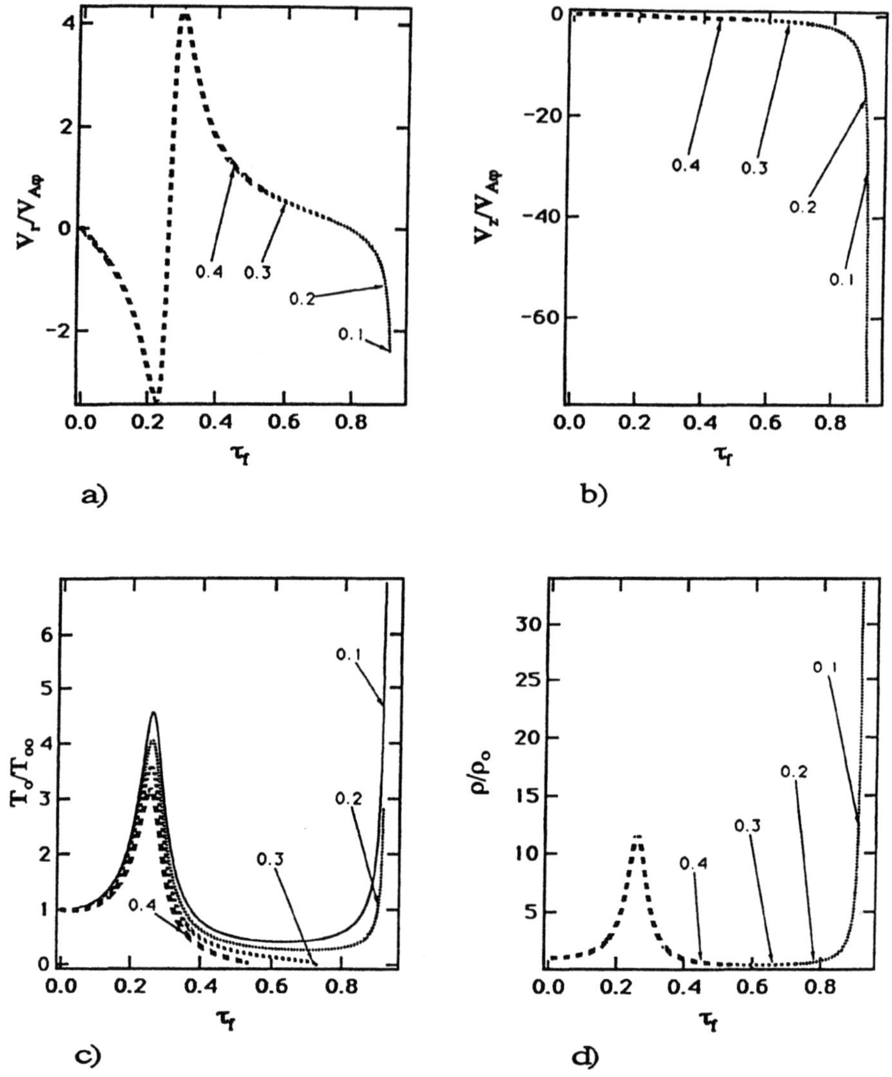

Fig. 4. Time histories of amplitudes of a) radial velocity, b) longitudinal velocity, c) temperature in the center and d) density for the initial conditions as: $T_{00} = 10^7$ K, $B_{r/\phi}$=0.5, A=2/3 $\beta = 0.01$ $R_{A/f}$=0.3 and $R_{A/C}$=0.1, 0.2, 0.3, 0.4. Numbers with arrows indicate values of $R_{A/C}$.

we obtain the basic equations describing the time evolution of the $a(t)$, $v\varphi(t)$, and $b(t)$ as follows:

$$\ddot{a} = \beta a^3 b p_r - \frac{4 + R_{Af}^2}{2ab} + a v_\varphi^2 \tag{19}$$

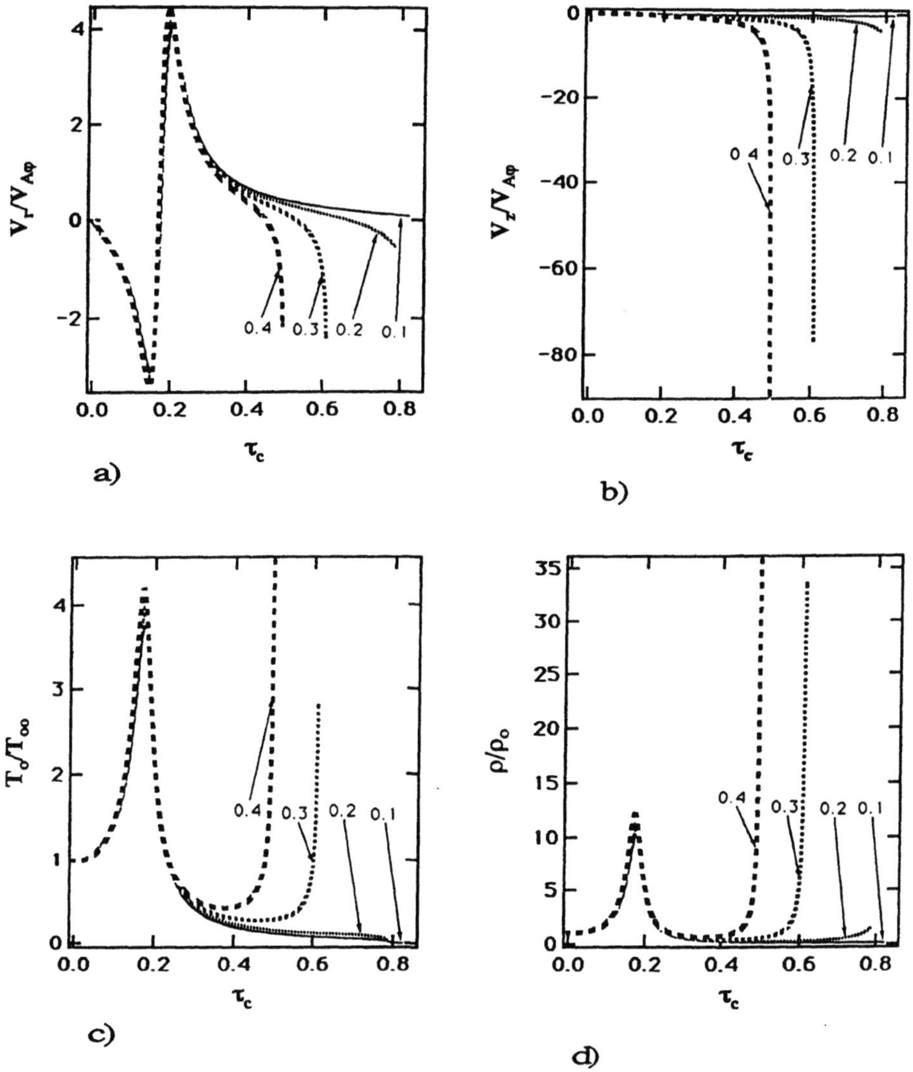

Fig. 5. Time histories of amplitudes of a) radial velocity, b) longitudinal velocity, c) temperature in the center and d) density for the initial conditions as: $T_{00} = 10^7$K $B_{r/\phi}$=0.5, A=2/3 $\beta = 0.01$ $R_{A/C}$=0.2 and $R_{A/f}$=0.1, 0.2, 0.3, 0.4. Numbers with arrows indicate values of $R_{A/f}$.

$$\dot{v}_\varphi = \frac{2B_1}{a^2b} - 2\frac{\dot{a}}{a}v_\varphi \tag{20}$$

$$\ddot{b} = \beta a^2 b^2 p_z - \frac{R_{Af}^2(1-A_r)}{a^2} \tag{21}$$

TABLE I
Parameter α versus Temperature.

$T_0 K$	α
$< 10^{4.6}$	0.0
$10^{4.6} - 10^{4.9}$	2.0
$10^{4.9} - 10^{5.4}$	0.0
$10^{5.4} - 10^{5.75}$	-0.2
$10^{5.75} - 10^{6.3}$	0.0
$10^{6.3} - 10^{7}$	-2/3
$> 10^{7}$	0.5

where the time, pressure and velocity are also normalized by τ_0, p_{00}, and $V_{A\varphi}$ respectively. The quantity $R_{Af} = \tau_A/\tau_f$ is the ratio between the Alfvén transit time and the free fall time $\tau_f = (4\pi\rho_0 G)^{-0.5}$.

From (13) we obtain the following expression for the normalized temperature

$$T = T_0 - \left(T_r \frac{r^2}{2\lambda^2} + T_z \frac{z^2}{2\lambda^2} \right) \tag{22}$$

where

$$T_0 = a^2 b P_0 \tag{23}$$

$$T_r = a^2 b p_r \tag{24}$$

$$T_z = a^2 b p_z \tag{25}$$

and temperatures are normalized by $T_{00} = c_s^2 m/k$.

So the Equations (9-25) are the full set of expressions governing a behavior of plasma in the collision region.

3 Results of Numerical Simulations

Here we present results of numerical simulations of Equations (9-25). We choose initial conditions according to the conditions in molecular and atomic clouds of galaxy corona and disc as follows: γ=5/3; $T_{00} = 10^7 K(a), 10^4 K(b)$ (a for corona, b for disc).

During the calculations the parameter α in the term of radiative losses of the optically thin plasma was changing according to Table I (Rosner et al. 1978) and χ was changing in the way to keep the continuous solutions when passing the margins of a different T_0 temperature range.

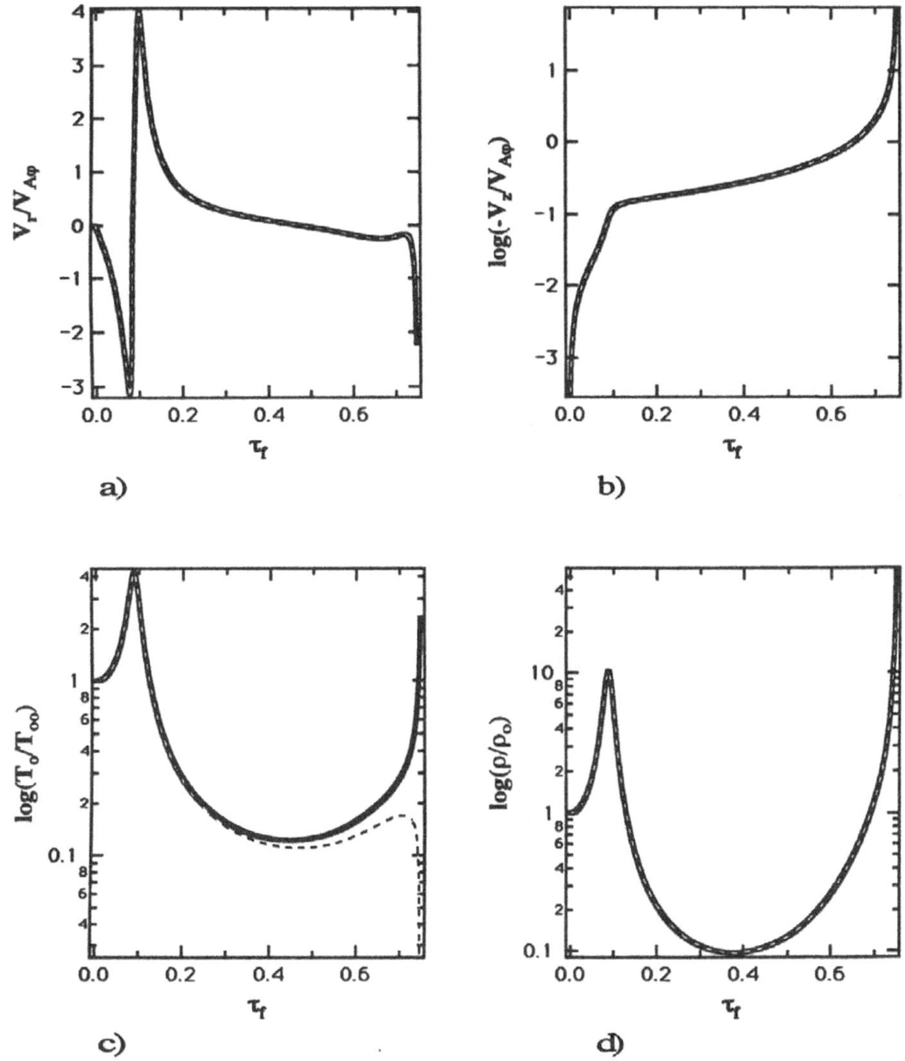

Fig. 6. Time histories of amplitudes of a) radial velocity, b) longitudinal velocity, c) temperature in the center and d) density for the initial conditions as: $T_{00} = 10^7$K, $B_{r/\phi}$=0.5, A=2/3,β=0.01. Dashed line corresponds to the case when $R_{A/C}$=0.11 and $R_{A/f}$=0.1 while solid line corresponds to the case when $R_{A/C}$=0.1 and $R_{A/f}$=0.1.

To avoid the effects of initial pressure gradient, we assume that initial pressure distribution near the RP is almost uniform so that initial pressure gradients in radial and longitudinal directions are vanishing small. Under these circumstances the only parameter that determines a configuration of the colliding filaments near the RP is A_r showing gravitational potential distribution as seen from equations (17) and (18). We consider three cases (Figure 1) when $A_r = 2/3$ (corresponds to a spheroid configuration) when $A_r = 1/3$ (rugby ball like ellipsoid with the axis of rotational symmetry parallel to Z axis) and $A_r = 5/6$ (disc like ellipsoid with the axis of rotational symmetry parallel to Z axis)

In general speaking the picture of process of current carrying filamentary cloud structures collision which was revealed by simulations is as follows. Initially one can see an inflow of plasma in radial direction with sudden change to outflow and finally collapse, while in the Z direction only inflow is observed. As for temperature and density profiles they show increase of these values, while radial velocity changes its sign and after some depression phase sudden increase of density with simultaneous cooling or heating of plasma (depending on parameters) is observed. Below we will use the following terminology to distinguish the three main phases of evolution of physical stage during the filamentary current carrying clouds collision. Increase of the temperature and the density during the changing of sign of radial velocity we will name by the preceding peak. Following phase of decrease of temperature and density will be called by the depression phase and final third phase we will call as fast condensation phase.

Figure 2 shows a dependence of time histories of the temperature in the center of the collision region on $R_{A/f}$ and $R_{A/C}$. As one can see when $R_{A/C}$ is less then $R_{A/f}$ final condensation shows a decrease of temperature whereas if $R_{A/f} > R_{A/C}$ one has a heating finally. As for the line $R_{A/f}/R_{A/C}$=1.0 (more details are shown on Figure 3) it shows an effect of the magnetic field and it is seen that when $R_{A/f}$=0.1 final fast condensation corresponds to a heating of plasma due to magnetic contractive force is dominant and cooling is not in time to begin act effectively, whereas when $R_{A/f} > 0.1$ or it means for a weaker magnetic fields, temperature profiles always show the final cooling of matter.

Figure 3 shows an influence of magnetic field strength on behavior of physical quantities. For the following initial conditions: $T_{00} = 10^7$K, $B_{r/\phi}$=0.5, A=2/3 and β=0.01 we set that initial free fall and cooling times are unchangeable and ratio $R_{A/f}/R_{A/C}$=1.0. Due to changing of an initial value of the magnetic field strength, $R_{A/C}$ takes the following values: $R_{A/C}$=0.1, 0.2, 0.3, 0.4, 0.5, 1.0. Simulations reveal that a decrease of magnetic field causes delay of preceding peaks of temperature and density and reduction of the depression phase duration while the fast condensation time is getting close to the one free fall time τ_f. And already for $R_{A/f}$=1.0 the preceding peak almost coincides with the final fast condensation. Another characteristic of this case is that when $R_{A/f}$=0.1 final fast condensation corresponds to a heating of plasma whereas when $R_{A/f}$>0.1 temperature profiles always show the final cooling of matter.

Figure 4 demonstrates a role of cooling effect. Changing $R_{A/C}$ as it is shown on the figure caption of Figure 3 and considering that Alfvén transit time is constant we get from simulations that decrease of cooling characteristic time τ_c, causes decrease of preceding peaks of the temperature profile and reduction of the depression phase duration. Most characterizing feature is that when $\tau_c < \tau_f$ then the fast condensation phase corresponds to a cooling of matter while if $\tau_c \leq \tau_f$ then finally a heating takes place.

Figure 5 demonstrates a role of gravity effect. Changing $R_{A/f}$ as it is shown on the figure caption of Figure 5 and considering that Alfvén transit time is constant we get from simulations that decrease of the free fall time τ_f, causes increase of preceding peaks of the temperature and density profiles and reduction of the depression phase duration. Similarly to the case of the Figure 4c Figure 5c shows that when $\tau_c < \tau_f$ then fast condensation phase corresponds to a cooling of matter while if $\tau_c \leq \tau_f$ then finally a heating takes place.

As it was seen from the Figures 4 and 5 there is a threshold so called thermal instability condition $\tau_c = \tau_f$ near which the final fast condensation phase crucially changes its character. Figure 5 represents the result of a small decline from the above equilibrium. As one can see from the Figure 6c even very small decline from the equilibrium, particularly small decrease of τ_c, causes the change of heating of a matter to a cooling during the final fast condensation phase.

Figure 7 shows an influence of current strength (parameter $B_{r/\phi}$) on the result of the interaction. As one can see increase of current strength causes increase of preceding peaks and shortening of depression time.

Figure 8 and Figure 9 show dependence of result on the geometry of colliding cloud filaments for different initial temperatures. Changing from the limit of infinitesimally flat disc to the string like collision does not influence on the preceding peaks but causes delay of final fast condensation and for the limit of string process becomes even stable. As it is seen from the Figure 8 for initial temperature 10^4 K all final fast condensation phases imply the final heating of matter while in the case when initial temperature is 10^7 K from the Figure 9 one can see that A_r=0.0, 1/3 cases show the final heating whereas A_r=2/3, 5/6 cases correspond to the final cooling.

4 Discussion

A model of protostar formation under two current carrying gas filaments collision is presented. The model implies MHD approach involving self-gravity and radiative cooling effects. We suppose that through the current carrying gas filament collision a magnetic field reconnection takes place. Using an appropriate self-consistent presentation for time and special dependences of physical quantities in MHD equations, we derive the full set of equations that describes time evolution of the physical quantities just after an occur-

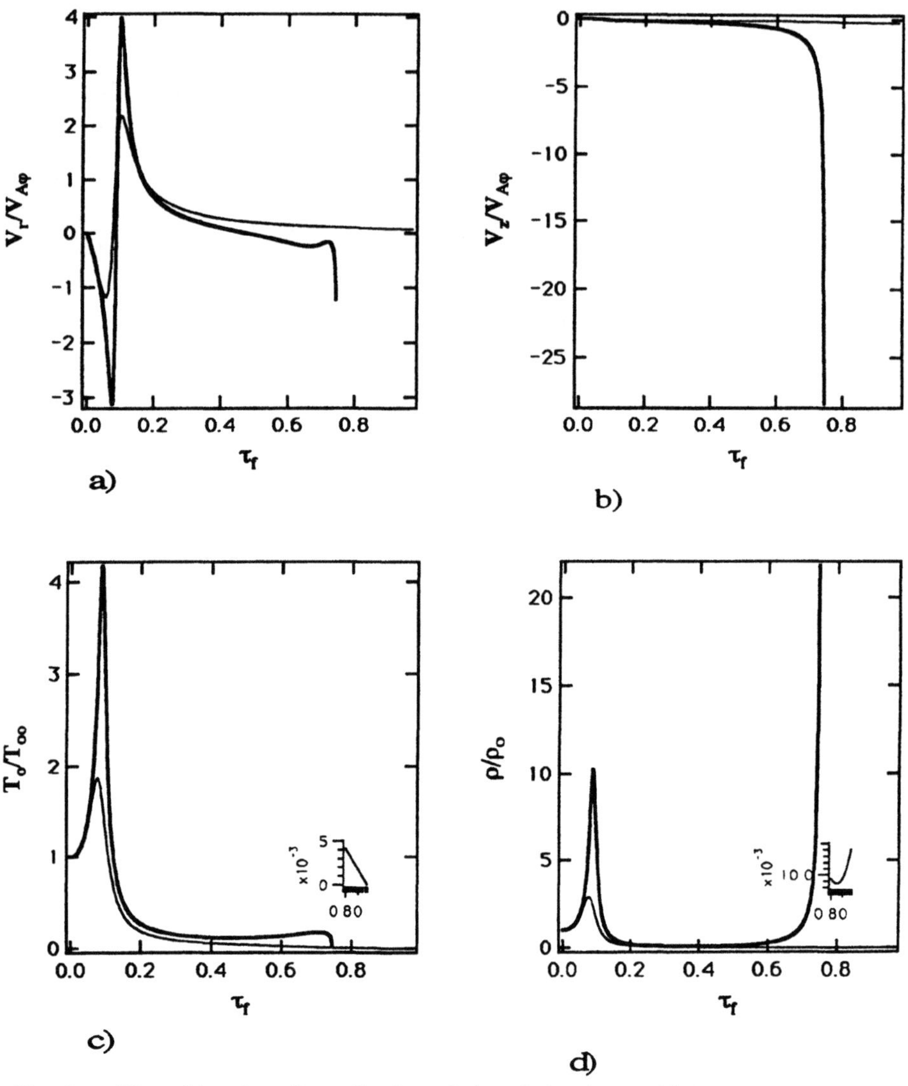

Fig. 7. Time histories of amplitudes of a) radial velocity, b) longitudinal velocity, c) temperature in the center and d) density for the initial conditions as: $T_{00} = 10^7$K, A=2/3, β=0.01, $R_{A/f}$=0.1, $R_{A/C}$=0.1. Solid line corresponds to the case when $B_{r/\phi}$=0.5 and dashed line to the case when $B_{r/\phi}$=1.0. Small graphics on the Figure c) and d) show details after the time τ_f=0.8.

rence of magnetic field reconnection. Numerical simulations reveal that the process consists of three main phases of evolution. The first is an appearance of preceding peaks in time profiles of density and temperature following by the next phase of depression of both temperature and density and the final

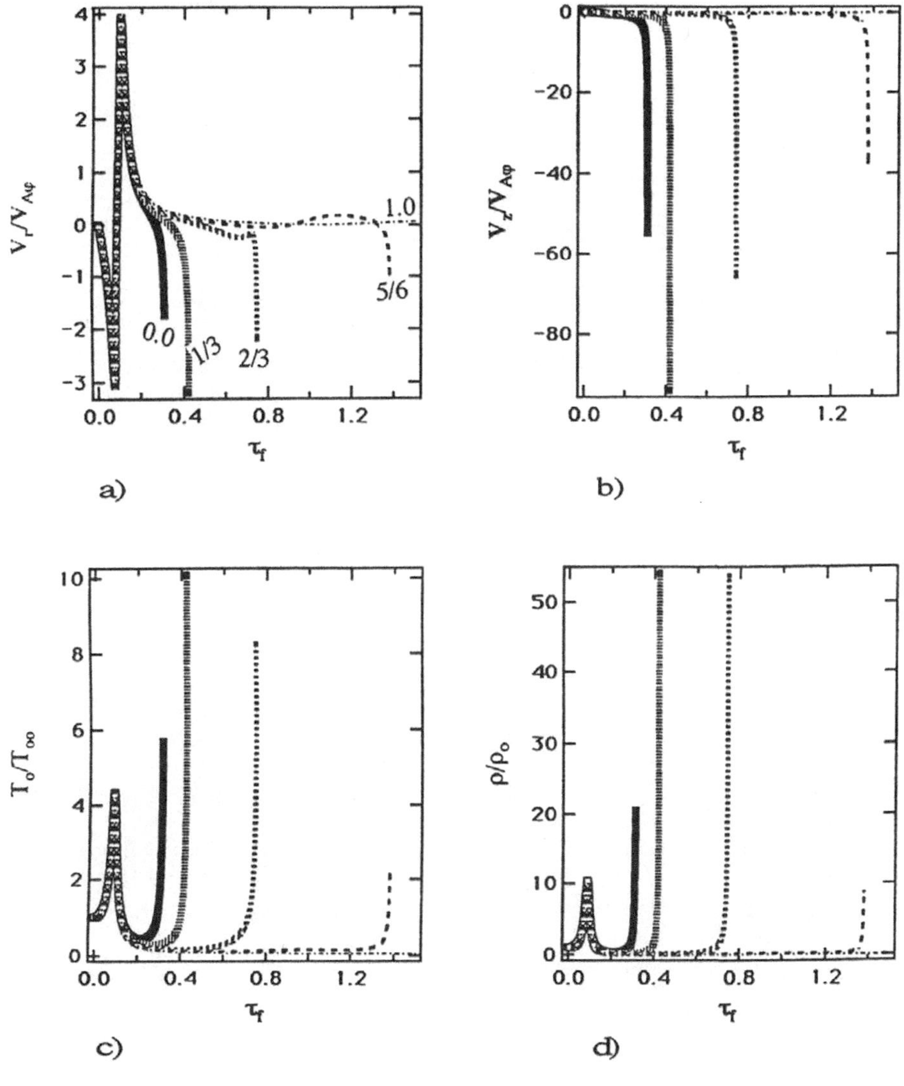

Fig. 8. Time histories of amplitudes of a) radial velocity, b) longitudinal velocity, c) temperature in the center and d) density for the initial conditions as: $T_{00} = 10^7$K, $B_{r/\phi}$=0.5, β =0.01, $R_{A/C}$=0.11, $R_{A//f}$=0.1, A_r=0.0, 1/3, 2/3, 5/6, 1.0. Numbers on the figure indicate values of A_r.

fast condensation phase with either cooling or heating of matter depending on initial parameters of problem. Effects of initial conditions like as magnetic field strength, current strength, initial gravity energy, cooling time and a geometry of collision are investigated. Main conclusion is that protostar formation takes place within the time interval less than one free fall time

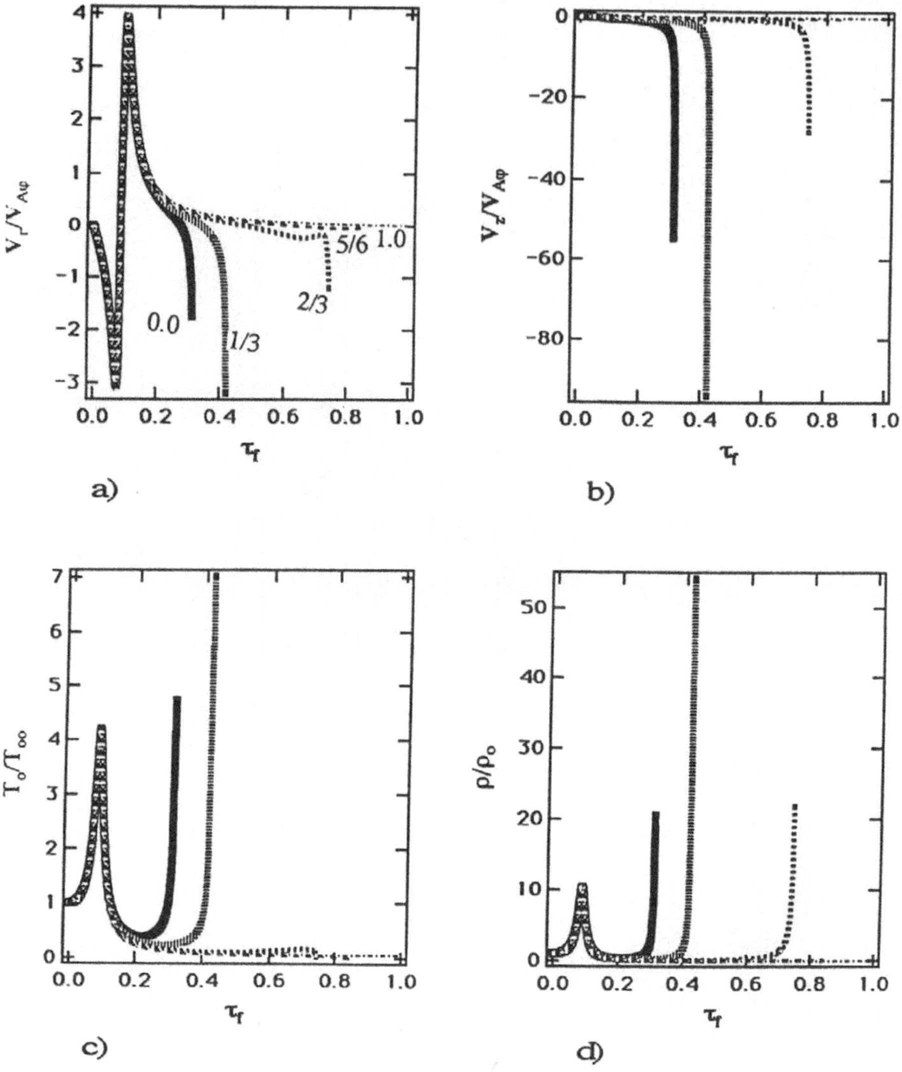

Fig. 9. Time histories of amplitudes of a) radial velocity, b) longitudinal velocity, c) temperature in the center and d) density for the initial conditions as: $T_{00} = 10^4$K, $B_{r/\phi}$=0.5, β =0.01, $R_{A/C}$=0.11, $R_{A/f}$=0.1, A_r=0.0, 1/3, 2/3, 5/6, 1.0. Numbers on the figure indicate values of A_r.

and it is preceded by the appearance of dense and hot matter with lifetime much less than free fall time. So we presume that the cores that are strung out along large filamentary structures of molecular clouds appear to be the actual sites of star formation. They may represent the preceding peak phase of protostar formation and in some sense are sort of embryos of future protostars. During the depression phase they are expected temporary to fade out and finally through the fast condensation to appear again as protostars. Final temperature of the protostar depends on the physical conditions of the colliding current carrying gas filaments. A similar scenario may take place in the atomic clouds of galactic corona if such an occasion as current carrying filamentary structure collision occur.

We would like to mention that our model does not concretizes parameters of the problem and does not pretend to perform quantitive characteristics of process but just figures out a qualitative picture of a such phenomena as protostar formation through the current carrying filamentary structures collision. The full 3D MHD code and particle simulations will reveal more details of problem and that is the subject of our nearest future study.

5 Acknowledgements

J. Sakai was partly supported by a Science Foundation from YKK and by the Foundation for the Promotion of Higher Education in Toyama Prefecture. B. Chargeishvili was supported from Japanese Government (Monbusho) Scholarship. We are grateful to Jie Zhao for his contribution in the computer simulations.

References

Carlberg R. G. and Pudritz R.E.: 1987, *Mon.Not.R.Astr.Soc.*Vol. no. **247**, pp. 353
Elmegreen B. G.: 1988, *Astrophysical J.*Vol. no. **326**, pp. 616
Nakano T. and Umebayashi T.: 1986, *Mon.Not.R.Astr.Soc.*Vol. no. **221**, pp. 319
Shu F.H., Adams F.C., and Lizano S.: 1987, *Ann.Rev.Astron.Astrophys.*Vol. no. **25**, pp. 23

EVOLUTION OF CURRENT LOOPS IN SPACE

S. J. GOLDSTEIN, JR.
University of Virginia
Charlottesville, VA 22903 USA

Abstract. Ampere's law requires that every magnetic field have an associated current. The analysis of magnetic fields in this paper begins with that current in a circular loop and calculates the forces that make the loop evolve. A circular current generates a dipole field; and a second-order, ordinary differential equation represents the evolving magetic field. The theory describes cases where the conductor shrinks as the loop increases in size. The temperature of the conducting ions and electrons then decreases. The theory also describes cases where the conductor grows as the loop grows. Then the conducting particles heat up.

Analysis shows that the magnetic clouds in the solar wind belong to the first type. In the provisional model adopted, the Klein–Burlaga clouds at one astronomical unit have a toroidal shape, centered on the sun, with a conductor radius of .125 au, and temperature (same for conducting electrons and protons) of 10^5 K. After 26 days the toroid has a radius of 7.1 au, the conductor radius is .025 au, and the temperature is 2600 K.

Key words: Current Loops, Solar Wind, Supernovae, Magnetic Fields, Solar Flares, Interstellar Media, Intergalactic Media

1 Introduction

Use of a current loop as the foundation for analyzing magnetic fields in space has several advantages. The electrodynamic forces on the loop and on the conductor may be calculated readily for a circular loop and an ordinary differential equation derived to represent the evolution of the loop. The large current that sets up the field and the local currents due to electrons and ions trapped in the field are easily distinguished. The magnetic field that results from a current loop is realizable, that is it does not rapidly evolve into a different form, and it rigorously obeys Maxwell's equations. The same cannot be said for many magnetic fields in the literature.

If a loop were not initially circular, there would be a stronger outward force where the conductors were closer, and it would evolve into a circle. An irregular distribution of trapped matter leads to a more complex shape, but searching for simplicity I consider circular current loops in this paper. One might assume that the simplest form for a current loop is a filament of infinitesimal thickness, but there is infinite energy in such a model because the magnetic flux density is unbounded near the filament. Instead, the loop is assumed here to have a uniform circular cross section.

Current loops are well established in astronomy. Menzel (1949) derived the loop current to sustain a large sunspot; Alfvén (1981) discussed the currents necessary for the heliosphere and for planetary magnetospheres; Greyber (1964; 1988) has made current loop models for extragalactic radio

Astrophysics and Space Science **227**: 217–228, 1995.

sources; Chen (1989) studied solar current loops in the solar corona which are anchored to the photosphere; Goldstein and Reed (1984) made a current loop model to explain the Faraday rotation toward the extragalactic source 3C 27; Goldstein (1992) interpreted the ring of line emission centered on supernova 1987A in terms of a current loop.

Since Ampere's law describes the only way of generating a magnetic field in classical electricity and magnetism, the well observed and stable magnetic fields measured by astronomers *require* closed current paths in space. It remains to be seen if circular loops can fit the observations.

To specify a current loop model one needs the radius of the loop r, and that of the conductor b, the current I, the temperature T, and the density n of the conducting electrons, and finally the velocity of expansion v, six quantities. I assume that there will be an identical density of conducting protons at the same temperature. To study the evolution of a loop, the derivatives of these quantities are required. The following relations are used: Newton's second law for the radial motion of the loop; conservation of magnetic flux; balance between the gas pressure in the conductor and the inward electrodynamic force; conservation of total energy; conservation of conductor mass. The calculations presented here omit the gravitational force of a massive object at the center of the loop, consider only radial motion and omit both the thermal energy of material trapped in the expanding field and the gas pressure difference between the trapped material and surrounding medium. The self inductance of the inductor is calculated from the assumption that the flux paths are circular.

An intricate quadratic equation results from the above assumptions whose roots are the parameter h, the ratio of the fractional time derivative of the radius of the conductor to that of the loop. In mathematical terms the quadratic equation gives two values of $h = (\dot{b}/b)/(\dot{r}/r)$. When h is negative, the conductor shrinks as the loop expands and the temperature of the current carrying electrons and protons decreases. This is the condition that applies to the magnetic clouds in the solar wind Klein and Burlaga (1982) used as an example in this paper. When the loop current is much stronger, h is positive and the conductor grows as the loop expands and the temperature increases. This mode may have applications to solar flares and to heating the interstellar and intergalactic media.

Having calculated h, I solve the differential equation for r numerically by Euler's method and evaluate the temperature, density, conductor radius, and radial velocity. The last step in the calculation is to test the theory and calculation by adding all the energies together. The results presented here conserve energy to about one part in a thousand.

2 Preliminary Calculations

Let the loop radius be a in this section and let r be the distance of some point in the plane of the loop from the center. Assume that the flux paths are circles centered on the loop and that a current I flows around the loop. The magnetic flux density perpendicular to the plane is

$$B(r) = \frac{\mu I A}{\pi(a^2 - r^2)}, \tag{1}$$

where $r < a$ and μ is the permeability of free space. It is convenient to use mks units to give I in amps, and the mks value for μ is $4\pi \cdot 10^{-7}$.

The flux linked by the loop is

$$\Phi = \int_0^{a-b} 2\pi B(r)\, r\, dr. \tag{2}$$

Then the inductance of the loop is

$$L = \frac{\Phi}{I} = \mu a \log\left(\frac{a^2}{2ab - b^2}\right). \tag{3}$$

To find the effective volume of the magnetic field and hence of the trapped matter, divide an expression for the total stored energy, $LI^2/2$, by the maximum energy density $B_1^2/2\mu$ where B_1 is the maximum flux density. (This procedure is analogous to finding the effective width of a filter.) B_1 can be found from equation 1 at $r = a - b$; but a more accurate value can be found by deriving a similar expression to equation 1 for $r > a$,

$$B'(r) = \frac{\mu I A}{\pi(r^2 - a^2)}, \tag{4}$$

and taking the average,

$$B_1 = \frac{B(a-b) + B'(a+b)}{2}. \tag{5}$$

Then

$$B_1 = \frac{2\mu I a^2}{\pi(4a^2 b - b^3)}, \tag{6}$$

and the effective volume is

$$V = V_c\left(2 - \frac{b^2}{a^2} + \frac{b^4}{8a^4}\right)\log\left(\frac{a^2}{2ab - b^2}\right), \tag{7}$$

where V_c is the volume of the conductor, $2\pi^2 ab^2$.

Suppose that both radii are subject to change and that over a narrow range

$$\frac{\dot{b}}{b} = h\frac{\dot{a}}{a}, \tag{8}$$

where h can be considered constant. The stored energy in an inductor is

$$E = \frac{\Phi^2}{2L}, \tag{9}$$

which for the current loop under discussion is

$$E = \frac{\Phi^2}{2\mu a \log\left[a^2/(2ab - b^2)\right]}. \tag{10}$$

The forces acting on the loop are obtained by differentiating equation 10 with respect to a and b, respectively, with Φ considered constant (i.e. flux is conserved). One obtains for the force on the loop

$$\frac{dE}{da} = \frac{\partial E}{\partial a} + \frac{\partial E}{\partial b}\frac{db}{da} = -\frac{I^2\mu}{2}\left(\log\frac{a^2}{2ab - b^2} - \frac{2(a-b)(1-h)}{2a-b}\right). \tag{11}$$

Similarly for the force on the conductor, the result is

$$\frac{dE}{db} = \frac{\partial E}{\partial b} + \frac{\partial E}{\partial a}\frac{da}{db} = \frac{I^2\mu}{2}\frac{a}{b}\left(\frac{2(a-b)(1-1/h)}{2a-b} - \frac{1}{h}\log\frac{a^2}{2ab - b^2}\right). \tag{12}$$

Suppose that the inward force on the conductor is balanced by the gas pressure of the electrons and protons that carry the current. Then for electron temperature equal to the proton temperature,

$$\frac{\mu I^2}{2} = \frac{8\pi^2 nkT\, b^2 h}{2(a-b)(h-1)/(2a-b) - \log\left[a^2/(2ab-b^2)\right]}. \tag{13}$$

Differentiating equation 13 and dividing by the original equation gives

$$\frac{\dot{I}}{I} = \frac{1}{2}\left[\frac{\dot{n}}{n} + \frac{\dot{T}}{T} + 2\frac{\dot{b}}{b} - 2\frac{\dot{a}}{a}\frac{-2a^2 + 2ab - b^2 + h(2a^2 - ab + b^2) - h^2(ab)}{(2a-b)^2\Delta}\right], \tag{14}$$

where

$$\Delta \equiv \frac{2(a-b)(h-1)}{2a-b} - \log\frac{a^2}{2ab - b^2}. \tag{15}$$

Let the mass of the conductor, since it is shielded by the magnetic field, be considered constant. Then $nab^2 = \text{const.}$ and

$$\frac{\dot{n}}{n} + \frac{\dot{a}}{a} + 2\frac{\dot{b}}{b} = 0. \tag{16}$$

Another relation for $\dot{I}/I$ results from the conservation of magnetic flux, $d/dt\ (LI) = 0$:

$$\frac{\dot{I}}{I} = -\frac{\dot{a}}{a}\left(-1 + \frac{\Delta}{\log\left[a^2/(2ab - b^2)\right]}\right). \tag{17}$$

Combining equations 14, 16, and 17 gives

$$\frac{\dot{T}}{T} = s\frac{\dot{a}}{a}, \tag{18}$$

with

$$s \equiv \frac{2(h-1)(2a^2-2b)}{(2a-b)\log\left[a^2/(2ab-b^2)\right]} + 2\frac{-2a^2+2ab-b^2+h(2a^2-ab+b^2)-h^2(ab)}{(2a-b)^2\Delta} - 1 \tag{19}$$

3 The Basic Differential Equation

Newton's second law for the current loop with a uniform density of trapped matter is

$$\frac{d}{dt}\left[(m_t + m_c)\dot{r}\right] + f_{ed} = 0, \tag{20}$$

where m_t is the mass of trapped matter, m_c is the mass of conducting matter, and f_{ed} is the outward electrodynamic force given by equation 11. In this section the radius of the loop is referred to as r, the terminology of the introduction. With the results and assumptions of the previous section equation 20 becomes

$$r\ddot{r}(x_1x_2\rho + \rho_c) + \dot{r}^2\rho\left[(1+2h)x_1x_2 + (1-h)(x_2x_3 + x_1x_4)\right] + 4nkTh = 0. \tag{21}$$

where ρ is the density of trapped matter, ρ_c that of the conductor, and

$$x_1 \equiv \log\frac{r^2}{2rb - b^2}, \qquad x_2 \equiv 2 - \frac{b^2}{r^2} + \frac{b^4}{8r^4}, \quad x_3 \equiv \frac{2(r-b)}{r-2b}, \qquad x_4 \equiv \frac{2b^2}{r^2} - \frac{b^4}{2r^4} \tag{22}$$

4 The Quadratic Equation for h

The energy of the model is assumed to consist of three terms: E_1, the stored magnetic energy; E_2, the kinetic energy of the evolving loop; and E_3, the thermal energy of the conductor.

Reader, you may be able to improve these calculations by adding other terms.

Summing the three terms and differentiating with respect to time and applying the results of section (2) and the symbols of section (3) gives

$$r\ddot{r}(x_1x_2\rho+\rho_c)+\frac{1}{2}\dot{r}^2\rho\left[(1+2h)x_1x_2+(1-h)(x_2x_3+x_1x_4)\right]+$$
$$\frac{3}{2}nkT\left[-1+(h-1)\frac{x_3}{x_1}+\frac{(x_6+x_7h+x_8h^2)}{x_3(h-1)-x_1}\right]+4nkTh \;=\; 0, \quad (23)$$

where

$$x_6\equiv\frac{2(-2r^2+2rb-b^2)}{(2r-b)^2},\qquad x_7\equiv\frac{2(2r^2-rb+b^2)}{(2r-b)^2},\qquad x_8\equiv-\frac{2rb}{(2r-b)^2}. \quad (24)$$

Subtracting equation 23 from equation 21 eliminates the acceleration term and gives a quadratic equation in h. Let $g\equiv\dot{r}^2\rho/(3nkT)$. The equation is

$$Ah^2+Bh+C=0, \quad (25)$$

and the coefficients are

$$A \;=\; x_3\left[g(2x_1x_2-x_2x_3-x_1x_4)-2\frac{x_3}{x_1}\right]-x_8 \quad (26)$$

$$B \;=\; x_3\left[g(x_1x_2+x_2x_3+x_1x_4)+1+2\frac{x_3}{x_1}\right]$$
$$-(x_1+x_3)\left[g(2x_1x_2-x_2x_3-x_1x_4)-2\frac{x_3}{x_1}\right]-x_7 \quad (27)$$

$$C \;=\; -(x_1+x_3)\left[g(x_1x_2+x_2x_3+x_1x_4)+1+2\frac{x_3}{x_1}\right]-x_6. \quad (28)$$

Equation 25 is referred to as the giant quadratic equation, GQE, in the remainder of this paper. To calculate the coefficients A, B, and C one must choose values for r, b, v, n, T, and ρ. If the quadratic equation yields real roots for h, the model can be applied. Return now to equation 21.

5 Numerical Solution of Differential Equations

The following equations apply:

$$dr \;=\; v\,dt, \quad (29)$$

$$db \;=\; \frac{hbv}{r}dt, \quad (30)$$

$$dT \;=\; \frac{sTv}{r}dt, \quad (31)$$

$$dn \;=\; -\frac{n(1+2h)v}{r}dt, \quad (32)$$

$$dv \;=\; \ddot{r}\,dt. \quad (33)$$

Here dt is the time step and s is defined by equations 18 and 19, while $\ddot{r}$ is the acceleration from equation 20. Euler's method suffices to solve this set of ordinary differential equations. It is indeed an advantage to trace the evolution of r, b, T, n, and v by such a simple method, when compared to a three–dimensional hydromagnetic formulation with partial differential equations.

6 Application to Magnetic Clouds in the Solar Wind

The magnetic clouds in the solar wind have been observed from spacecraft by measurements of magnetic flux and of plasma density, thus it is interesting to apply the theory of the previous section to these objects. The gravitational force of the sun is not part of the theory, but the average solar wind velocity (I adopt 400 km/sec) is much larger than the escape velocity at 1 au (42 km/sec), so the correction to equation 20 can be neglected. Further, it may be shown that a gravitational term has no effect at all on the GQE and its roots. Burlaga, *et al.* (1981) and Klein and Burlaga (1982) have identified a species of magnetic clouds whose dimension along the line to the sun (determined from their duration and the solar wind velocity) is about .25 au.

Burlaga (1991) has interpreted the form of these clouds as rooted to the sun in two places and extending past the earth. He states "It is unlikely that the magnetic clouds have the topology of a torus, since that would imply that two passages through the magnetic cloud would be observed when the axis (center lines) of the magnetic cloud is near the ecliptic, whereas no such double–passages have been reported. Nevertheless, it might be appropriate to approximate the local geometry of a magnetic cloud as that of a torus when the radius of curvature of the axis is comparable to that of the flux tube."

I will assume that a magnetic cloud is a heliocentric current loop, a particular torus, and free to expand in the solar wind. It is not easy to understand how such a loop is created, but it is also not easy to comprehend how a torus fixed to the sun extends indefinitely into space. The currents required to drive this latter structure have not been described.

The ionized interstellar matter seem by Jacobsen, *et al.* (1991) around the supernova SN1987A is an elliptical ring and has been interpreted by Goldstein (1992) as containing a current loop. The synchrotron emitting ring inside of Tycho's supernova reported by Wood, *et al.* (1992) is apparently another current loop. In this case the polarization measurements corrected for Faraday rotation show that the magnetic field is transverse to the loop as Ampere's law requires. The ejecta from Tycho's supernova have passed through this current loop without disrupting it. The current loops in our star may be able to merge into a large one such as those observed in the supernova explosions. At least having observed two astrocentric loops, the assumed existence of heliocentric loops seems plausible.

The simple picture of a uniform conductor surrounded by another uniform volume of trapped matter is not in good agreement with the observations of Burlaga, *et al.* (1981) The profiles of the magnetic field as the cloud moves by the spacecraft have a roughly constant component that rotates through nearly 180 deg. If the spacecraft were moving through trapped matter about a uniform conductor, Burlaga, *et al.* (1981) would have observed a stronger field at closest approach to the conductor and not a roughly constant one. Thus the trapped matter and the current seem intermingled, and the conductor is not supported by the gas pressure as assumed in the model. Nevertheless, I use the model to give some provisional results on the evolution of a magnetic cloud.

The following initial conditions are adopted: $r = 1$ au, $b = r/8$, $v = 400$ km/sec, and $T = 10^5$ K. The electron density of the conductor n was adjusted to agree with the 12 nT average field at the edge of the conductor given by Klein and Burlaga (1982) ($n = 60$ cm^{-3}). The density of trapped matter is taken to be 10 H atoms/cm^3. Integration steps are .01 days. The calculation includes 26 days in which the radius of the loop increases from 1 au to 7.1 au and the loop current decreases from $1.51 \cdot 10^9$ to $3.64 \cdot 10^7$ amps. Figure 1 gives the results. In figure 1 the radius of the loop increases nearly uniformly with time, but there is a small acceleration, most visible near $t = 0$. Thus the velocity of the solar wind controls the radius of the loop. The radius of the conductor decreases sharply from .125 to .0249 au, an electrodynamic effect. The program that produced the data for Figure 1 will also integrate backwards in time. At $r = .772$ au the adopted value of $h = -4.097$ gives $b = .207$ au, $T = 360,000$ K, and $n = 28.3$ cm^3 The value of h subsequently becomes too large numerically for the current to have a real value.

In figure 2 the forward calculation is repeated with the positive value for h from the GQE. The same initial conditions are used but the steps are ten times smaller. The loop current is $3.37 \cdot 10^{10}$ amps initially and $2.20 \cdot 10^{10}$ amps at the last step. The magnetic field in this case is initially 20 times greater than with the negative value for h and there is a correspondingly stronger magnetic field than the average value given by Klein and Burlaga (1982). Now the outward velocity decelerates noticeably and the radius of the conductor increases from .125 au to .207 au. The integration stops before the current turns imaginary. The temperature of the conductor electrons and protons increases sharply while the electron density decreases.

7 Allowable Ranges for Model Constants

A major constraint on the constants is that they give real roots to the GQE. Consequently the discriminant $B^2 - 4AC$ must be positive. Because the x_i

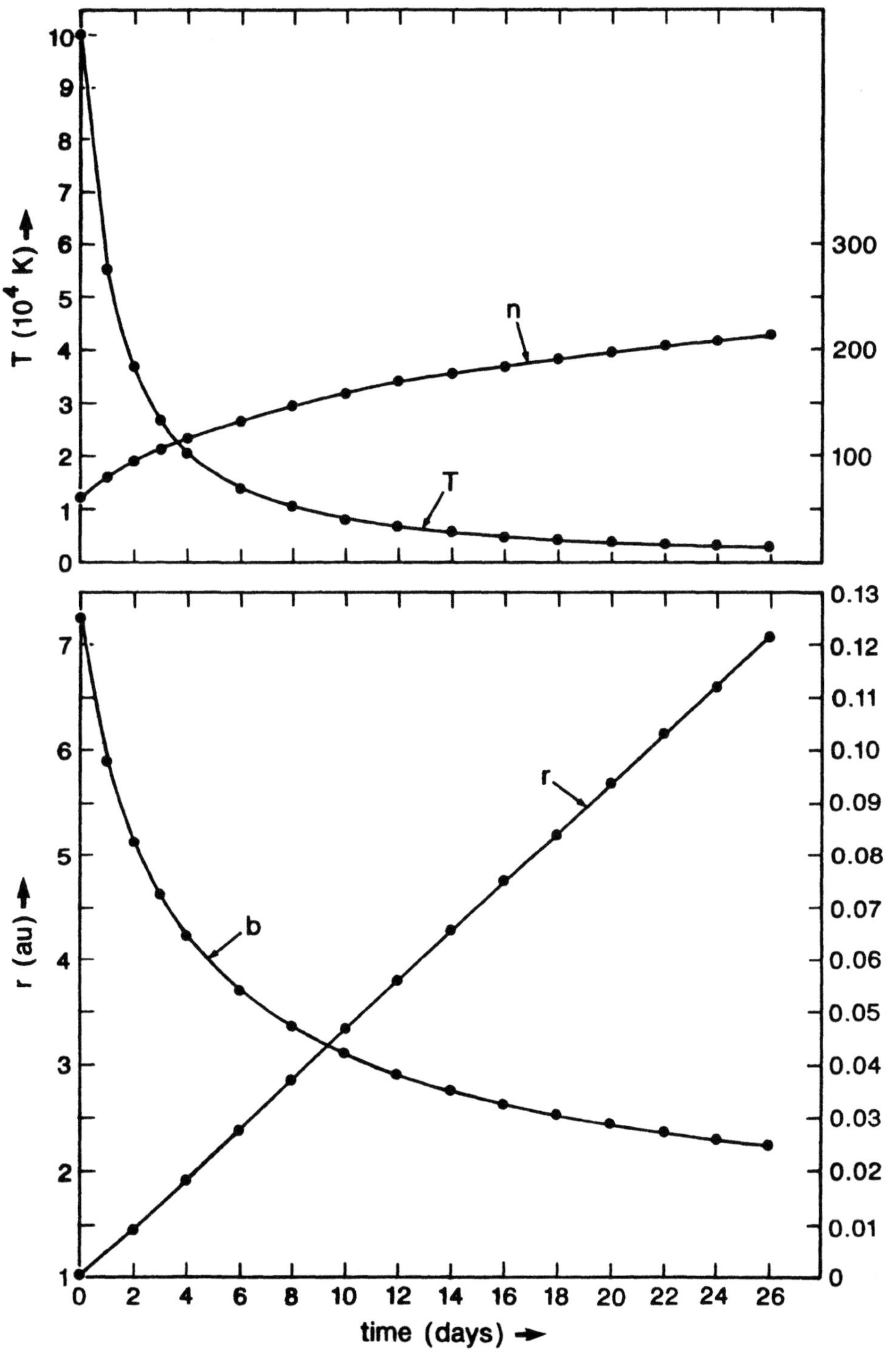

Fig. 1. Evolution of a current loop model, fitted to the average magnetized cloud described by Klein Burlaga (1981). Negative h.

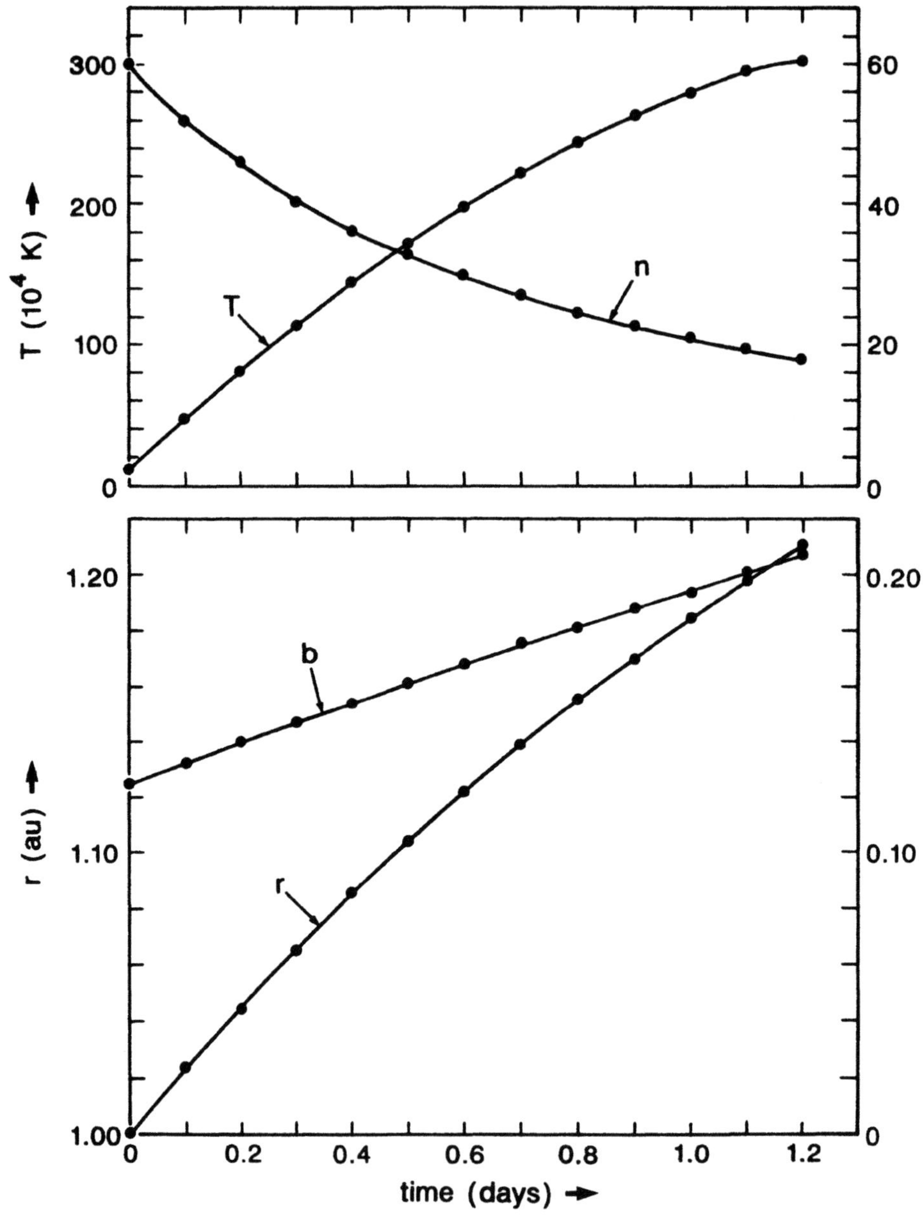

Fig. 2. Evolution of a current loop model. Same as Figure 1 except that the positive value of h was selected. The magnetic field for this model would be 20 times stronger than that for Figure 1.

TABLE I
Lower limits to g

r/b	lowest value of g for real roots
8	.218
20	.0931
100	.0344
1000	.0137

that appear in the GQE are all functions or r/b it is possible to establish the boundary for which $B^2 - 4AC = 0$ by a simple empirical study in terms of the constant g, which is the only other adjustable quantity in the GQE. Table 1 gives the lower limit to g in terms of r/g.

Since $g^2 = v^2\rho/(3nkT)$ we may use the values in the table to select allowable values for v, ρ, n, and T. There is a minimum value for v and ρ and a maximum value for n and T when the other parameters have been chosen. Note that the size of the loop does not enter into finding h.

The second major constraint on the model constants comes from the balance between the electrodynamic and gas pressure forces from equation 13. The right hand side of equation 13 must be positive. All negative values of h give a positive right hand side and thus allow the model to be applied. For positive h, however, h must be greater than $1 + x_1/x_3$, and too large a value of h makes it impossible for the right hand side of equation 13 to match the left side. The positive values of h lead to rather critical constants in the model.

8 Conclusions

The dynamics of magnetic fields in space can be described in terms that rigorously obey Maxwell's equations. This is not possible within the framework of fluid dynamics because a fluid has interactions only between adjoining cells, whereas magnetism has interactions between charged particles and the exciting current at large distances. I don't doubt that there are some fluid dynamic corrections that would improve the accuracy of equation 20, however.

The negative h model matches the magnetic field measured near one astronomical unit and thus seems able to serve as an interim description of the magnetized clouds in the solar wind.

The positive h model, though critical to its constant, attracts interest because it might heat up the interstellar matter about supernovae and the intergalactic matter about clusters of galaxies.

The origin of current loops has not been treated here. The most likely sources are individual stars. But the process for generating them and that for detaching them from the star remain unsolved astrophysical problems. A possible improvement to the theory presented here would be a more realistic method of supporting the conductor against the inward electrodynamic force.

References

Alfvén, H.: 1981,**Cosmic Plasma** (Reidel:Dordrecht), ch. 3.

Burlaga, L. , Sittler, E., Mariani, F., and Schwenn, R.: 1981, *J. Geophys. Res.* **Vol. 86**, pp. 6673–6684.

Burlaga, L.F.: 1991 in **Physics of the Inner Heliosphere**, ed. Schwenn, R. and Marsch, E. (Springer–Verlag: Berlin) **Vol. II**, pp. 1–22.

Chen, J.: 1989, *Astrophys. J.* **338**, pp. 453–470.

Goldstein, S.J. and Reed, J.A.: 1984, *Astrophys. J.* **283**, pp. 540–545.

Goldstein, S.J.: 1992, *IEEE Trans. Plasma Sci.* **20**, pp. 874–876.

Greyber, H.D.: 1964, *Astron. J.* **69**, p. 542.

Greyber, H.D.: 1988, in **Supermassive Black Holes**, ed. M. Kafatos (Cambridge Univ. Press: Cambridge).

Jacobsen, P., *et al.*: 1991, *Astrophys. J.* **369**, pp. L63–L66.

Klein, L.W. and Burlaga, L.F.: 1982, *J. Geophys. Res.* **87**, pp. 613–624.

Menzel, D.H.: 1949, **Our Sun** (Blakiston: Philadelphia), p. 120.

Wood, L.A., Mufson, S.L., and Dickel, J.R.: 1992, *Astron. J.* **103**,pp. 1338–1345.

RADIATION PROPERTIES OF PULSAR MAGNETOSPHERES: OBSERVATION, THEORY, AND EXPERIMENT

KEVIN R. HEALY
Very Large Array Operations Center
National Radio Astronomy Observatory
Socorro, New Mexico, USA

and

ANTHONY L. PERATT
Physics Division
Los Alamos National Laboratory
Los Alamos, New Mexico, USA

Abstract. In the three decades since their discovery, the accumulated body of observational data from pulsar sources puts constraints on models that seek to explain their periodic radiation. This paper reviews this data, reports on a VLA search for haloes predicted by an early model, and reinvestigates a magnetospheric disk–field–aligned–current transmission line system as the origin of the observed radiation, with external wave excitation by as yet an unexplained source. Three dimensional, fully electromagnetic particle–in–cell simulations of the pulsar surface and magnetosphere are used to explore the waveshape and polarization properties of the observed radiation.

Key words: Pulsars, Radio Astronomy, Plasma Astrophysics

1 Introduction

Unexpectedly in 1967, J. Bell and A. Hewish detected a highly stable train of pulses of radio emission from the constellation Vulpecula during an investigation of extragalactic radio sources (Burnell 1983, Backer 1988). This train had a periodicity of 1.337 s.

The low galactic altitudes of this first 'pulsar', and soon thereafter, several other 'pulsars', suggested that these radiation regions were imbedded within our Galaxy. Furthermore, millisecond intensity fluctuations of the radio signals implied that the emitting regions had dimensions less than 300 km.

Because of the stability of the pulse period, Thomas Gold (1968) concluded that pulsars must be highly magnetized, rapidly rotating neutron stars. This model of a pulsar's emission qualitatively describes it as high energy plasma beams emitted from the poles of a spinning neutron star and accelerating along the stars's dipole magnetic field, emitting curvature radiation as pulses towards the observer. In the three decades since their discovery, most new observations have been able to be described in terms of the standard, search-light, model.

Astrophysics and Space Science **227**: 229–253, 1995.

In spite of the successes of the standard model, there yet exists no self-consistent theory to describe the pulsar electrodynamics. The basic radiation mechanism, the acceleration of relativistic electrons along a curvature producing force, is beyond doubt. Nevertheless, how the relativistic electrons beams are produced and why they possess the observed polarizations are major problems any model must answer (Verschuur & Kellerman 1988).

In particular, any model must address the following known pulsar features:

1. Radio luminosities estimated to lie in a range $10^{27} - 10^{30}$erg/s ($10^{20} - 10^{23}$W);
2. Pulse periods within a range of a millisecond to several seconds;
3. Stable mean pulse profile having one, two, or three components;
4. Monotonic rotation of linear polarization vector with a signature that depends on the component structure;
5. Slow decrease of the pulse width with increasing radio frequency, and with increasing period;
6. Slow increase of periodicity, usually with dP/dt in a range between 10^{-14}– 10^{-16} sec/sec, but for millisecond pulsars as small as 10^{-19} sec/sec;
7. *Glitches*, a jump-like *decrease* of the period P;
8. Interpulse components midway between the main pulses;
9. A continuum, low-power background;
10. Linear polarization angle modes spaced by 90°;
11. A circularly polarized component that reverses polarity between peaks in the main pulse;
12. Memory in the pulse–to–pulse fluctuations which in some pulsars takes the form of periodically spaced subpulses drifting slowly in time;
13. Erratic pulse–to–pulse amplitude fluctuations;
14. Micropulses within single pulses that are sometimes periodically spaced.

Other problems a model must address include:

15. The extreme bandwidth of radiation observed, in some pulsars from radio frequencies to gamma ray energies;
16. A very high Doppler-interpreted velocity, up to 500 km/s; and finally:
17. The global pulsar system and topology which, for the Crab Nebula, encompasses a fibrous and filamentary region 2×10^{16} m beyond the pulsar; 'wisps', 'jets', and other features symmetrically oriented 10^{15} m beyond the active region (Hester *et al.* 1994).

A number of suggestions have been offered to address many of the seventeen items listed above. For example, the orientation of core and conal components of beam emission regions with respect to the axis of rotation of a neutron star, its magnetic axis, and the location of the observer are postulated to answer items 3, 4, 8, and 10. Naik and Kulkarni (1994) suggest the existence of field–aligned currents and utilize the Carlqvist Relation (Peratt 1992a) to explain item 9.

A fundamental problem in any pulsar model, the actual emission mechanism for relativistic charged particle beams, is investigated in some detail by Beskin *et al.* (1993). They describe the generation of electron-positron plasma with relativistic parameter $\gamma \geq 10^6 - 10^7$ in a relativistic double layer ($V_{||} \approx 10^{12} - 10^{13}$V) between the surface of the neutron star and its magnetosphere.

Glitches, item 7, once described entirely in terms of 'starquakes,' has more recently been attributed to fluid motions within a stellar core.

Several radiation loss mechanisms have been suggested for Item 6. For example, gravitational quadrapole radiation had been cited as a cause for periodicity lengthening. However, the discovery of millisecond pulsars do not show a decay consistent with this theory (Backer *et al.*).

The content of this paper is ordered as follows. In Section 2 we describe the historical search for pulsar haloes predicted by the standard model that led to the writing of this paper. In Section 3 we report current results with VLA observations of halo-candidate pulsars PSR0809+74, PSR0942-13, and PSR0950+08. Section 4 is a reexamination of the disk-magnetosphere model and a theory of radio emission in a magnetospheric transmission line model. Section 5 reports the measured properties of pulses in a model magnetosphere experimental setup. We present our conclusions in Section 6.

2 The Search for Haloes

In the standard model, what a beam of plasma does after it has left the inner part of the magnetic field is not commonly described in the literature. For example, if the plasma leaves the pulsar system and interacts with the interstellar medium (ISM), the effects of the interaction might be visible to telescopes on Earth.

Twenty years ago, Blandford *et al.* proposed a theory of 'ghost remnants' surrounding old pulsars. While considerable effort has been spent trying to confirm or refute this theory, little progress has been made. Today, the original Blandford theory is outdated, but the observational search for pulsar–related nebulae continues. Past observations of ancient pulsars above 1 GHz have had no success, while searches below 1 GHz have had detections. Recent detections by the Very Large Array and optical observatories of 'pulsar wind nebulae' around fast–moving pulsars have shown the detectability of pulsar wind/interstellar medium interactions. Further possibilities have been opened by recent Hubble Space Telescope observations of the Crab pulsar.

Observations made after Blandford *et al.* (1973) have disproved the swept–shell model, but relativistic pulsar winds as a driving mechanism in the formation of pulsar synchrotron nebulae is still a popular notion. The interpretation of the 'pulsar wind nebulae' is a bow shock formed by ram–

pressure enhanced pulsar wind (Frail and Kulkarni, 1991), while the optical haloes seen around the Crab pulsar have a very striking symmetry related to the pulsar's beams of relativistic particles (J. J. Hester, 1994).

The original theory of Blandford *et al.* (1973) describes a radio halo in terms of a relativistic wind blowing away from the pulsar and interacting with the ISM. This wind would travel through the medium surrounding the pulsar at velocities faster than the Alfvén speed in the medium, expelling the medium and the ambient magnetic field outward. Balancing the flux of the rotational energy lost by the pulsar with the pressure of the interstellar medium provides a way to predict the size of the halo. That is,

$$\frac{\dot{E}_R}{12\pi R^2 V_A} \sim p_{ism}$$

where $\dot{E}_R$ is the rotational energy loss rate, R is the radius of the wind/ISM boundary at equilibrium, V_A is the Alfvén speed in the ISM, and p_{ism} is the pressure of the interstellar medium. For the values provided in Blandford *et al.* (1973) of $\dot{E}_R \sim 10^{32}$ erg s^{-1}, $V_A \sim 10$ km s^{-1} (for $n_e \sim 1$ cm^{-3}, $B \sim 5 \times 10^{-6}$ Gauss), and $p_{ism} \sim 10^{-12}$ dyne cm^{-2}, this gives a value for R $\sim$ 1 pc (see Figure 1a). A halo this size would form in $T \sim R/V_A \sim 10^5$ years. This time is much greater than the age of a supernova remnant, so a halo could easily be distinguished from the original remnant (Blandford *et al.*, 1973).

However, the reports of the first positive detections of a halo around PSR 0950+08 (Gopal–Krishna 1978, Glushak, Pynzar, and Udal'tsov 1981) stated that the object was ~1 arcmin (1') across, only 0.04 pc (8000 AU) in diameter. Even taking into account the modern value of $\dot{E}_R$ for PSR0950+08 is closer to 10^{30} erg s^{-1}, this only lowers the expected diameter of the halo to ~0.1 pc. Obviously another mechanism is at work here, for the wind must be confined by a pressure 10^4 times higher than the pressure of the interstellar medium (see Figure 1b).

2.1 Past Observations

Although the first detection of a possible pulsar radio halo was seen around PSR1919+21, the case for a halo around PSR0950+08 is stronger. It is now thought that PSR1919+21, a pulsar with a proper motion of ~200 km s^{-1}, is superimposed on a more distant supernova remnant, while there appears to be no confusing source behind PSR0950+08.

Shortly after Blandford *et al.* (1973), Weiler, Goss and Schwarz reported on observations done at the Westerbork Radio Synthesis Telescope at 1415 MHz. The purpose of their observations was two–fold: to search for new supernova remnants in the vicinity of known pulsars, and to search for pulsar 'ghosts', the newly theorized haloes.

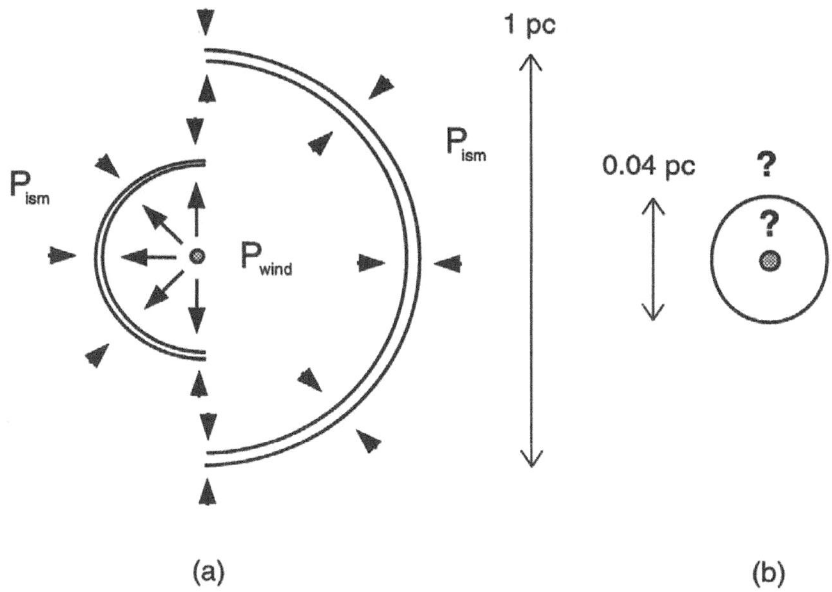

Fig. 1. (a) The Blandford *et al.* model describes a pulsar halo as the interaction of the pulsar's relativistic wind and the interstellar medium, (b) The revised model is still under construction.

To the sensitivity of Westerbork, no supernova remnants or haloes were seen (Weiler, Goss, and Schwarz, 1974). But in light of the later measurements of the size of the PSR0950+08 halo (1' in diameter) and the synthesized beamwidth of the Westerbork array (23 × 196 arcsec in the PSR0950+08 field), it is unlikely that the unpulsed halo emission could have been resolved as seperate from the pulsar's emission.

In 1973 and 1978, Gopal–Krishna, using the proximity of PSR0950+08 to the ecliptic, observed the pulsar during two lunar occultations to search for unpulsed emission. He collected four scans of the pulsar, two immersions and two emersions, each with a different position angle, using the Ooty radio telescope at 327.5 MHz. Analyzing the immersion data from the 1973 occultation and the emersion data from the 1978 occultation seperately, Gopal–Krishna saw evidence of a "halo of continuous emission" with a flux density of 0.2±0.07 Jy at 327.5 MHz and a size of ~1' (Gopal–Krishna, 1978). In addition, he used the null detection of a halo around PSR0950+08 by Schönhardt (1973) at 2.7 GHz to determine a spectral index, α, (the flux,

S, at frequency, ν, is given by S $\propto \nu^{\alpha}$) for the halo observed. His value of a steep index, -0.6, agreed well with the synchrotron emission model for haloes (Gopal–Krishna, 1978). The next report of a positive detection of a radio halo came to a similar conclusion.

Four years later, Glushak, Pynzar', and Udal'tsov reported that they had confirmed a halo around PSR0950+08. Their observations were made at 102.5 MHz with two different telescopes at Pushchino. The BSA, a multi-beam meridional telescope, and the BRT, a interferometer using the BSA and the DKR–1000 telescope, had resolutions of 35' and 16', respectively, in the PSR0950+08 field (Glushak, Pynzar, and Udal'tsov, 1981).

After observing their candidates several times, Glushak, Pynzar', and Udal'tsov found that the PSR0950+08 halo had a 102.5 MHz flux of 2$\pm$1 Jy. Using past positive and null detections of the halo, they determined the spectral average for the emission to be -2 ± 1, which was in good agreement with Gopal–Krishna's value (Glushak, Pynzar, and Udal'tsov, 1981).

2.2 Wind Nebulae and Optical Nebulae

Glushak, Pynzar', and Udal'tsov also predicted that high–velocity pulsars would have no visible haloes because the pulsar's relativistic wind would be dispersed over the path of the pulsar, keeping the local injection rate low (Glushak, Pynzar, and Udal'tsov, 1981). In this, they were half right; synchrotron haloes have been seen near fast pulsars only in x–rays to date, but Hα– and radio–emitting shocks have also been detected (Kulkarni and Hester 1988, Frail and Kulkarni 1991, and Cordes, Romani, and Lundgren 1993). These three shock detections all fall into a loose catagory known as 'pulsar wind nebulae'.

Two other observations appear to fall into another catagory of shock–powered structures around relatively slow pulsars. A Very Large Array observation of the Vela pulsar has been made to detect a standing shock in the pulsar's relativistic wind. This 'wisp' is believed to be part of a toroidal shock formed by the pulsar's wind interacting with the interior of the Vela supernova remnant (see Figure 2a) (Bietenholz, Frail and Hankins, 1991). A recent Hubble Space Telescope image of the vicinity of the Crab pulsar shows a pair of circular haloes concentric with what is interpreted as the pulsar's spin axis. These haloes are then thought to be some type of interaction of the pulsar's relativistic beams of particles with the circumstellar medium (J. J. Hester, 1994).

3 VLA Observations of PSR0809+74, PSR0942-13, and PSR0950+08

In March 1993, T. H. Hankins of NRAO and one of the authors, K. R. Healy, observed three nearby pulsars with the VLA at 327.5 MHz. We used a real-

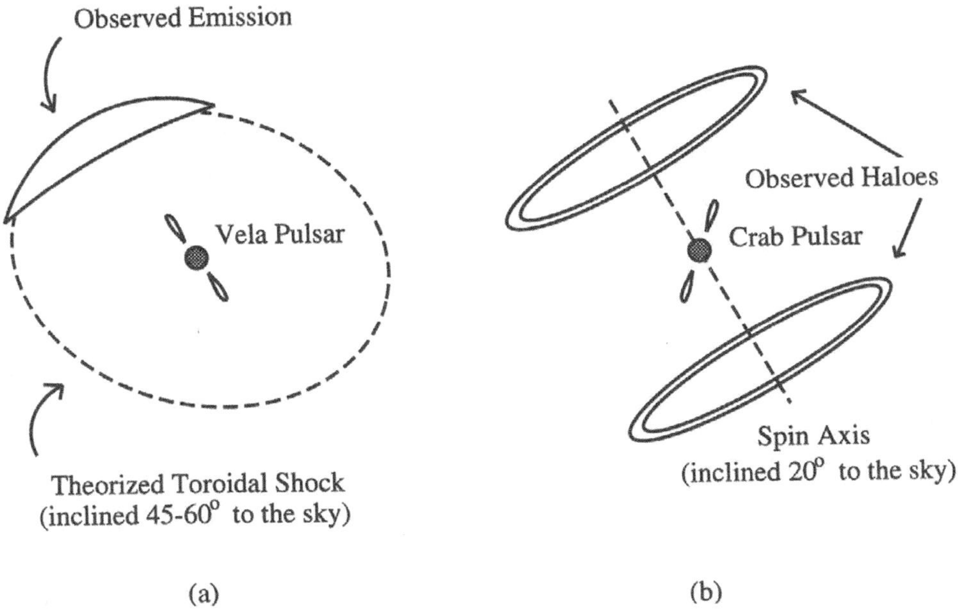

Fig. 2. (a) The 'wisp' detected by Bietenholz, Frail, and Hankins is thought to be part of a larger toroidal shock, (b) while the haloes seen by Hester *et al.* are thought to be the interaction of the pulsar's relativistic particle beams with the circumstellar medium.

time gating system which T. H. Hankins had developed to study pulsars with high time–resolution. The system can produce high time–resolution pulse profiles of pulsars and can split the incoming pulsar signal synchronously with the pulsar's period into two signals; the 'on–pulse' emission and the 'off–pulse' emission. This approach had been successfully used by Bietenholz, Frail and Hankins in 1991 to produce the very high dynamic range maps of the wisp near the Vela pulsar (Bietenholz, Frail and Hankins, 1991). We hoped to use this filtering technique to remove the bright pulsar from the data and make high sensitivity maps of the area around our three targets.

Our targets were selected on several criteria. Our first choice was PSR0950 +08, the target source for Gopal–Krishna's occultation experiment, and the Glushak, Pynzar', and Udal'tsov positive detection. From this candidate, we determined our selection criteria to find other pulsars with similar properties. PSR0950+08 is relatively stationary in space, with a transverse velocity of $\sim$15 km s^{-1}, and it is very nearby, 130 pc. It also appears to be a fairly ancient pulsar, having an energy loss rate of $\sim 10^{32.75}$ erg s^{-1}, well below the canonical $\dot{E}_R$ of 10^{36} erg s^{-1} for the newest pulsars. The properties of our other candidates are summarized in Table I.

TABLE I
Properties of VLA Pulsar Halo Search Candidates

1950 Name	Period (s)	Distance (pc)	v_t (km s^{-1})	$\log(\dot{E}_R)$ (erg s^{-1})
PSR0809+74	1.29224	320	41	30.49
PSR0942−13	0.57026	710	48	31.00
PSR0950+08	0.25307	130	15	32.75

Each of these pulsars was observed in B Array (beamwidth of 15") in Phased Array Spectral Line mode for one hour. This allows a theoretical noise level for the map of 1 Jy/beam. In addition, the Phased Array was required to ensure all antenna signals would arrive at the gating system in phase, and Spectral Line mode was used to allow data that were corrupted by terrestrial interference to be removed.

A very clean map of PSR0809+74 shows no evidence of structure like that seen by Gopal–Krishna; maps of the other two pulsars are noiser and are currently being further processed.

We now turn to the task of modeling a pulsar. Our research is divided into three tasks: i) the total pulsar system and environment, with dimensions reaching 10^{16} m, ii) the pulsar magnetosphere, whose dimensions may be of order 10^8 m, and iii) the active radiation region, whose dimensions may range from hundreds of kilometers to 10^7 m. The remainder of this paper deals only with the pulsar magnetosphere.

4 Magnetospheric Transmission Line Model

Michel and Dessler (Michel & Dessler, 1981; Michel, 1982) suggested the presence of a fossil disk in all pulsars. The purpose of the disk hypothesis was to solve a problem concerning the balance of charges in a corotating magnetosphere. It was soon found that one of the other attributes of a disk was a realistic mechanism for increasing P. For example, in the Michel-Dessler model, the rate of energy loss varies as P^{-3}. This was consistent with the slow rate of energy loss by millisecond pulsars in the presence of a smaller magnetic field, when these were discovered.

If the disk consists of matter in the plasma state, then it is likely that this material is electromagnetically coupled to the magnetosphere via field-aligned currents. To model this coupling, we employ a transmission line mod-

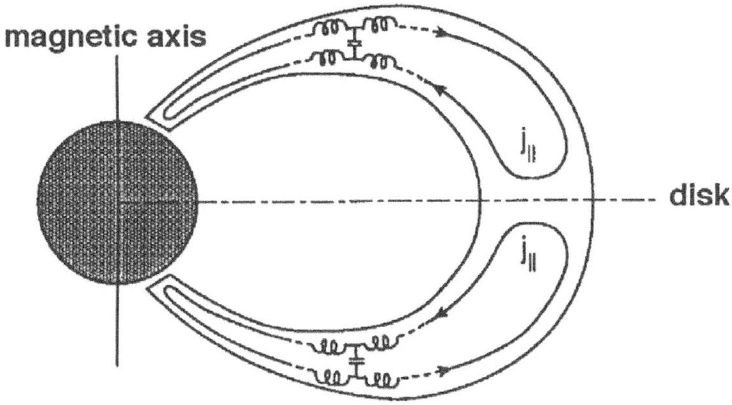

Fig. 3. Transmission line model for magnetic field-aligned currents in a magnetospheric plasma configuration. The north and south pole transmission lines need not be symmetric. The disk may lie between the north and south pole currents, on the equator, or may extend to higher latitudes.

el similiar to that used to describe planetary magnetospheres (Sato 1987)[1]. Figure 3 illustrates a magnetospheric transmission line model for field-aligned currents $j_{||}$ that couples a source region to a load (radiation) region.

The following sections investigate the properties of synchrotron radiating, field-aligned current pulses propagating in a pulsar's magnetosphere.

4.1 Characteristics of a Transmission–Line Pulse

Any disturbance on a transmission line, such as an arc-discharge or an interruption of steady-state conditions, results in the initiation of traveling waves, which propagate towards the ends of the line, where they may be reflected

[1] The similarity between the pulsar phenomenon and the Jovian decametric emissions were noted almost immediately (Michel 1982). Jovian decametric bursts originate from relativistic electrons ($\sim$ 30 MeV; comparable to the 170 MeV pole–to–equator potential drop) spiralling around Jupiter's dipole magnetic field ($\sim$ 4 G) and in the Io-Jupiter flux tube, not from planetary-satellite rotation. The highest (decametric) frequencies come from regions closest to the planet while the lower (kilometric) frequencies are emitted much farther away.

and modified, and attenuated and distorted by any losses until they die out.

The mathematical description of the pair of voltage ϕ and current i traveling waves, along direction z, are given by :

$$\begin{aligned} \phi(z,t) &= \phi^{+}(z-ut)+\phi^{-}(z+ut) \\ i(z,t) &= \tfrac{1}{Z_c}\left[\phi^{+}(z-ut)-\phi^{-}(z+ut)\right] \end{aligned} \quad (1)$$

where $Z_c = (L/C)^{1/2}$ is the characteristic impedance of a transmission line with inductance L and capacitance C, and $u = (LC)^{-1/2}$ is the propagation constant of the line. The left and right traveling voltage components are (Peratt 1992b):

$$\begin{aligned} \phi^{+} &= \tfrac{Z_c}{Z_g+Z_c}\sum_{n=0}^{\infty}\Gamma_g^n\Gamma_l^n\phi\left\{t-\tfrac{1}{u}\left[z+2l\ n\right]\right\} \\ \phi^{-} &= \tfrac{Z_c}{Z_g+Z_c}\sum_{n=0}^{\infty}\Gamma_g^n\Gamma_l^{n+1}\phi\left\{t-\tfrac{1}{u}\left[-z+2l\ \ (n+1)\right]\right\} \end{aligned} \quad (2)$$

The voltage reflection coefficients are:

$$\Gamma_g = \frac{Z_g - Z_c}{Z_g + Z_c},\ \Gamma_l = \frac{Z_l - Z_c}{Z_l + Z_c} \quad (3)$$

for line surface and disk resistances Z_g and Z_l, respectively.

Consider now a particular case of a lossless 0.4 Ω impedance transmission line of length $l = 10^8$ m (Peratt 1992c). We launch a 20 ms, 10^{13-16}V voltage pulse ($\gamma = 2 \times 10^{7-10}$) from the surface. [Either surface or disk launching is permissible; both lead to the same result]. This voltage is a value associated with relativistic double layers (Peratt 1992d). We suppose that highly ionized plasma of high conductivity terminates both ends of the line so that, to first approximation, $Z_g = Z_l = 0$ and $\Gamma_g = \Gamma_l = -1$. On a lossless line with perfect end conductors the pulses propagate back and forth, forever. In reality, kinetic energy of the relativistic currents flowing along the guide magnetic field is lost in synchrotron radiation.

It is of interest to measure the currents and voltages in the line near the surface, where the current curvature is strongest. These waveforms are shown in Figure 4 for probes located 3×10^5 m (300 km) and 10×10^6 m (10^4 km), respectively, from the surface. As shown, after time 0, each waveform consists of two components, an incident (left-traveling) wave and a reflected (right-traveling) wave. Together, these produce the following features seen in more detail in Figure 5:

- Either a double-hump or single-hump current profile with periodicity 0.65 s;

- A concomitant voltage pulse that reverses polarity over the current profile;

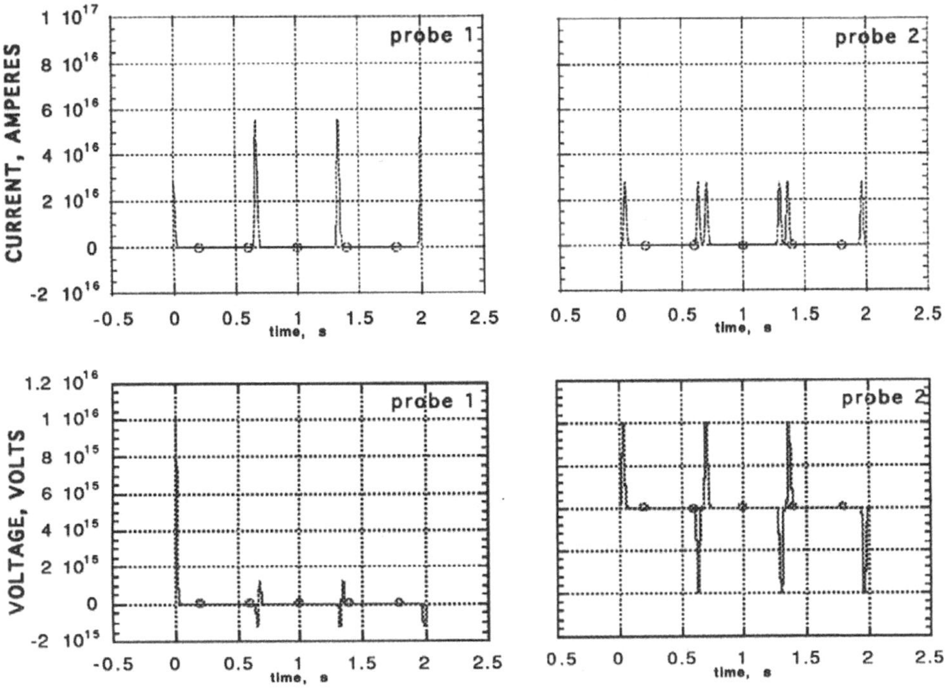

Fig. 4. Current and voltage pulses near the pulsar surface. (top) Current pulses at 3×10^5m and 10×10^6m. (bottom) Voltage pulses.

The relativistic 10^{13-16} A current flowing at the surface is thought responsible for the pulsar emission. The spatial extent of the current is determined by the depth to which the incident wave propagates. This, in turn, is determined by the response of the lower pulsar magnetosphere to an incident broad-band electromagnetic pulse. The problem then turns to the analysis of electromagnetic waves incident on a bounded plasma or wave resonator region (Lysak, 1993; Thompson & Lysak, 1995).

Figure 6 illustrates the electromagnetic waves associated with traveling waves on a transmission line. Depending on how the transmission line intersects the conducting surface or lower magnetosphere, either **E** or **B** may be perpendicular to the plane of incidence. In either case, the reflected electric field is 180° out of phase with the incident electric field. [The change in polarity of the voltage pulses in Figure 5 means that the electric field also reverses polarity].

By convention, the relativistic electrons flow in a direction opposite to that of the high-γ current, and the radiation is beamed in the direction of electron propagation. This direction remains unchanged at the plane of

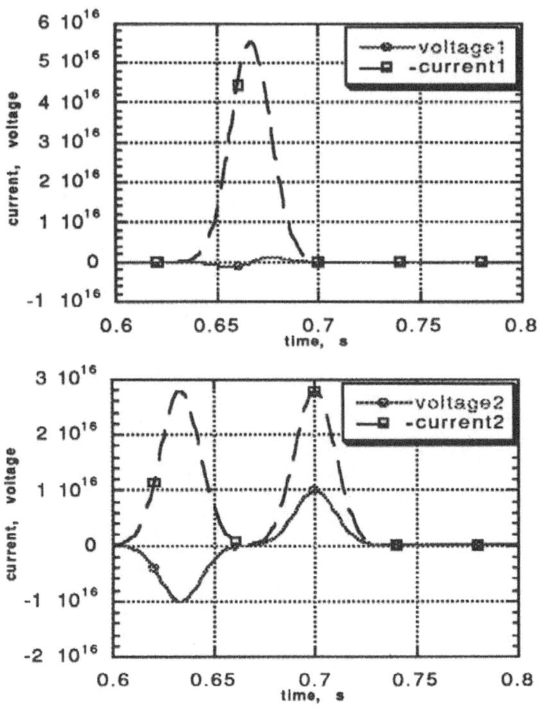

Fig. 5. Magnified current and voltage pulses from figure 4 at (top) 3×10^5m, (bottom) 10×10^6m.

incidence but the field polarization changes by 180°. This is observation is consistent with D and S_t class pulsars that show either a double or single pulse, but a polarization change across the entire pulse profile.

The property of total reflection has been used to produce circularly polarized waves. If the incident wave is linearly polarized in a direction which is neither parallel to nor perpendicular to the plane of incidence, the reflected wave is, in general, elliptically polarized. The degree of circular/elliptical polarization in pulsars can be 10–30% of the total intensity in pulsars. It is signficant that the circular component changes polarity at the maximum of a S_t pulse, characteristic of an oblique reflection off a conducting surface.

The reflection coefficient R for the field is:

$$R \approx \frac{\cos\theta - \frac{1}{2}(1+i)\,\gamma_\omega}{\cos\theta + \frac{1}{2}(1+i)\,\gamma_\omega} \tag{4}$$

where $\gamma_\omega = \sqrt{2\omega\varepsilon_1\mu_2/\mu_1\sigma_2}$.

At the interface $E = (1+R)$ and the properties of the emitted radiation are frequency dependent. Any pulse component at frequency ω prop-

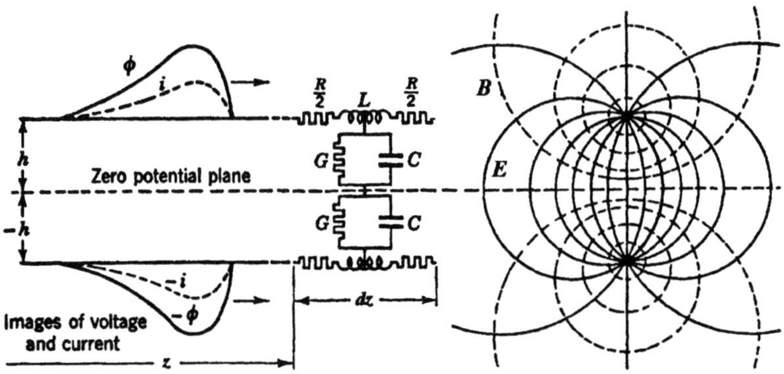

Fig. 6. Traveling waves $\phi(z,t)$ and $i(z,t)$ on a transmission line and associated electromagnetic fields **E** and **B**. The parameters L, C, R, and G are the self-inductance, capacitance, resistance, and transverse conductance, per unit length, respectively. The electromagnetic fields shown are those for a two-filament transmission line (intense field-aligned sheet currents generally undergo filamentation into current strands).

agates with spatial dependency $\exp(-z/\lambda_E)$ to the *collisionless skin depth*, $\lambda_E = c/\sqrt{\omega_{pe}^2 - \omega^2}$, where c is the speed of light and ω_{pe} is the angular plasma frequency. Therefore, in order to calculate the collisionless skin depth, some estimate must be made regarding the lower magnetosphere density and distribution.

Regardless of density profile, if $\omega >> \omega_{pe}$, λ_E is imaginary and the wave propagates freely. Hence, high frequencies propagate freely until a reflection condition $\omega = \omega_{pe}$ is encountered. The radiating currents are thin surface currents.

For frequencies $\omega << \omega_{pe}$, $\lambda_E \approx c/\omega_{pe}$ and the current may be spatially distributed over great distances within the magnetosphere. Table II lists collisionless skin depth versus plasma density for a number of cosmic plasmas.

For the readers convenience, we list some electromagnetic frequency/wavelength bands in Table III. Comparison of λ_E to λ shows that the extent of

TABLE II
Collisionless Skin Depths Versus Plasma Density

Plasma	Density, cm^{-3}	λ_E
stellar wind	0.1	16.8 km
current sheet (Earth)	1	5.3 km
magnetosheath	5	2.4 km
solar corona	10^7	1.68 m
solar surface	10^{15}	0.168 mm
solar center	10^{26}	0.53 nm

the emission region may be extreme depending on frequency of observation.

TABLE III
Electromagnetic Frequency/Wavelength Bands

Band	f-lower	f-upper	λ-lower	λ-upper
MF	300 kHz	3 MHz	100 m	1 km
HF	3 MHz	30 MHz	10 m	100 m
VHF	30 MHz	300 MHz	1 m	10 m
UHF	300 MHz	3 GHz	10 cm	1 m
SHF	3 GHz	30 GHz	1 cm	10 cm
EHF	30 MHz	300 GHz	1 mm	1 cm
Submillimeter	300 GHz	3 THz	100 μm	1 mm
Infrared	3 THz	430 THz	700 nm	100 μm
Visible	430 THz	750 THz	400 nm	700 nm
Ultraviolet	750 THz	30 PHz	10 nm	400 nm
X Ray	30 PHz	3 EHz	100 pm	10 nm
Gamma Ray	3 EHz			100 pm

4.2 Simulation of Pulses in a Magnetospheric transmission Line

We choose to simulate the pulse carrying magnetosphere with a three-dimensional particle-in-cell model. This approach allows us to:

- Insert dipole magnetic fields and double layer electric fields along the line.

– Investigate the pulse characteristics over a wide range of frequencies. For example PSR 0531+21 (Crab) is observed from radio frequencies (430 MHz) to gamma ray energies (greater than 50 MeV).

– Include surface and line curvature and associated field gradients.

– Include nonuniform plasma densities.

– View directly the relativistic currents in the high-curvature, synchrotron emitting regions.

– Investigate glitches when high voltages cause particle leakages across the line.

Figure 7 illustrates the geometry at hand. Field-aligned current pulses propagate at high latitude along the dipole magnetic field. Two regions of high-density plasma terminate this magnetospheric transmission line: plasma near the pulsar surface and plasma at the disk. The transmission line may also contain another active element, a double layer, whose field orientation is also depicted in Figure 7. This figure is axially symmetric about the magnetic axis so that the complete figure of revolution is a torus and the relativistic current pulses propagate within a toroid waveguide. The emission region is a cone of radiation centered on the magnetic axis. No other constraints are placed upon the orientation or rotation of the model; the observer need only be looking in the direction of the instantaneous velocity of the relativistic electrons.

The pulsar magnetosphere transmission line is modelled by two current conducting boundaries that represent the inward and outward field-aligned currents. These boundaries should generally follow a dipole magnetic field profile but, in the simulation model are simply taken to be coaxial. Thus the waveguide geometry is cylindrical in the (z,r,θ) coordinate frame. The line is terminated by two conductors: the pulsar surface and disk. In this scoping study both conductivities are set to $\sigma = \infty$. In reality, cosmic plasmas have finite conductivities that are frequency dependent.

Again, a voltage pulse is launched from the pulsar surface along a line of arbitrary length 100. [Either disk or surface launching of waves is permissible in the simulation]. To measure the electrical constituitive parameters within the line, we place clusters of probes along the transmission line.

Figure 8 shows contours of the propagating radial E_r field at arbitrary computer times 200 (traveling leftwards towards the surface) and 280 (propagating rightwards after reflecting off the surface). Because of the conducting boundary, the polarity of E_r has reversed 180° while the polarization of the azimuthal magnetic field B_θ remains the same (Figure 9).

The simulation analog of Figure 5, the time histories of current and voltage probes placed within the high curvature region, are shown in Figure 10. We

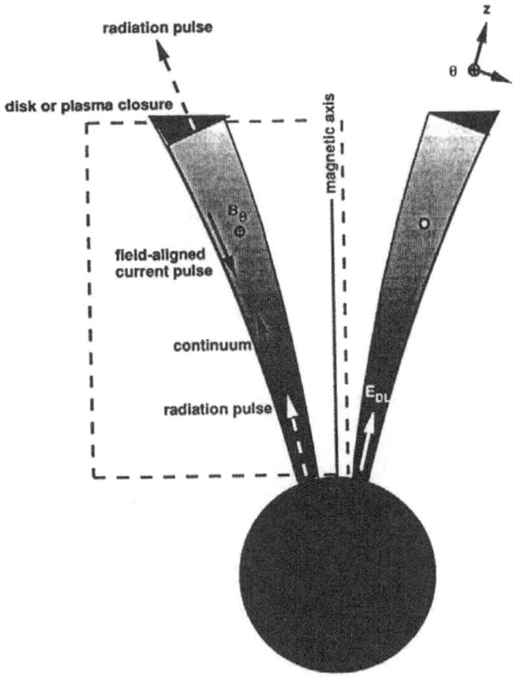

Fig. 7. Simulation geometry. Relativistic current pulses propagate along the dipole magnetic field lines at high latitude and are reflected off high–density plasma near the pulsar surface and magnetosphere disk. The disk may be either located at the equator (long transmission line case) or, as in this figure, at a higher latitude (short transmission line case). The figure is axially symmetric about the magnetic axis and forms a toroidal waveguide. Three emission regions are shown: near the surface, at the disk, and a continuum along the entire transmission line. The dashed–box outlines the simulation region.

again measure either a single or double pulse profile. Figures 8, 9, and 10 pertain to a relativistic pulse traveling on a transmission line in the absence of plasma. When plasma is allowed to flow between the current boundaries a number of important modifications are observed in the simulated waveforms.

As the electric field propagates along the line (e.g., Figure 8), electrons accelerate away from the boundary in the $-E_r$ direction. However, because of the intense B_θ field associated with the current pulse (Figure 7), they are constrained to flow along the boundary in $\mathbf{E} \times \mathbf{B}$ drift rather than across the line.

When scaled to magnetosphere transmission line dimensions, the entire 10^8 m line acts a continuous source of of low-level radiation and noise. [All pulsars show background radiation, *i.e.*, noise between the pulses with amplitudes between several millijansky to a few hundreds of mjy. For example,

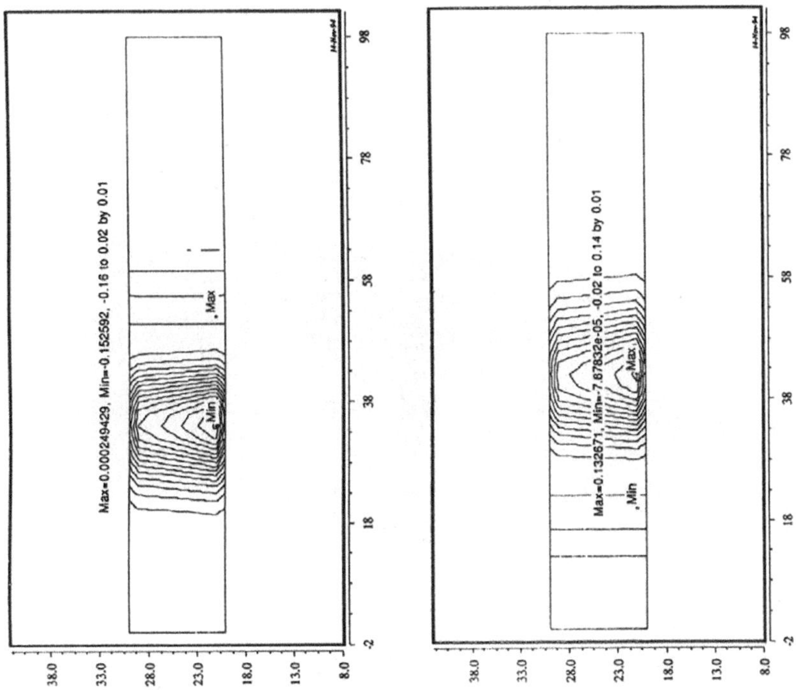

Fig. 8. Contours of the propagating radial E_r field. (Left) At computer time 200 (traveling downwards towards the surface). (Right) At time 280 (propagating upwards after reflecting off the surface).

the PSR0950+08 never 'turns off' as might be expected].

In the vicinity of the surface the B_θ magnetic insulation effect is lost and electrons accelerate across the line. During a 20 ms pulse duration, highly relativistic electrons can propagate as far as 6000 km and can modulate and slice the radiating current. Figure 11 depicts the electron flow in the scaled simulation. This flow is short-lived, lasting only as long as the negative potential is at the conductor surface. Figure 12 shows the associated current pulse. In contrast to the idealized cases shown in Figures 5b and 10b, the current in plasma is noisy and shows loss structure.

5 Pulsar Transmission Line Experiment

The pulsar magnetosphere was modelled with a 1 m long, pulse-power, coaxial transmission line. Electric energy in the form of a 1 ns, 2 MV/cm field was switched into the line. The geometry of the line was such that both ends, input and load, were terminated by conducting plates. The relative

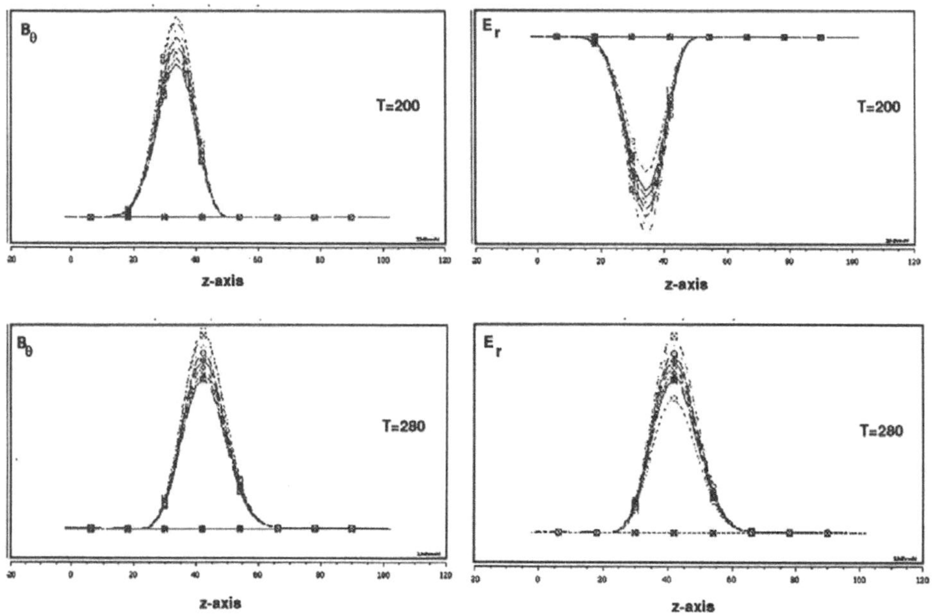

Fig. 9. (Top) Slices of B_θ and E_r at time 200. (Bottom) Slices of B_θ and E_r at time 280. The abscissa is in simulation spatial units.

permittivity of the dielectric medium was $\varepsilon = 1.998$. The pulse velocity is $v = c/\sqrt{\varepsilon} = 2.12 \times 10^8$ m/s.

The observed periodicity was 11.3 ns from voltage and current probes placed 20 cm from the input, as shown in Figure 13. For comparison, $i(t)$ and $v(t)$ from Eq.(1) are plotted on top of the data. While the calculated waveforms follow closely the observed waveforms, a number of differences

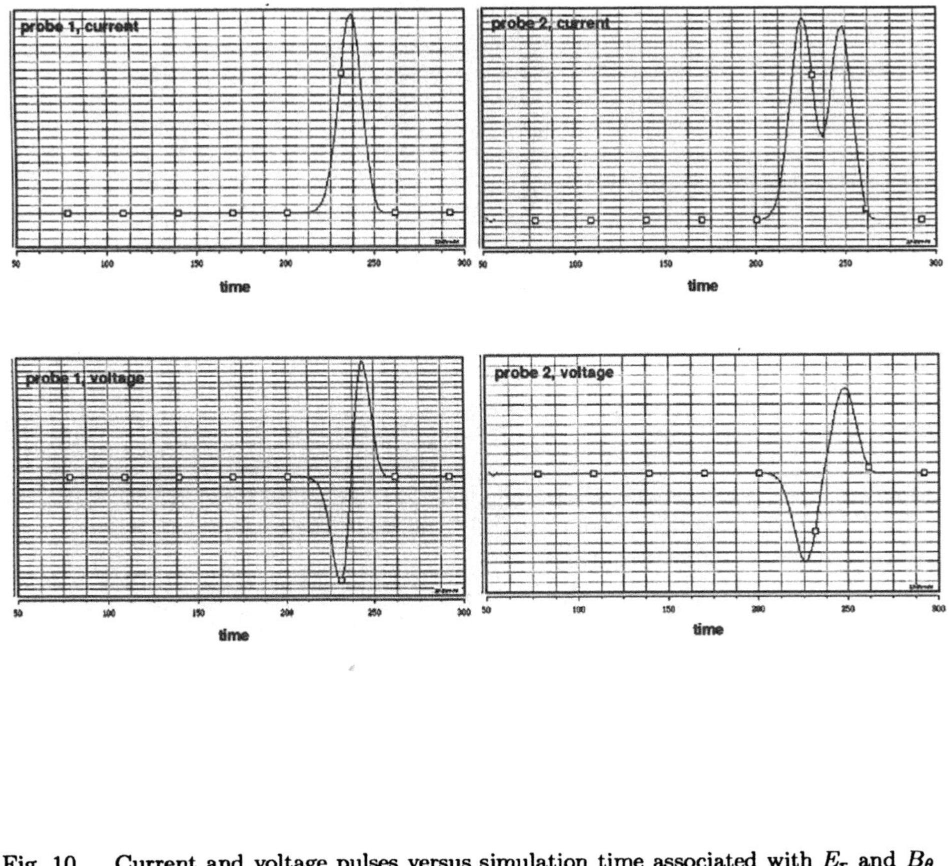

Fig. 10. Current and voltage pulses versus simulation time associated with E_r and B_θ of Figure 9. (Top) Current pulses at a probe located in the simulation region at (z, r, θ) locations (2,22,0) and (12,22,0), respectively. (Bottom) Voltage pulses at the same probe locations.

were noted. As mentioned, the entire line is a source of background radiation. Additionally, microstructure, noise, and low frequency drifts were seen on the observed data.

The experiment was instrumented with a number of x-ray detectors (XRD's) and the current waveform verifies relativistic transmission line currents as

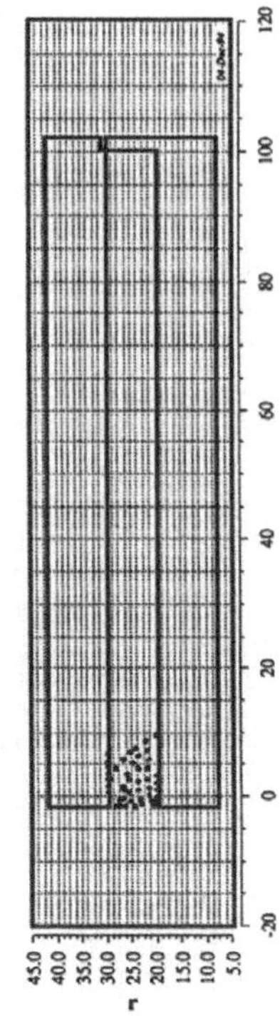

Fig. 11. Electrons flowing across the boundary near the pulsar surface at simulation time 240. The predominant flow is from the left to right boundary.

the source of the radiation observed. The voltage waveform verifies the linear polarization properties of the radiation and its 180° reversal over a pulse. Surprisingly, because of the relative shortness of the line, the detectors picked up radiation signals from both the input conducting plate and the short itself. While no probe was present at the short, the current waveform can be reconstructed and Figure 14 shows the current pulse-interpulse that might be expected from the 10^8 m magnetospheric transmission line, if the surface and disk terminating conductors are nearly parallel.

Because the currents at the surface/disk share the same polarity, if they are parallel, the radiation is in the same direction and both 'main' pulse and 'interpulse' components are observed. This phenomena can only occur for shorter lines; in a curved magnetosphere transmission line, the surface and disk currents may no longer be parallel and the high-γ radiation patterns will point in different directions.

It is signficant that the highly stressed line produced glitches when the magnetic insulation failed and electrical breakdown occurred. In all cases the glitches at high in the magnetosphere produce a line shortening and a

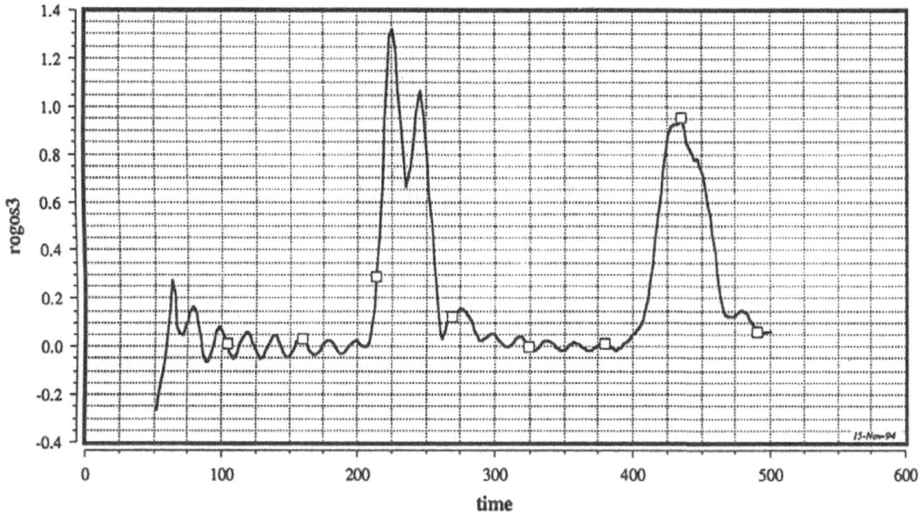

Fig. 12. Current pulses modified by the flow of electrons across the boundaries near the surface (Figure 11).

concomitant shortening in periodicity. This experimental observation is in agreement with pulsar observations.

6 Conclusions

VLA observations of the halo candidate pulsars PSR0809+74, PSR0942-13, and PSR0950+08 were made at 327.5 MHz. Each of these pulsars was observed in B Array (beamwidth of 15") in Phased Array Spectral Line mode for one hour. A very clean map of PSR0809+74 shows no evidence of a halo; maps of the other two pulsars are noiser and are the in process of further analysis.

On the theoretical side, our research involves the modeling of the global pulsar system, the pulsar magnetosphere, and the synchrotron radiation region, respectively. This paper reports only our magnetosphere transmission line simulations. We hope to publish the large scale and small scale simulations elsewhere. The models utilize transmission line equations and full three-dimensional, fully electromagnetic, particle-in-cell simulations to

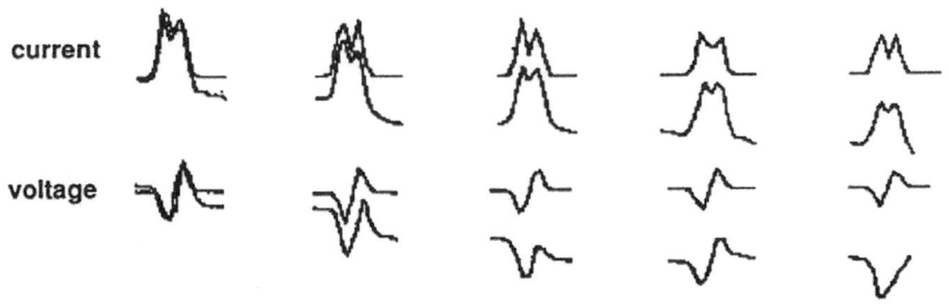

Fig. 13. Individual experimental current and voltage waveforms. The periodicity between waveforms is 11.3 ns. (Top) Current pulses. (Bottom) Voltage pulses. The calculated waveforms (upper waveforms) lie on a horizontal baseline while the measured waveforms lie on a downwards low-frequency component. The full-width-half-maximum duration of the pulses is about 5.3 ns.

determine pulse characteristics.

The simulated model produced a train of 10^{13-16} ampere pulses with periodicity 0.65 s. These $\gamma \sim 10^{7-10}$ currents are thought to be the source of the synchrotron radiation observed. The polarization properties of the model are consistent with observation: a high degree of linear polarization that changes 180 degrees over the pulse width, and circular or elliptical components. Whether or not the radiated pulse has a single or double pulse profile depends on the extent of the radiation region at the pulsar surface. If the radiating current is confined to a shallow region instead of an extended region, a single pulse profile will replace the double pulse. As shown below, this effect is frequency dependent: lower frequencies are associated with fields that are extended in the emission region and are more likely to show pulse structure. This is consistent with (Rankin) S_d class pulsars (Hankins & Rickett 1986).

The simulation results were verified with a high-voltage, transmission line

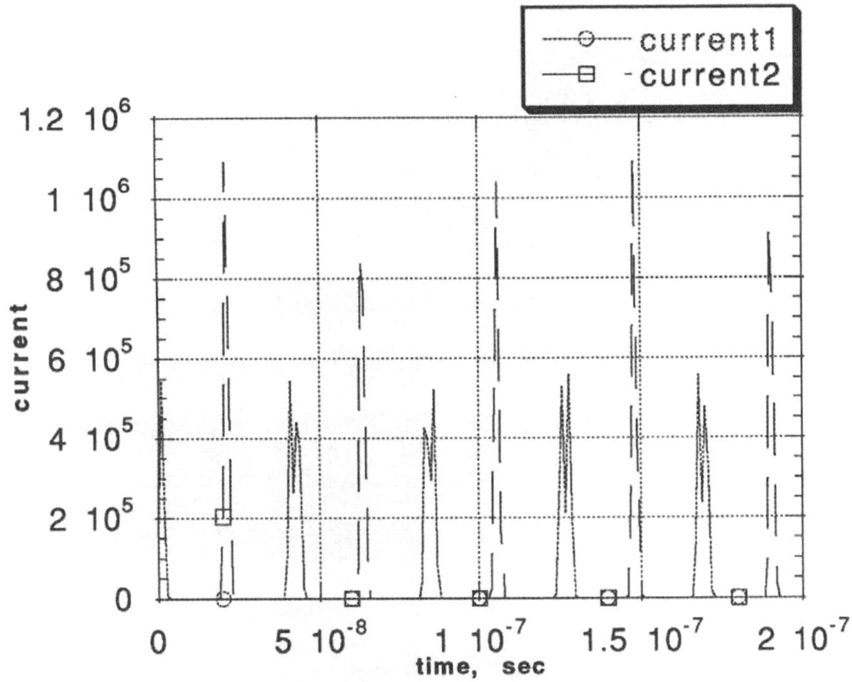

Fig. 14. Pulse and interpulses from the surface and disk locations.

experiment. Because of losses in the dielectric media and in synchrotron emission, the periodicity of the propagating pulses increases. However, the experiment dramatically showed that glitches, the flow of electron flux across the magnetosphere, can shorten the line and concomitantly the period.

The fractional frequency stability scaling versus measurement interval up to about 3×10^7 s ($\sim$ 1 year) for pulsars is nearly identical to that for trapped–ion clocks. This supports the pulsar surface–magnetosphere relativistic double layer model; itself a trapped-ion mechanism.

Both simulation and experiment suggest that micropulses and subpulses are produced particle–wave interactions in nonuniform plasma irradiated by an electromagnetic wave. This effect is produced when the magnetically insulated voltage pulse reaches the pulsar surface. Because of the curvature, magnetic insulation is lost, and plasma flows across this region. This tends to produce a resonating or modulation component on the the proper current pulse. We are currently investigating this phenomena with simulations of the emission region. If this effect is verified, then the source of the radiation energy may not be contained within the pulsar, but may instead derive from either the pulsar's interaction with its environment or by energy delivered by an external circuit (Alfvén 1981)[2]. This hypothesis is consistent with both the long-term memory effect (provided either by the transmission line parameters or the double layer) of the time-averaged pulse and the occurance of nulling, when no subpulses are observed. As noted earlier, our results support the 'planetary magnetosphere' view (Michel 1982) where the extent of the magnetosphere, not emission points on a rotating surface, determines the pulsar emission. We are currently investigating this phenomena with simulations of the global environment.

7 Acknowledgements

One of the authors (KRH) thanks T. H. Hankins for advice on all past and ongoing parts of this project.

The other author (ALP) thanks Galen R. Gisler and Edward R. Harrison for suggesting the experimental approach to modeling pulsars.

The National Radio Astronomy Observatory is operated by Associated Universities, Inc., under cooperative agreement with the National Science Foundation.

References

Alfvén, H. : 1981 *Cosmic Plasma*, D. Reidel, Dordrecht, Holland

Backer, D. C. : 1988 "Pulsars", in *Galactic and Extragalactic Radio Astronomy*, G. L. Verschuur and K. I. Kellerman, eds., Springer–Verlag, New York, Chap. 11

Bietenholz, M. F., Frail, D. A., and Hankins, T. H. : 1991, "Does the Vela Pulsar have "Wisps"?", *ApJ*, **Vol. no. 376**, pp. L41–L44

Blandford, R. D., Ostriker, J. P., Pacini, F., and Rees, M. J. : 1973, "Radio Haloes around Old Pulsars — Ghost Supernova Remnants", *Astron. & Astrophys.*, **Vol. no. 23**, pp. 145–146

Burnell, J. B. : 1983 "The Discovery of Pulsars", in *Serendipitous Discoveries in Radio Astronomy*, K. Kellerman and B. Sheets, eds., National Radio Astronomy Observatory, Green Bank, West Virginia, pp. 160–170

Cordes, J. M., Romani, R. W., and Lundgren, S. C. : 1993 "The Guitar nebula: a bow shock from a slow–spin, high-velocity neutron star", *Nature*, **Vol. no. 362**, pp. 133–135

Frail, D. A., and Kulkarni, S.R. : 1991 "Unusual interaction of the high–velocity pulsar PSR1757−24 with the supernova remnant G5.4−1.2", *Nature*, **Vol. no. 352**, pp. 785–787

Glushak, A. P., Pynzar', A. V., and Udal'tsov, V. A. : 1981 "Pulsar Radio Haloes", *Sov. Astron.*, **Vol. no. 25(2)**, pp. 182–187

Gold, T. : 1968 *Nature*, **Vol. no. 218**, p. 731

Gopal–Krishna : 1978, "A ghost supernova remnant around PSR 0950+08", *MNRAS*, **Vol. no. 185**, pp. 521–525

Hester, J. J. : 1994 private communication

Jones, D. S. : 1964, *The Theory of Electromagnetism*, Macmillan, New York

Kennel, C. F., Fujimura, F. S., and Pellat, R. : 1979 "Pulsar Magnetospheres", *Space Science Reviews*, **Vol. no. 24**, pp. 407–436

[2] A double layer, set up by an electrostatic instability in a particle beam and fed from an external plasma source, seems to offer the long term P stability demanded by observation.

Kulkarni, S. R., and Hester, J. J. : 1988 "Discovery of a nebula around PSR1957+20", *Nature*, Vol. no. **335**, pp. 801–803
Lysak, R. L.: 1993 "Generalized Model of the Ionospheric Resonant Cavity", in *Auroral Plasma Dynamics, Geophysical Monograph Series*, Vol. no. **80**, R. L. Lysak, ed. (American Geophysical Union, Washington, DC), p.121
Michel, F. C. and Dessler, A. J. : 1981 *Astrophys. J.*, Vol. no. **251**, pp. 654–664
Michel, F. C. : 1982 "Theory of the Pulsar Magnetospheres", *Rev. Mod. Phys.*, Vol. no. **54**, pp. 1–66
Naik, P. K., and Kulkarni, V. H. : 1994 "Emission of Electromagntic Waves in Magnetized Plasma in the Presence of Field Aligned Currents", *Astrophys. Space Sci.*, Vol. no. **218**, pp. 13–21
Peratt, A. L. : 1992a, *Physics of the Plasma Universe*, Springer-Verlag, New York. P.58
Peratt, A. L. : 1992b, *Ibid.*, Appendix A
Peratt, A. L. : 1992c, *Ibid.*, P.314
Peratt, A. L. : 1992d, *Ibid.*, Chap. 5
Thompson, B. & Lysak, R. L. : 1995 "The Effect of Static Electric Fields in the Alfvén Wave Model of Auroral Acceleration", *unpublished*
Verschuur, G. L. and Kellermann, K. I. : 1988, *Galactic and Extragalactic Radio Astronomy*, Springer-Verlag, New York.
Weiler, K. W., Goss, W. M., and Schwarz, U. J. : 1974, "A Search for Supernova Remnants in the Vicinity of Pulsars", *Astron. & Astrophys.*, Vol. no. **35**, pp. 473–477

TITIUS-BODE SERIES
GALAXY GROUP RED SHIFT DIFFERENCES
CALCULATED FROM ROOTS OF THE BESSEL EQUATION

DANIEL R. WELLS
Department of Physics
University of Miami
Coral Gables, Florida, USA
wells@phyvax.ir.miami.edu

Abstract. It has been shown in detail by Wells (Wells, 1986) that the Bode numbers and measured velocity ratios of the planets are accurately predicted by the eigenvalues of the Euler-Lagrange equations resulting from the variation of the free energy of the generic plasma that formed the Sun and planets. This theory is reviewed and extended to show that the equations make accurate predictions for all the major planets out to and including Pluto. The semimajor axes and velocity ratios of Pluto and Neptune are predicted exactly. The Bode numbers are shown in Table I to correspond to the roots of the first-order Bessel functions. The extrema of the roots of the zeroth-order Bessel function predict the ratios of the measured planetary velocities almost without error for the outer planets. Both sets of roots correspond to the same eigenvalue solution of the forcefree equation. The eigenvalues are set by the initial energy input to the plasma nebula. Both the Titius-Bode series and Kepler's harmonic law are predicted by the "relaxed state solution" of the free-energy equation for the generic plasma that formed the Sun and planets. Newton's law of gravitation is not used in the calculations. The solution makes exact predictions for the outer planets where the Titius-Bode series fails completely.

The work of Arp (Arp, 1985) adds to the growing body of observable evidence of objects which appear to be attached to galaxies or galaxy systems but display red shifts, sometimes quite large, differing from those of the associated galaxies. Adding to the mystery and confusion are a series of objects that have quantized red shifts.

It is widely recognized that the history of these objects involves extremely high energy processes. The solutions of the equations of the relaxed state of the resulting high energy plasmas is discussed and it is shown that the predictions of red shift frequencies are quantized and agree numerically with many of the quantized shifts reported by Arp and Sulentic.

Key words: Titius-Bode, plasma

1 Introduction

For more than 30 years it has been understood that the lowest energy state of a finite volume of plasma subject to the appropriate boundary condition corresponds to a forcefree collinear flow structure (Chandrasekhar, 1958). This early work was applied to laboratory systems by Wells (Wells, 1964) and later by Taylor (Taylor, 1976) and Siemin (Siemin, 1985). The author considered the problem of applying these methods to the morphology of the solar system (Wells, 1986). He was able to predict the Bode numbers and show that they corresponded to the roots of the Bessel function solutions of

Astrophysics and Space Science **227**: 255–263, 1995.

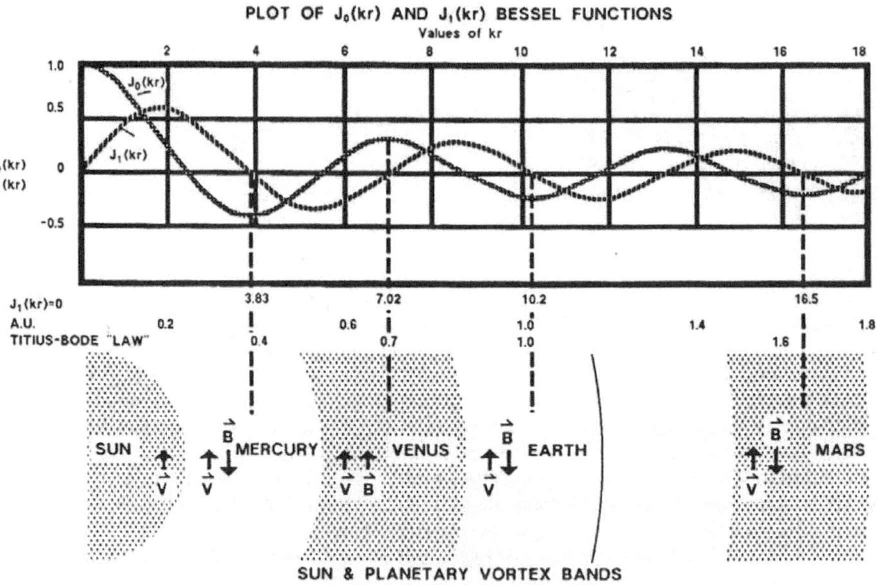

Fig. 1. Agreement of Bessel functions with Titius-Bode and measured planetary distances. A plot of the J_1 and J_n Bessel functions is scaled here to the geometry of the solar system The values here are also summarized in Table I Corotational vortex bands (planets) must alternate with contrarotational ones. In the first case, orbital motion (vector **B**) are aligned; in the latter, they are opposed. The missing planet just beyond Earth (indicated by a curved line) would have to have been contrarotational. Note that the orbital motion of all planets, and also the rotational motion of the Sun, are in the same direction, as they are in fact.

the appropriate forcefree equation. We outline the theory in Section II. In Section III the theory is extended to a prediction of the semimajor axes and velocity ratios of the outer planets not given in the earlier papers. These same methods are applied to the formation of groups of galaxies.

2 How was the Morphology of the Solar System Set?

Assume a major disturbance in the primordial gas of the galaxy. The compressed gas forms a cylindrical volume of plasma which is moving through the background plasma and rotating with a finite angular velocity (Figure 2). Assume that there is a finite magnetic field present during the formation of the plasmoid. As this mass of plasma propagates through surrounding gas, it loses energy by accelerating the surrounding plasma. The cylinder will

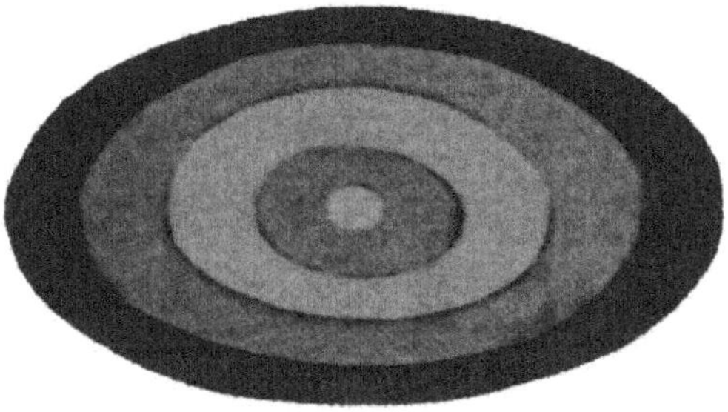

Fig. 2. Planetary vortex bands. At this stage the concentric vortex cylinders have collapsed into a disc of concentric bands. The matter in each band will subsequently form a ring and then a planet

lose energy and settle down to a minimum-energy "relaxed" state, forcefree collinear cylindrical structure. It is shown in detail elsewhere (Wells, 1987; Wells, 1986; Wells, 1988; Wells, 1964) that the resulting "field equations" for the flow are given by

$$\nabla \times \mathbf{B} = k\mathbf{B} \tag{1}$$

$$\mathbf{v} = \pm \left[\frac{\gamma - 1}{\gamma - 2}\right] \frac{\mathbf{B}}{(\mu_0 \rho)^{\frac{1}{2}}} \tag{2}$$

$\rho \stackrel{\text{def}}{=}$ fluid density

$\mathbf{B} \stackrel{\text{def}}{=}$ Magnetic Induction Field

$\mathbf{v} \stackrel{\text{def}}{=}$ velocity of the center of mass of a fluid element

$\gamma \stackrel{\text{def}}{=}$ ratio of specific heats of the gas

A **pseudoplane** solution to the forcefree equation (1) is given by Bjorgum and Godal and is shown to be (Bjorgum, 1993; Chandrasekhar, 1958)

$$\begin{aligned}
B_r &= -k^2 \tilde{a} \frac{J_1(kr)}{kr} sin\theta \\
B_\theta &= k^2 \tilde{a} \left\{ \frac{J_1(kr)}{kr} - J_0(kr) \right\} \cos\theta \\
B_z &= k^2 \tilde{a} J_1(kr) \cos\theta \\
B^2 &= \tilde{a}^2 k^4 \left\{ \left(\frac{J_1}{kr}\right)^2 - \frac{2J_0 J_1}{kr} \cos^2\theta + (J_0^2 + J_1^1) \cos^2\theta \right\}
\end{aligned}$$

where B_r, B_θ, and B_z are the magnetic induction components in the striated rings of the gas cylinder, and

$$\begin{aligned} \tilde{a} &= l\theta + k_2 z \\ k^2 &= k_1^2 + k_2^2 \end{aligned}$$

where ℓ, k_1 and k_2 are constants supplied by the boundary conditions (Bjorgum, 1993). Figure 1 shows a plot of J_1 and J_0 with the functions scaled to the geometry of the solar system. We observe that for $J_1 = 0$

$$\begin{aligned} B^2 &\sim J_0^2 \\ B_\theta &\sim J_0. \end{aligned}$$

This maximizes the magnetic and kinetic energy at the origin. In the cylindrical structure formed by the supernova explosion, the first root corresponds to the structure of the star at the center of the hypothetical solar system, the second root corresponds to a ring of gas just outside the star. The corresponding flow velocities in the rings is given by equation (2). The signs reverse for every other ring (corotational and contrarotational) so that the azimuthal velocities are all prograde. The azimuthal velocity of the gas in each ring has a direct relationship to the velocities of the planets as they exist today. An examination of Table I shows that the Bode numbers of the Planets are predicted by the roots of the equations describing the "relaxed state" of the primordial gas. Comparison of the measured velocity ratios with the ratios of the extrema of $J_0(kr)$, show very close agreement.

3 The Outer Planets

For the outer planets, the Bode series fails completely for Neptune and Pluto but the plasma solutions, the Bessel function roots, give exact predictions. The author's early papers on this subject took the Bessel function solutions out to and including the planet Jupiter (Wells, 1986; Wells, 1988). In a letter to the editor of 21st Century Science and Technology (Allen, 1986), William D. Allen observed that if the asymptotic expansions of $J_1(kr)$ and $J_0(kr)$ were carried out, the theory could be checked all the way out to and including Pluto. Beyond the 50th root, $J_1(kv)$ increases by approximately 3.14 between roots. $J_0(kr)$ is given by

$$J_0(kr) \simeq \left(\frac{2}{(\pi(kr)}\right)^{\frac{1}{2}}$$

The corresponding normalized (kr) values are shown in Table I. The predicted ratios of the successive peak velocities of the gas in the rings checks the

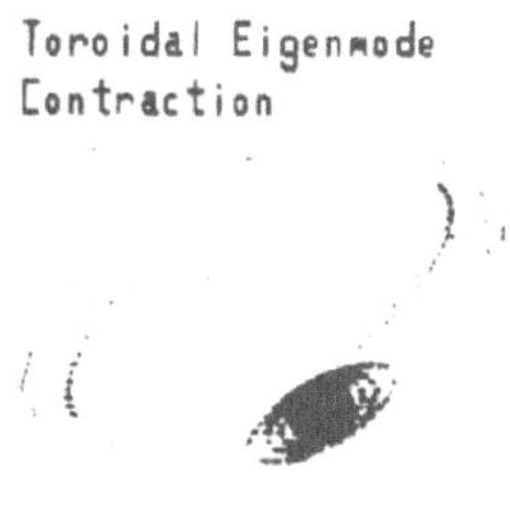

Fig. 3. Toroidal eigenmode contractions.

measured velocity ratios of the inner planets within a few percent. The velocity ratios for Uranus, Neptune, and to Pluto are exact. The relaxed state of the generic plasma predicts both the Bode number series and Kepler's harmonic law,

$$p^2 = a^3$$

where

$$\begin{aligned} p &\stackrel{\text{def}}{=} \text{period of the planet} \\ a &= \text{average radius of planet} \end{aligned}$$

It is suggested that the rings of gas in the planet structure "roll up" azimuthally to form balls of gas that eventually evolve into the planets (Figure 3) (Wells, 1986). The roll up of vortex rings to form balls of gas is a well known phenomena that has been observed in laboratory experiments (Wells, 1986). This is illustrated in Figure 3.

The theory developed in this paper not only accurately predicts the semimajor axes of the nine major planets and their velocity ratios without invoking Newton's law of gravitation (inverse square law), but also hints at other phenomena associated with the morphology of the system. These may be listed as follows:

1. A planet is predicted at 1.3 A.U. No such planet exists today. It is suggested that the missing planet suffered a catastrophe either in the birthing process or at a later time and the residue is our moon.

2. There are eleven rings possible between Mars and Jupiter. The Asteroid belt is located in this region. Table I indicates that the average predicted semimajor axes of these rings corresponds to the average semimajor axis of the principle asteroids to an accuracy of -8.1%.

TABLE I
Bode Numbers

Planet	Root n	Bode numbers	Normalized (kr_n) values	Measured AU	% Error	Extremum $J_0(kr_n)$	Velocity Ratio	Extremum Ratio	% Error
Mercury	1	0.4	0.38	0.39	-2.6	0.4028			
							1.37	1.34	-2.2
Venus	2	0.7	0.69	0.72	-4.2	0.3001			
							1.18	1.20	+1.7
Earth	3	1.0	1.0	1.0	—	0.2497			
							1.24	1.27	+2.4
Mars	5	1.6	1.62	1.52	-6.6	0.1965			
							1.86	1.76	-5.4
Jupiter	16	5.2	5.02	5.20	-3.5	0.1117			
							1.34	1.39	+3.7
Saturn	31	10.2	9.62	9.54	+0.8	0.0805			
							1.42	1.41	-0.7
Uranus	62	19.6	19.17	19.2	-0.16	0.0571			
							1.25	1.25	0.0
Neptune	97	38.8	29.95	30.1	-0.49	0.0457			
							1.15	1.15	0.0
Pluto	128	77.2	39.49	39.5	-0.03	0.0398			
Asteroids	11		3.08 (Avg)	3.35 (Avg)	-8.1				

3. The errors in the predicted semimajor axes of Mars and Jupiter and the velocity ratio of Mars to Jupiter is large compared to the values for the outer planets. This may be caused by a disruption due to whatever mechanism violently formed the asteroids.

4. Both the predicted semimajor axes and velocity ratios are almost exact for the outer planets (Uranus, Neptune and Pluto).

5. The exact match for the predicted semimajor axes of Uranus, The exact match for the predicted semimajor axes of Uranus, Neptune and Pluto is especially interesting. For a fixed k (see equation 1), i.e. a fixed eigenvalue for the forcefree equation, the spacing of the roots as a function of radius is fixed. If one now normalizes the system so that the third root has an argument equal to 1 A.U. (the semimajor axes of earth) then there is a high order root (62 for Uranus, 97 for Neptune, 128 for Pluto), corresponding **exactly** to the semimajor axes of Uranus, Neptune and Pluto. The extrema of $J_0(kr)$ also accurately predict the measured velocity ratios of these planets.

6. It is suggested that the rings corresponding to the roots that do not match any existing planets were either pulled into the planetary rings or formed gas balls that were not stable. The gas was then lost to the system or later formed planetesimals that account for the high level of meteoritic activity in the early solar system. These same methods can now be applied to the formation of groups of galaxies.

Careful measurements of the red shifts of major galaxies along with the shifts of their companion satellite galaxies (Arp, 1985), indicate that the companion galaxies are red shifted by different amounts than the principal nearby galaxies. If, again, the concept of galaxy group formation by the production of forcefree minimum energy state plasma structures is invoked, it is found that the measured red shifts are duplicated, i.e. if the Z component of the velocity for succeeding roots of the Bessel equation describing the main Galaxy and the companion Galaxies is compared to the measured values for the same quantities, the agreement is excellent, (See Table II). The most interesting result, in this case, is that the red shifts are quantized in agreement with the quantized red shifts measured by the major observatories. The corresponding errors are very small, 5%.

4 Discussion

It has been demonstrated in other papers (Wells, 1989; Wells, 1992) that a "virtual plasma" theory of physical forces unifies all forces as "fluid" or "Magnus" forces generated by vortex structures (particles) in the virtual

Table II
Galaxy Group Red Shifts

Galaxy Group	Main Velocity (1)	Companion Velocity (2)	(1)+(2)	[(1)+(2)]/2	Extremum Ratio	% Error
NGC672						
A20	592	56	648	1.10	1.15	4.5
NGC1023						
UGC2080	763	311	1074	1.41	1.41	0
NGC1637						
NGC1507	639	171	810	1.27	1.27	0
NGC2146						
UGC3439	1108	578	1686	1.52	1.41	7
NGC2146						
UGC3384	1108	157	1265	1.14	1.15	0.009
NGC2146		-126		1.11		3
A130,00039	1108	-105	1234	1.09	1.15	5
NGC2336						
UGC3794	2386	449	2835	1.2	1.25	4
NGC2336						
UGC3834	2386	-161	2547	1.07	1.15	7
NGC2655						
UGC4646	1629	193	1822	1.2	1.25	4
NGC3507						
NGC3501	879	161	1040	1.18	1.15	2.5
NGC3507						
UGC6181	879	201	1080	1.23	1.25	1.6
NGC3507						
NGC3457	879	173	1052	1.2	1.25	4
NGC3655						
NGC3659	1384	184	1568	1.13	1.15	2
NGC5457						
NGC5474	379	44	423	1.12	1.15	3
NGC5457						
NGC5585	379	87	466	4.23	1.25	2
NGC5713						
UGC5740	1871	313	2184	1.17	1.15	2
NGC5713						
NGC5705	1871	134	2005	1.07	1.15	7
NGC5921						
UGC9830	1508	346	1854	1.23	1.25	2
NGC5005						
NGC5002	1000	142	1442	1.14	1.15	1
NGC5005						
NGC5014	1000	90	1090	1.09	1.15	6
NGC5951						
NGC5954	1860	218	2078	1.12	1.15	3
UGC3742						
UGC3743	3822	1257	5079	1.33	1.34	1
UGC3742						
UGC3752	3822	8081	4703	1.23	1.25	2

plasma gas. The theory generates gravitational fields that conform to the Schwartzshield field in the proper limits. It generates the electrostatic field in the atom, "explains" the form and action of the Schrödinger equation. The general form of the strong nuclear forces is also generated by the theory. An extension of the concept of planetary formation to this virtual plasma also predicts the mass spectrum for the leptons and hadrons. Finally, the quantized red shifts observed by Arp and Sulentic are also explained by the theory.

References

Allen, W. D.: 1986, "Letter to the editor", *21st Century Science and Technology* **September**

Arp, H. and J. W. Sulentic: 1985, "Accurate red shifts", *"Appl. Phys. J.* **Vol. 291**, p. 94

Bjorgum O. and T. Godal: 1952,"On Beltrami fields and flows – the case where Ω is constant in space", *Naturvitenskapelig Rekke* Yearbook, Univ. of Bergen, Norway, **N.13**

Chandrasekhar, S.: 1958, "On the equilibrium configuration of an incompressible fluid", *Proc. Nata. Acad. Sci U.S.* **Vol. 42**, pp. 273-279

Siemin, R. E.: 1985, "Review of the Los Alamos FRC experiments", *Los Alamos Nat. Lab Report* **LA-UR-85-935**

Taylor, J. B.: 1976, "Relaxation of toroidal plasma and generation of reverse magnetic fields", *Phys. Rev. Lett.* **Vol. 33**, pp. 1139-1143

Wells, D. R. and L. Hawkins: 1987, "Containment forces in low energy states of plasmoids", *J. Plasma Phys.* **Vol. 38**, pp. 263-274

Wells, D. R.: 1986, "Titius Bode and the helicity connection", *IEEE Trans. Plasma Sci.* **Vol. PS-14**, pp. 865-873

Wells, D. R.: 1988, "How the solar system was formed", *21st Cen. Sci. and Techn.* **July-Aug**

Wells, D. R.: 1964, "Axially symmetric forcefree plasmoids", *Phys. Fluids* **Vol. 7**, pp. 826-831

Wells, D. R.: 1989, "Quantization effects in the plasma universe", *IEEE Trans. Plasma Sci.* **Vol. 17**, pp. 270-281

Wells, D. R.: 1989, "Unifications of gravitational, electrical and strong forces by a virtual plasma theory", *IEEE Trans. Plasma Sci.* **Vol. 20**, pp. 939-943

SOLAR WIND-MAGNETOSPHERE INTERACTION AS SIMULATED BY A 3-D EM PARTICLE CODE

KEN-ICHI NISHIKAWA
Department of Physics and Astronomy, The University of Iowa, Iowa City, Iowa 52242, USA

TORSTEN NEUBERT
Space Physics Research Laboratory, The University of Michigan, Ann Arbor, Michigan 48109, USA

and

OSCAR BUNEMAN*
STAR Laboratory, Department of Electrical Engineering, Stanford University, Stanford, California 94305, USA

Abstract. We have studied the solar wind-magnetosphere interaction using a 3-D electromagnetic particle code. The results for an unmagnetized solar wind plasma streaming past a dipole magnetic field show the formation of a magnetopause and a magnetotail, the penetration of energetic particles into cusps and radiation belt and dawn-dusk asymmetries. The effects of interplanetary magnetic field (IMF) have been investigated in a similar way as done by MHD simulations. The simulation results with a southward IMF show the shrunk magnetosphere with great particle entry into the cusps and nightside magnetosphere. This is a signature of a magnetic reconnection at the dayside magnetopause. After a quasi-stable state is established with an unmagnetized solar wind we switched on a solar wind with an northward IMF. In this case the significant changes take place in the magnetotail. The waving motion was seen in the magnetotail and its length was shortened. This phenomena are consistent with the reconnections which occur at the high latitude magnetopause. In our simulations kinetic effects will determine the self-consistent anomalous resistivity in the magnetopause that causes reconnections.

Key words: Solar wind, Magnetosphere, IMF, Magnetic Reconnection

1 Introduction

The solar wind interaction with the Earth's magnetic field gives rise to a number of important and intriguing phenomena many of which are only partially understood. These include reconnection between the solar wind magnetic field at the dayside magnetopause and flux transfer events, the drag of the solar wind exerted on the magnetotail and associated instabilities at the magnetopause, plasma convection in the magnetosphere/ionosphere, and the generation of field-aligned current systems. Reviews of these and other phenomena can be found in (Haerendel and Paschmann, 1982; Elphic, 1987; Huang, 1987; Mauk and Zanetti, 1987). The range of physical processes

* Deceased January 24, 1993; R. Buneman *et al.* 1993.

Astrophysics and Space Science **227**: 265–276, 1995.

involved in the solar wind-magnetosphere system is quite overwhelming, and consequently a wide array of methods have been used to study these, ranging from detailed studies of select phenomena with the assumption of a specific field geometry or set of boundary conditions, to fully three dimensional simulations with less spatial resolution but self-consistent particle or field geometries and with no (little) influence of the boundary conditions.

The first 3-dimensional (3-D), global magnetohydrodynamic (MHD) simulations of the solar wind-magnetosphere system was reported by (Brecht *et al.* 1981; Leboeuf *et al.* 1981; Wu *et al.* 1981). Since then, global 3-D MHD simulations have been used with some success to study a range of processes including reconnection in the tail (Brecht *et al.*, 1982; Walker *et al.*, 1987), reconnection at the day-side magnetopause (Sato *et al.*, 1986), the dependence on the interplanetary magnetic field (IMF) orientation of magnetospheric convection and field-aligned currents (Ogino *et al.*, 1985, 1986, 1994; Ogino, 1986; Walker *et al.*, 1993), the solar wind-magnetosphere-ionosphere current-voltage relationship (Fedder and Lyon, 1987), the self-excitation of auroral arcs (Watanabe and Sato, 1988), and the reconnection voltage between the closed geomagnetic field and the IMF as a function of the IMF clock angle (Fedder *et al.*, 1991).

In MHD-codes the micro-processes are represented by statistical (macroscopic) constants such as diffusion coefficients, anomalous resistivity, viscosity, temperature, and the adiabatic constant. Recently, Winglee (1994) has developed magnetoplasma dynamics (MPD) including the surface current due to the ambipolar electric field set up by the different penetration between electrons and ions.

To include particle dynamics explicitly, a number of simulations have been performed using hybrid codes (fluid electrons, particle ions) and full particle codes in one and two dimensions. With such codes local problems have been investigated such as shocks formed in the solar wind as it encounters the magnetosheath (Quest, 1988; Burgess, 1989; Omidi *et al.*, 1990; Thomas *et al.*, 1990; Thomas and Winske, 1990; Winske *et al.*, 1990) and the current layer at the dayside magnetopause (Gary and Sgro, 1990; Berchem and Okuda, 1990; Okuda, 1991,1992,1993). Recently also global problems have been investigated in three dimensions with hybrid codes, namely the solar wind interaction with the dayside of Venus (Moore *et al.*, 1991) and the solar wind interaction with Mars and Venus (Brecht and Ferrante, 1991, 1993).

With the model presented here, we intend to take the full step, namely the global simulation of the solar wind interaction with a planetary magnetic field using a particle code which contains the complete particle dynamics. As will be shown below, the advantage is that the model contains the complete physics. The price to be paid is that some scaling of plasma parameters must be done at the present time.

2 A 3-D EM Particle Simulation Model

Our code is a successor to the TRISTAN code (Buneman *et al.*, 1980; Peratt *et al.*, 1980; Peratt, 1986). Its new features (Buneman, 1994) are: (1) Poisson's equation and Fourier transforms have been eliminated by updating the fields locally from the curl equations and depositing the particle currents according to charge-conserving formulas (Villasenor and Buneman, 1992), (2) radiating boundary conditions are applied to the fields using a first order Lindman approximation (Lindman, 1975), (3) filtering is done locally, (4) localisation makes the code ideally suited to modern parallel machines which call for minimising data paths, (5) the code is in Fortran and fully transportable: modest versions run on PCs and on workstations. In the past, the TRISTAN code has successfully simulated large scale space plasma phenomena such as the formation of systems of galaxies (Peratt, 1986). The new version of the code (Buneman, 1994) has been applied to the study of the dynamics of low-β plasma clouds (Neubert *et al.*, 1992) and the whistler wave driven by Spacelab 2 electron beam (Nishikawa *et al.*, 1994a), electron-positron plasmas (Zhao *et al.* 1994a,b) and current loop coalescence (Nishikawa *et al.*, 1994b: Sakai *et al.*, 1994).

For the simulation of solar wind-magnetosphere interaction the following boundary conditions were used for the particles: (1) Fresh particles representing the incoming solar wind (in our test run unmagnetized) are continuously injected across the yz-plane at $x = x_{min}$ with a thermal velocity plus a bulk velocity in the $+x$ direction, (2) thermal solar particle flux is also injected across the sides of our rectangular computation domain, (3) escaping particles are arrested in a buffer zone and redistributed there more uniformly by making the zone conducting in order to simulate their escape to infinity. They are then written off.

For the fields, boundary conditions were imposed just outside these zones: radiation is prevented from being reflected back inward, following Lindman's ideas (Lindman, 1975). The lowest order Lindman approximation was found adequate: radiation at glancing angles was no problem. However, special attention was given to conditions on the edges of the computational box.

In order to bring naturally disparate time- and space-scales closer together in this simulation of phenomena dominated by ion inertia and magnetic field interaction, the natural electron mass was raised to 1/16 of the ion mass and the velocity of light was lowered to twice the incoming solar wind velocity. This means that charge separation and anomalous resistivity phenomena are accounted for qualitatively but perhaps not with quantitative certainty. Likewise, radiation related phenomena (e.g. whistler modes) are covered qualitatively only.

3 Simulation Results

A first test exploring the solar wind-magnetosphere interaction was run on the CRAY-YMP at NCAR using a modest 105 by 55 by 55 grid and only 200,000 electron-ion pairs (Buneman *et al.*, 1992). We also have reported on our second test run on the CRAY-2 at NCSA using a larger 215 by 95 by 95 grid and about 1,000,000 electron-ion pairs for unmagnetized solar wind (Buneman *et al.*, 1994). Initially, these fill the entire box uniformly and drift with a velocity $v_D = 0.5c$ in the $+x$ direction, representing the solar wind. The electron thermal velocity is $v_{th} = 0.02c$ while the magnetic field is initially zero. A circular current generating the dipole magnetic field is increased smoothly from 0 to a maximum value reached at time step 65 and kept constant at that value for the rest of the simulation. The center of the current loop is located at (70.5Δ, 47.5Δ, 48Δ) with the current in the xy-plane and the axis in the z-direction. The initial expansion of the magnetic field cavity is found to expel a large fraction of the initial plasma. The solar wind density is about 0.7 electron-ion pairs per cell, the mass ratio is $m_i/m_e = 16$, and $\omega_{pe}\Delta t = 0.84$.

In this report we concentrate on the simulation results with the solar wind with southward or northward IMF. Figure 1 is shown the electron density in the center xz-layer containing the dipole center at time step 768. (Data were recorded at 64 step intervals and while no fully steady state was approached, no major change was seen between steps 768 and 1024 (Buneman *et al.*, 1992, 1994).) The electron density is color coded and the magnetic field component in the plane is shown with arrows at every third grid point. The magnitude of the field has been scaled in order to see the field direction for weak fields. Thus the length of the vectors is not a true representation of the field magnitude. The plasma is flowing through the simulation domain from left to right (low to high x-values). In the process the dipolar field is compressed at the side facing the plasma wind and is extended to a long tail on the down-wind side, just as the Earth's magnetic field in the solar wind.

Some particles have entered the cusp regions and, on the down-wind side (night side), particles have gathered on closed field lines in what could be equivalent to the radiation belt. The particles at the up-wind magnetopause are forming a bow shock. The temperature of these particles is much elevated from the background temperature. Also seen up-wind from the magnetopause is what could be a second shock or a foreshock. The magnitude of both shocks has a minimum at the sub-solar point and is increasing to a maximum at some distance from the sun-Earth line.

The system becomes quasi-steady around time step 768 shown in Fig. 1. At this point in time a southward IMF is introduced gradually with the injection of solar wind particles from the left boundary. The value of the southward IMF is set as slightly larger than the noise level outside the magne-

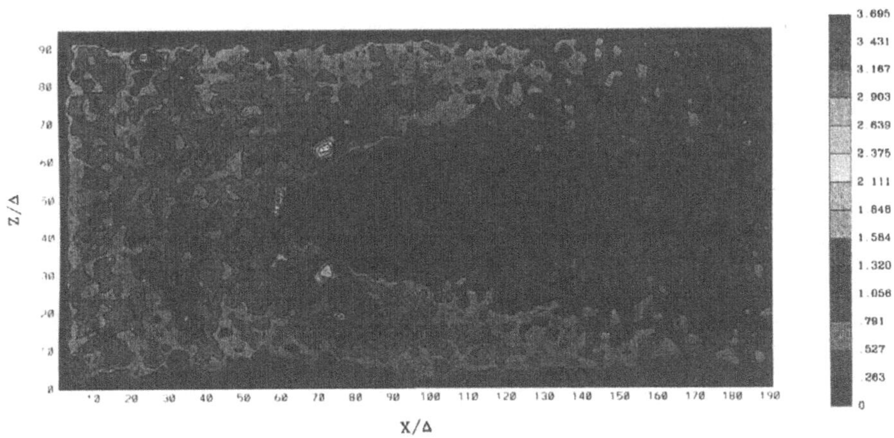

Fig. 1. Electron density in the center xz-layer containing the dipole center at step 768. "Relative amplitude" on color bar signifies simulation units.

topause ($B_z = -0.4$). The induced electric field ($E_y = -v_D \times cB_z$) is applied at the boundary with the form of $0.5*(1-\tanh((10-(step-768))/5))$ $step >$ 768. After about 20 steps the applied IMF becomes constant. After 394 steps the front of the solar wind with the southward IMF reaches $x/\Delta = 100$, if it is not slowdown by the magnetopause and the magnetosphere. Figure 2 shows the electron density at time step 1152.

Even the solar wind with the constant southward IMF is attacking at the magnetopause only for about 150 steps, the more electrons are penetrated through the cusps and the magnetopause as shown in Fig. 2. Consequently the magnetosphere is shrunk and more electrons trapped in the nightside. Ion density also shows the similar change as the electron density. Figure 3 shows the electron and ion fluxes along the z-direction at the noon-midnight meridian cross-section ($y = 48\Delta$) near the earth. The x point is located around $x = 58\Delta$, $z = 51\Delta$. The close examination reveals that the shape and intensities of fluxes are difference between the electrons and ions. The electrons and ions are move out from the x point.

Based on these simulation results we believe a magnetic reconnection

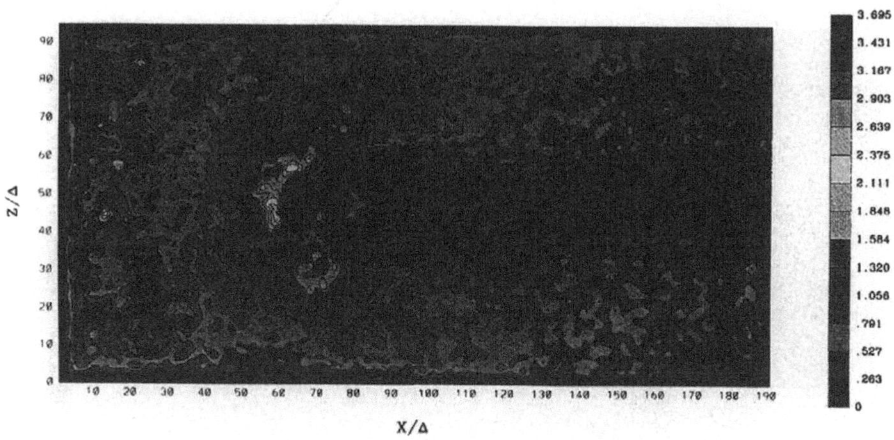

Fig. 2. Electron density in the center xz-layer containing the dipole center with a switched southward IMF at step 1152.

takes place in the dayside magnetopause as in MHD simulations. In our 3-D EM particle simulations the (anomalous) resistivity is in principle self-consistently determined by the particles and the electric and magnetic fields (waves) in the dayside magnetopause.

At step 768 the northward IMF is switched on. The magnetotail is shifted toward south and slightly shortened as shown in Fig. 4. This may be due to the reconnection at the higher magnetopause. The more particles are penetrated into the magnetotail near the earth. Obviously, we need a wider system to avoid the influences of the boundaries. The north-south asymmetry is caused by the different rates of particle penetrations due to the reconnection at the high latitude mantle.

4 Discussion

The results presented in Figures 1 through 4 show that even with the modest grid-size of 215 by 95 by 95 cells, our 3-D fully kinetic model is able to generate the complete magnetosphere with its basic characteristics with

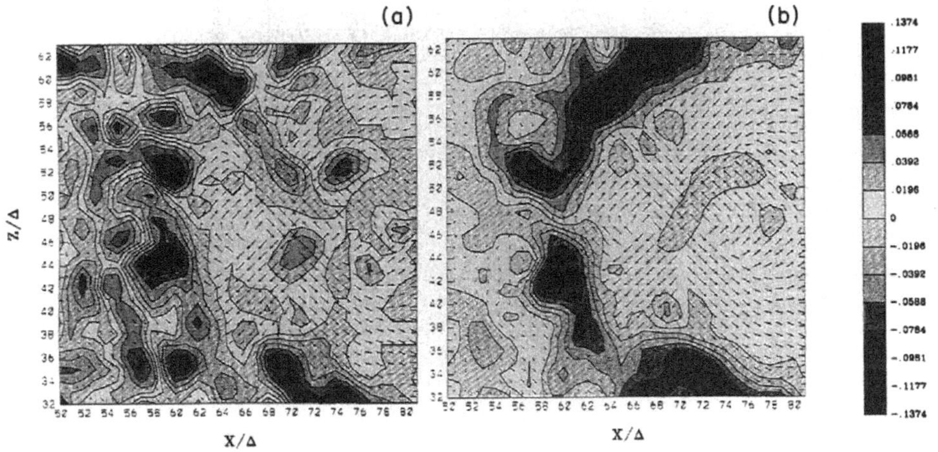

Fig. 3. **Particle flux near the magnetopause with magnetic fields by the arrows at step 1152: (a) electron flux, (b) ion flux.**

IMFs. The simulation with a southward interplanetary magnetic field (IMF) has been performed. After the solar wind with the IMF reaches the dayside magnetopause, more particles are diffused into the magnetosphere than without an IMF. The magnetosphere is shrunk and more particles are trapped in the nightside magnetosphere. This may be a manifestation of the magnetic reconnection at the magnetopause. In our 3-D EM particle simulations the resistivity in the magnetopause is self-consistently determined by the particles and the electric and magnetic fields. The detailed analysis with the close up diagnosis near the dayside magnetopause show the clear reconnection with jet-like flow from the x point.

On the contrary, in the case of solar wind with northward IMF, the magnetotail was shifted toward south and its length is shortened. This result is consistent with the occurrence of the magnetic reconnection at the high latitude mantle. Due to the more effective reconnection at the north side the more plasma move into from the north side, therefore, the distant magnetotail move toward south shortening its length.

These simulation results encourages us to continue our work using fin-

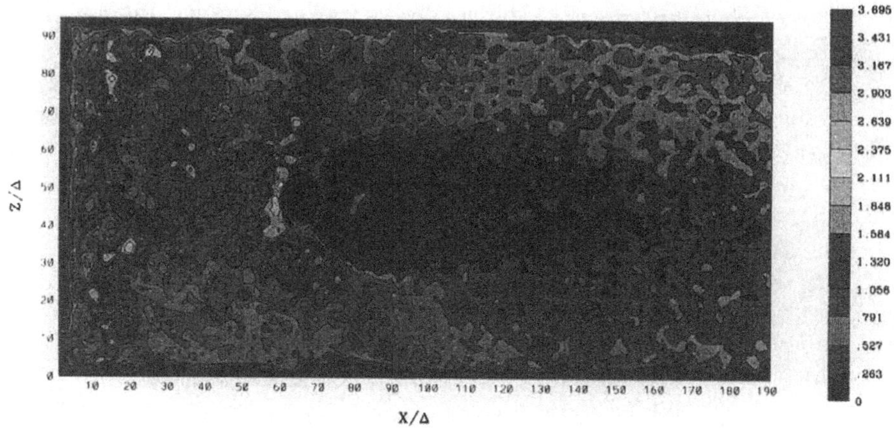

Fig. 4. Electron density in the center xz-layer containing the dipole center with a northward IMF at step 1408.

er grids (larger array sizes) and including such features as time-dependent IMFs, a tilt of the magnetic axis, and an inner conducting surface simulating the ionosphere. Larger dimensions will allow finer resolution of fields. The number of particles should also be increased. Longer runs would allow us to enhance realism by using smaller m_e/m_i and larger c/v_D. On presently available single processor computers the grid resolution could be increased by a factor of 2 in all dimensions. Multi-processor machines in the future will allow even greater improvement, since the code allows simultaneous updating of many cells.

Some important issues that with advantage can be addressed with a kinetic code are:

1. In MHD simulations reconnection phenomena are sometimes studied using artificial numerical dissipation to drive the process. Furthermore, artificially high diffusion coefficients are often used in order to achieve a stable solution. On the contrary a full kinetic code can handle reconnection processes self-consistently without such problems.

2. The motions of individual particles are tracked. In many regions of the magnetosphere, such as the magnetopause and the tail, gradients in the magnetic field are of such magnitude that the magnetic field changes significantly during a particle gyration. This gives rise to non-isotropic, non-Maxwellian distribution functions which drives the physics of those regions. The recent research into chaotic particle orbits in the magnetotail (Chen and Palmadesso, 1986; Burkhart and Chen, 1991; Ashour-Abdalla *et al.*, 1994) is examples. The complex particle motion gives rise to such phenomena as collisionless electron viscosity Dungey, 1988; Sonnerup, 1988; Lyons and Pridmore-Brown, 1990) and collisionless conductivity (Horton and Tajima, 1991). Furthermore, a particle code calculates the particle dynamics in self-consistent rather than model magnetic fields.

3. The complete wave dynamics from low frequency magnetohydrodynamic modes (Kelvin-Helmholtz instabilities at the magnetopause and ULF pulsations) to electron waves (Langmuir oscillations) is included. An example where electron kinetics is important is illustrated by recent studies of the dynamics of the magnetopause current layer (Berchem and Okuda, 1990; Okuda, 1991,1992,1993). It is thought that a generalized lower hybrid drift instability is generated, which in the case of the magnetopause has the effect of broadening the current layer and causing filamentation of the current.

4. Any complicated magnetic field configuration can be simulated by a simple modification of the source-current. This makes it straight-forward to simulate not only other planetary magnetospheres such as that of Uranus, but also the effect of the diurnal rotation of the Earth's dipole axis with respect to the direction to the sun.

However, some form of scaling is usually needed in particle simulations, even in one and two dimensions, and still such simulations are able to reveal much of the physics behind natural phenomena. Similar problems are also encountered, and to some degree overcome, in laboratory plasma simulations of macro-scale space plasma phenomena. Examples include the study of magnetic field line reconnection (Gekelman and Stenzel, 1984; Gekelman and Pfister, 1988), and the interaction of a magnetized plasma flow with a dipole magnetic field (Rahman *et al.*, 1991). We propose, therefore, that a kinetic particle model will be a useful tool for the study of the Earth's and other planetary magnetospheres in the solar wind, by supplementing present models that do not contain the complete particle dynamics.

5 Acknowledgements

Support for this work was provided by NASA contract NAS8-38772, NASA grant NAGW-2350, and NSF grants ATM-9106639, ATM-9119814, and ATM-9121116. The development of the simulation code was performed at the National Center for Supercomputing Applications, University of Illinois at Urbana-Champaign and test runs were performed at National Center for Atmospheric Research. Both centers are supported by the National Science Foundation.

References

Ashour-Abdalla. M. Zelenyi, L. M. Perroomian, V. and Richard, R. L.: 1991, "Consequences of magnetotail ion dynamics," *J. Geophys. Res.*, **Vol. 99**, pp. 14,891-14,916

Berchem, J. and Okuda, H.: 1990, "A two-dimensional particle simulation of the magnetopause current layer," *J. Geophys. Res.*, **Vol. 95**, pp. 8133-8147

Brecht S. H. and Ferrante, J. R.: 1991, "Global hybrid simulation of unmagnetized planets: Comparison of Venus and Mars," *J. Geophys Res.*, **Vol. 96**, pp. 11,209-11,220

Brecht, S. H. J. Lyon, S. H., Fedder, J. A. and Hain, K.: 1981, "A simulation study of East-West IMF effects on the magnetosphere," *Geophys. Res. Lett.*, **Vol. 8**, pp. 397-400

Brecht, S. H., Lyon, J. G. Fedder, J. A. and Hain, K.: 1982, "A time-dependent three-dimensional simulation of the Earth's magnetosphere: Reconnection events," *J. Geophys. Res.*, **Vol. 87**, pp. 6098-8108

Brecht, S. H. Ferrante, J. R. and Luhmann, J. G.: 1993, "Three-dimensional simulation of the solar wind interaction with Mars," *J. Geophys Res.*, **Vol. 98**, pp. 1345-1357

Buneman, O.: 1994, "TRISTAN: The 3-D, E-M Particle Code," *Computer Space Plasma Physics, Simulation Techniques and Software*, edited by H. Matsumoto and Y. Omura, Terra Scientific, Tokyo, pp. 67-84

Buneman, O., Barnes, C. W. Green, J. C. and Nielsen, D. E.: 1980, "Review: Principles and capabilities of 3-d, E-M particle simulations," *J. Comput. Phys.*, **Vol. 38**, pp. 1-44

Buneman, O. Neubert, T. and Nishikawa, K.-I.: 1992, "Solar wind-magnetosphere interaction as simulated by a 3D, EM particle code," *IEEE Trans. Plasma Sci.*, **Vol. 20**, pp. 810-816

Buneman, O. Nishikawa, K.-I. and Neubert, T.: 1994, "Solar wind-magnetosphere interaction as simulated by a 3D EM particle code," *Micro and Meso Scale Phenomena in Space Plasmas*, AGU Monograph, in press

Buneman, R. Barker, R. J., Peratt, A. L., Brecht, S. H., Langdon, A. B., Lewis, H. R.: 1994, "A Tribute to Oscar Buneman—Pioneer of Plasma Simulation," *IEEE Trans. Plasma Sci.*, **Vol. 22**, pp. 22-30

Burgess, D.: 1989, "Cyclic behavior of quasi-parallel collisionless shocks," *Geophys. Res. Lett.*, **Vol. 16**, pp. 345-348

Burkhart G. R. and Chen, J.: 1991, "Differential memory in the earth's magnetotail," *J. Geophys. Res.*, **Vol. 96**, pp. 14,033-14,049

Chen J. and Palmadesso, P.: 1986, "Chaos and non-linear dynamics of single particle orbits in a magnetotail-like magnetic field," *J. Geophys. Res.*, **Vol. 91**, pp. 1499-1508

Dungey, J. W.: 1988, "Noise-free neutral sheets," *Proceedings of an International Workshop in Space Plasma*, held in Potsdam, GDR, European Space Agency Special Report, *ESA SP-285* **Vol. II**, 15-

Elphic, R. C.: 1987, "The bowshock and magnetopause," *Rev. Geophys.*, **Vol. 25**, pp. 510-522

Fedder, J. A. and Lyon, J. G.: 1987, "The solar wind-magnetosphere-ionosphere current-voltage relationship," *Geophys. Res. Lett.*, **Vol. 14**, pp. 880-882

Fedder, J. A., Mobarry, C. M. and Lyon, J. G.: 1991, "Reconnection voltage as a function of IMF clock angle," *Geophys. Res. Lett.*, **Vol. 18**, pp. 1047-1050
Gary, S. and Sgro, A. G.: 1990, "The lower hybrid drift instability at the magnetopause," *Geophys. Res. Lett.*, **Vol. 17**, pp. 909-912
Gekelman, W. and Stenzel, R. L.: 1984, "Magnetic field line reconnection experiments 6. Magnetic turbulence," *J. Geophys. Res.*, **Vol. 89**, pp. 2715-2733
Gekelman, W. and Pfister, H.: 1988, "Experimental observations of the tearing of an electron current sheet," *Phys. Fluids*, **Vol. 31**, pp. 2017-2025
Haerendel G. and Paschmann, G.: 1982, "Interaction of the solar wind with the dayside magnetosphere," *Magnetospheric Plasma Physics*, edited by A. Nishida, Reidel Publishing Company, Dordrecht, Boston, London, pp. 49–142
Horton W. and Tajima, T.: 1991, "Collisionless conductivity and stochastic heating of the plasma sheet in the geomagnetic tail," *J. Geophys. Res.*, **Vol. 96**, pp. 15,811-15,829
Huang, C. Y.: 1987, "Quadrennial review of the magnetotail," *Rev. Geophys.*, **Vol. 25**, pp. 529-540
Leboeuf, J. N. Tajima, T. kennel, C. F. and Dawson, J. M.: 1981, "Global simulation of the three-dimensional magnetosphere," *Geophys. Res. Lett,*, **Vol. 8**, pp. 257-260
Lindman, E. L.: 1975, "Free-space boundary conditions for the time dependent wave equation," *J. Comp. Phys.*, **Vol. 18**, pp. 66-78
Lyons L. R. and Pridmore-Brown, D. C.: 1990, "Force balance near an X-line in a collisionless plasma," *J. Geophys. Res.*, **Vol. 95**, pp. 20,903-20,909
Mauk B. H. and Zanetti, L. J.: 1987, "Magnetospheric electric fields and currents," *Rev. Geophys.*, **Vol. 25**, pp. 541-554
Moore, K. R. Thomas, V. A. and McComas, D. J.: 1991, "Global hybrid simulation of the solar wind interaction with the dayside of Venus," *J. Geophys. Res.*, **Vol. 96**, pp. 7779-7791
Neubert, T. Miller, R. H. Buneman, O. and Nishikawa, K.-I.: 1992, "The dynamics of low-β plasma clouds as simulated by a 3-dimensional, electromagnetic particle code," *J. Geophys. Res.*, **Vol. 97**, pp. 12,057-12,072
Nishikawa, K.-I. Buneman, O. and Neubert, T.: 1994a, "New aspects of whistler waves driven by an electron beam as studied by a 3-D EM particle code," *Geophys. Res. Lett.*, **Vol. 21**, pp. 1019-1022
Nishikawa, K.-I. Sakai, J. I. Zhao, J. Neubert, T. and Buneman, O.: 1994b, "Coalescence of two current loops with a kink instability simulated by a 3-D EM particle code," *Ap. J.*, in press
Ogino, T.: 1986, "A three-dimensional simulation of the interaction of the solar wind with the Earth's magnetosphere: the generation of field-aligned currents," *J. Geophys. Res.*, **Vol. 91**, pp. 6791-6806
Ogino, T., Walker, R. J. Ashour-Abdalla, M. and Dawson, J. M.: 1985, "An MHD simulation of the By-dependant magnetospheric convection and field-aligned current during northward IMF," *J. Geophys. Res.*, **Vol. 90**, pp. 10,835-10,842
Ogino, T., Walker, R. J. Ashour-Abdalla, M. and Dawson, J. M.: 1986, "An MHD simulation of the effects of the interplanetary magnetic field By component on the interaction of the solar wind with the Earth's magnetosphere during southward interplanetary magnetic field," *J. Geophys. Res.*, **Vol. 91**, pp. 10,029-10,045
Ogino, T. Walker, R. J. and Ashour-Abdalla, M.: 1994, "A global magnetohydrodynamic simulation of the response of the magnetosphere to a northward turning of interplanetary magnetic field," *J. Geophys. Res.*, **Vol. 99**, pp. 11,027-11,042
Okuda, H.: 1991, "Numerical simulations on the magnetopause current layer," *Modeling Magnetospheric Plasma Processes*, American Geophysical Union, Geophysical Monograph 62, pp. 9-16
Okuda, H.: 1992, "Structure of the magnetopause current layer at the subsolar point," *J. Geophys. Res.*, **Vol. 97**, pp. 1389-1395
Okuda, H.: 1993, "Numerical simulation of the subsolar magnetopause current layer in the Sun-Earth meridian plane," *J. Geophys. Res.*, **Vol. 98**, pp. 3953-3962

Omidi, N., Quest, K. B. and Winske, D.: 1990, "Low Mach number parallel and quasi-parallel shocks," *J. Geophys. Res.*, **Vol. 95**, pp. 20,717-20,730

Peratt, A. L., Green, J. and Nielsen, D.: 1980, "Evolution of colliding plasmas," *Phys. Rev. Lett.*, **Vol. 44**, pp. 1767-1770

Peratt, A. L.: 1986, "Evolution of the plasma universe: II. The formation of systems of galaxies," *IEEE Trans. Plasma Sci.*, **Vol. PS-14**, pp. 763-778

Quest, K. B. 1988, "Theory and simulation of collisionless parallel shocks," *J. Geophys. Res.*, **Vol. 93**, pp. 9649-9680

Rahman, H. U., Yur, G. White, R. S. Birn, J. and Wessel, F. J.: 1991, "On the influence of the magnetization of a model solar wind on a laboratory magnetosphere," *J. Geophys. Res.*, **Vol. 96**, pp. 7823-7829

Sakai, J. I. Zhao, J. and Nishikawa, K.-I.: 1994, "Loops heating by D.C. electric current and electro- magnetic wave emissions simulated by a 3-D EM particle code," *Solar Phys.*, in press

Sato, T., Shimada, T. Tanaka, M. Hayashi, T. and Watanabe, K.: 1986, "Formation of twisting flux tubes on the magnetopause and solar wind particle entry into the magnetosphere," *Geophys. Res. Lett.*, **Vol. 13**, pp. 801-804

Sonnerup, B. U. O.: 1988, "On the theory of steady-state reconnection," *Comput. Phys. Commun.*, **Vol. 49**, 143-159

Thomas, V. A. and Winske, D.: 1990, "Two dimensional hybrid simulations of a curved bow shock," *Geophys. Res. Lett.*, **Vol. 17**, pp. 1247-1250

Thomas, V. A., Winske, D. and Omidi, N.: 1990, "Re-forming supercritical quasi-parallel shocks 1. One- and two-dimensional simulations," *J. Geophys. Res.*, **Vol. 95**, pp. 18,809-18,819

Villasenor J. and Buneman, O.: 1992, "Rigorous charge conservation for local electromagnetic field solvers," *Comp. Phys. Comm.*, **Vol. 69**, pp. 306-316

Walker, R. J., Ogino, T. and Ashour-Abdalla, M.: 1987, "A magnetohydrodynamic simulation of reconnection in the magnetotail during intervals of with southward interplanetary magnetic field," *Magnetotail Physics*, Edited by A. T. Y Lui, Johns Hopkins University Press, Baltimore, pp. 183-190

Walker, R. J., Ogino, T. Raeder, J., and Ashour-Abdalla, M.: 1993, "A global magnetohydrodynamic simulation of the magnetosphere when the interplanetary magnetic field is southward: The onset of magnetotail reconnection," *J. Geophys. Res.*, **Vol. 98**, pp. 17,235-17,249

Watanabe, K. and Sato, T.: 1988, "Self-excitation of auroral arcs in a three-dimensionally coupled magnetosphere-ionosphere system," *Geophys. Res. Lett.*, **Vol. 15**, pp. 717-720

Watanabe, K. and Sato, T.: 1990, "Global simulation of the solar wind-magneto- sphere interaction: The importance of its numerical validity," *J. Geophys. Res.*, **Vol. 95**, pp. 75-88

Winglee, R. W.: 1994, "Non-MHD influences on the magnetospheric current system," *J. Geophys. Res.*, **Vol. 99**, pp. 13,437-13,454

Winske, D., Omidi, N. Quest, K. B. and Thomas, V. A.: 1990, "Re-forming supercritical quasi-parallel shocks 2. Mechanism for wave generation and front re-formation," *J. Geophys. Res.*, **Vol. 95**, pp. 18,821-18,832

Wu, C. C. Walker, R. J. and Dawson, J. M.: 1981, "A three-dimensional MHD model of the Earth's magnetosphere," *Geophys. Res. Lett.*, **Vol. 8**, pp. 523-526

Zhao, J. Nishikawa, K.-I. Sakai, J. I. and Neubert, T.: 1994a, "Study of nonlinear Alfvén waves in an electron-positron plasma with 3-D EM particle code," *Phys. Plasmas*, **Vol. 1**, pp. 103-108

Zhao, J. Sakai, J. I. Nishikawa, K.-I. and Neubert, T.: 1994b, "Relativistic particle acceleration in an electron-positron plasma with a relativistic electron beam," *Phys. Plasmas*, in press

RECOLLECTIONS OF MAX BORN

EMIL WOLF*
Department of Physics and Astronomy
University of Rochester
Rochester, NY 14627 USA

Abstract. This article is essentially the text of lectures presented September 7, 1982 at the Max Born Centenary Conference held in Edinburgh, Scotland and October 21, 1982 at the Max Born Symposium held during the Annual Meeting of the Optical Society of America.

Key words: History of Science

1 Introduction

The invitation to address this commemorative meeting has given me the rare opportunity to set aside my customary activities and try to recall a period of my life several decades ago when I had the great fortune of being able to collaborate with Max Born. As the title of my talk suggests, this will be a rather personal account, but I will do my best to present a true image of a scientist who has contributed in a decisive way to modern physics in general and to optics in particular; it will also present glimpses of a man who, under a somewhat brusque exterior, was a very humane and kind person and who in the words of Bertrand Russell was brilliant, humble, and completely without fear in public utterances.

The early part of my story is closely interwoven with another great scientist, Dennis Gabor, through whose friendship I became acquainted with Born.

I completed my graduate studies in 1948 at Bristol University. My PhD thesis supervisor was E. H. Linfoot, who at just about that time was appointed Assistant Director of the Cambridge University Observatory. He offered me, and I accepted, a position as his assistant in Cambridge. During the next two years while I worked in Cambridge I frequently traveled to London to attend the meetings of the Optical Group of the British Physical Society. They were usually held at Imperial College and were often attended by Gabor, whose office was in the same complex of buildings. From time to time I presented short papers at these meetings. At the end of some of the meetings Gabor would invite me to his office for a chat. He would comment

* Emil Wolf is Wilson Professor of Optical Physics at the University of Rochester. Reprinted from *Optics News*, November/December 1983 with permisssion from the author and the Optical Society of America

Fig. 1. Max Born at his desk, ca 1950. (Credit: AIP Niels Bohr Library).

on the talks, make suggestions regarding my work, and speak about his own researches. Gabor liked young people, and he always offered encouragement to them. He knew Born from Germany, and he had great admiration for him.

Through Gabor I learned in 1950 that Born was thinking of preparing a new book on optics, somewhat along the lines of his earlier German book Optik, published in 1933, but modernized to include accounts of the more important developments that had taken place in the nearly 20 years that had gone by since then. At that time Born was the Tait Professor of Natural Philosophy at the University of Edinburgh, a post he had held since 1936, and in 1950 he was 67 years old, close to his retirement. He wanted to find some scientists who specialized in modern optics and who would be willing to collaborate with him in this project. Born approached Gabor for advice, and at first it was planned that the book would be written jointly by him, Gabor, and H. H. Hopkins. The book was to include a few contributed sections on some specialized topics, and Gabor invited me to write a section on diffraction theory of aberrations, a topic I was particularly interested in at that time. Later it turned out that Hopkins felt he could not devote adequate time to the project, and in October of 1950, Gabor, with Born's agreement, wrote to Linfoot and me asking if either of us, or both, would be willing to take Hopkins' place. After some lengthy negotiations it was agreed that Born, Gabor, and I would co-author the book.

Fig. 2 Dennis Gabor

2 The start of collaboration

I was, of course, delighted with this opportunity, but there was the problem of my finding the necessary time to work on this project while holding a full-time appointment with Linfoot at Cambridge. I mentioned this to Gabor, and I told him that if there were any possibility of obtaining an appointment with Born, which would allow me to spend most of my time working on the book, I would gladly leave Cambridge and go to Edinburgh.

Gabor took up the matter with Born, who was interested. Toward the end of November 1950, Gabor wrote me that Born would be in London a few days later and that he (Gabor) was arranging for the three of us to meet the following weekend. It was agreed that I would come to Gabor's office at Imperial College on the following Saturday morning, December 2, 1950, and that we would then go to his home in South Kensington, within walking distance of Imperial College. Born was to come directly to Gabor's home from his London hotel, and the three of us and Mrs. Gabor would have lunch there.

I arrived at Gabor's office just before lunch, and I have a vivid recollection of that meeting. There was a long staircase leading to the entrance hall of the building. As we were walking down the staircase, Gabor suddenly became somewhat apprehensive. He knew that our luncheon meeting might lead to an appointment for me with Born, and he said to me, "Wolf, if you let me down, I will never forgive you. Do you know who Born's last assistant was?

Fig. 3. The building on Drummond Street in Edinburgh that housed Max Born's Department of Applied Mathematics

Heisenberg!" This statement was not accurate. Born had other assistants after Heisenberg, but the remark shows how nervous Gabor was on that particular occasion. Fortunately, all turned out well, and Gabor certainly seemed in later years well satisfied with the consequences of our luncheon with Born.

During that meeting Born asked me a few questions, mainly about my scientific interests, and before the lunch was over he invited me to become his assistant in Edinburgh, subject to the approval of Edinburgh University. It seemed to me remarkable that Born should have made up his mind so quickly, without asking for even a single letter of reference, especially since I had published only a few papers by that time and was quite unknown to the scientific community.

Later, when I got to know Born better, I realized that his quick decision was very much in line with one trait of his personality; he greatly trusted the judgment of his friends. Since Gabor recommended me, Born considered further inquiries about me to be superfluous. Unfortunately, as I also learned later, Born's implicit trust in people whom he considered to be his friends was occasionally misplaced and sometimes created problems for him.

A few days after our meeting I received a telegram from Born inviting me to a formal interview at Edinburgh University. The interview took place about two weeks later, and the next day Born wrote me saying that the committee which interviewed me recommended my appointment as his private assistant, beginning January 23, 1951. I resigned my post in Cambridge and took up the new appointment. Later I learned that committee approval was not really needed because my salary was to be paid from an industrial grant that was entirely at Born's disposal. However, on this occasion Born was

Fig. 4. David Hilbert, 1912.

careful, because some time earlier he had had on his staff Klaus Fuchs, who turned out to be a spy for the Russians, and Born got rather bad publicity from that.

Now, the name Fuchs means fox in German, and before inviting me to Edinburgh, Born apparently wrote to Sir Edward Appleton, the Principal of Edinburgh University at that time, saying that he felt the decision about this particular appointment should not be made by him alone; since he would like to appoint a Wolf after a Fox!

3 Arrival at Edinburgh

I arrived in Edinburgh toward the end of January 1951, eager to start on our project. Born's Department of Applied Mathematics was located in the basement of an old building on Drummond Street. I was surprised by the small size of the department. Physically it consisted of Born's office; an adjacent large room for all of his scientific collaborators, about five at that time; a small office for Mrs. Chester, his secretary; two rooms for the two permanent members of his academic staff, Robert Schlapp, a senior lecturer, and Andrew Nisbet, a lecturer; and one lecture room. The rest of the building was occupied by experimental physicists under the direction of Professor Norman Feather. In earlier days the building housed a hospital, in which Lord Lister, a famous surgeon known particularly for his work on antiseptics, also worked.

In spite of his advanced age Born was very active and, as throughout all his adult life, a prolific writer. He had a definite work routine. After coming to his office he would dictate to his secretary answers to the letters that arrived in large numbers almost daily. Afterward he would go to the adjacent room where all his collaborators were seated around a large U-shaped table. He would start at one end of it, stop opposite each person in turn, and ask the same question: "What have you done since yesterday?" After listening to the answer he would discuss the particular research activity and make suggestions. Not everyone, however, was happy with this procedure. I remember a physicist in this group who became visibly nervous each day as Born approached to ask his usual question, and one day he told me that he found the strain too much and that he would leave as soon as he could find another position. He indeed did so a few months later. At first I too was not entirely comfortable with Born's question, since obviously when one is doing research and writing there are sometimes periods of low productivity. One day when Born stood opposite me at the U-shaped table and asked, "Wolf, what have you done since yesterday?" I said simply, "Nothing!" Born seemed a bit startled, but he did not complain and just moved on to the next person, asking the same kind of question again.

Born was always direct in expressing his views and feelings, but he did not mind if others did the same, as this small incident indicates. There will be more examples of this later.

4 Work at Edinburgh

We started working on the optics book as soon as I came to Edinburgh. It was understood right from the beginning that Born's main contribution would consist of making material available from his German *Optik*, but he was to take part in the planning of the new book, make suggestions, and provide general advice. Most of the actual writing was to be done by Gabor and me and a few contributors. However, like Hopkins earlier on, Gabor soon found it difficult to devote the necessary time to the project, and it was mutually agreed that he would not be a co-author after all, but would just contribute a section on electron optics. So in the end it became my task to do most of the actual writing. Fortunately I was rather young then, and so I had the energy needed for what turned out to be a very large project. I was in fact 40 years younger than Born. This large age gap is undoubtedly responsible for a question I am sometimes asked, whether I am a son of the Emil Wolf who co-authored *Principles of Optics* with Max Born!

Although I did most of the writing, Born read the manuscript and made suggestions for improvements. We signed a contract with the publishers about a year after I came to Edinburgh, and we hoped to complete the

Fig. 5. Max Born as *Privatdozent* in Göttingen. (Reproduced from *Hilbert* by Constance Reid, 1970).

manuscript by the time Born was to retire, one-and-a-half years later. However, we were much too optimistic. The writing of the book took about eight years altogether.

Throughout his life Born was a quick, prolific writer, publishing more than 300 scientific papers, about 31 books (not counting different editions and translations), apart from numerous articles on nonscientific topics.[1] In spite of my relative youth I could not compete with the speed with which Born wrote, even at his advanced age, and it soon became clear to me that he was not too pleased with my slow progress.

One day when I started writing an Appendix on Calculus of Variations, Born said that the best treatment of that subject he knew of was in his notes of lectures given by the great mathematician David Hilbert in Göttingen in the early years of this century. Born suggested that he dictate the Appendix to me, following in the main Hilbert's presentation, and that we acknowledge this in the preface to our book. So we started in this way. After each dictating session I was to rewrite the notes and give them to Born the next day for his comments. But we did not get very far this way. After about two dictating

[1] A bibliography of Born's scientific publications is given in "Max Born," by N. Kemmer and R. Schlapp in Biographical Memoirs of the Royal Society, 17, London: the Royal Society, 1971, pp. 17-52. Born's autobiography (Born, 1978) was published posthumously, first in German in 1975 and is, therefore, not included in that bibliography.

Fig. 6. Max Born in the 1920s.

sessions Born said he could prepare the whole Appendix himself much faster without my help, which he then did. It is essentially in this version, written by Born, that the Appendix on Calculus of Variations appears in our book.

5 Born's revered teacher

Incidentally, David Hilbert, whose presentation Born closely followed, was one of Born's great heroes. To physicists Hilbert is mainly known in connection with the concept of the Hilbert space and as co-author of the classic text *Methods of Mathematical Physics*, referred to generally as "Courant-Hilbert." But Hilbert contributed in a fundamental way to many branches of mathematics and was generally considered to have been the greatest mathematician of his time. Born became acquainted with Hilbert soon after coming to Göttingen in 1905, later becoming Hilbert's private assistant. In one of his later writings Born refers to Hilbert as his "revered teacher and friend," and in a biography of Hilbert by Constance Reid (Reid, 1970), Born is quoted as saying that his job with Hilbert was to him "precious beyond description because it enabled [him] to see and talk to him every day."

Born had an encyclopedic knowledge of physics and whatever problem one brought to him, he was able to offer a useful insight or suggest a pertinent reference. He also knew personally all the leading physicists of his time and would often recall interesting stories about them.

Optics in those days–remember we are talking about optics in pre-laser days–was not a subject that most physicists would consider exciting; in fact, relatively little advanced optics was taught at universities in those days. The

Fig. 7. Mrs. Hedwig Born, 1961.

Fig. 8. The house of Max and Hedwig Born in Edinburgh, at 84 Grange Loan.

fashion then was nuclear physics, particle physics, high energy physics, and solid state physics. Born was quite different in this respect from most of his colleagues. To him all physics was important, and rather than distinguish between "fashionable" and "unfashionable" physics he would only distinguish between good and bad physics research.

Born was equally broad-minded about the techniques used by physicists in their research. For example, when we were writing a section on certain math-

Fig. 9. Albert Einstein in the 1920s. (Credit: AIP Niels Bohr Library).

ematical methods needed to evaluate the performance of optical systems, we found that although the results given in a basic paper on this subject were correct, the derivation contained serious flaws. I was rather indignant about this, but Born just said something like, "In pioneering work everything is allowed, as long as one gets the right answer. Real justification can come later."

One of the earliest occasions when many physics students encounter Born's name comes when they start studying quantum theory of scattering. Here they soon learn about the *Born approximation*. This term also occurs in many of the papers on potential scattering that have been published in the more than half a century that has gone by since Born wrote a basic paper on this subject. Yet Born was rather irritated when the Born approximation was mentioned. He once said to me, "I developed in that paper the whole perturbation expansion for the scattered field, valid to all orders, yet I am only given credit for the first term in that series!"

6 Resistance to new discoveries

It was not always easy for Born's collaborators to convince him quickly of new discoveries. Let me illustrate this by an example from my own experience. In the early 1950s I became very interested in problems of partial coherence. One day I found a result in this area of optics that seemed to me remarkable.

I phoned Born from my home one morning, told him I had an exciting new result, and asked him for an appointment to discuss it. We arranged to have lunch together that day.

When I came to his office just before lunch, Born wanted to know straight away what the excitement was all about. I told him I had found that not only an optical field, but also its coherence properties, characterized by an appropriate correlation function (now known as the mutual coherence function), are propagated in the form of waves. Born looked at me rather skeptically, put his arm on my shoulder and said, "Wolf, you have always been such a sensible fellow, but now you have become completely crazy!" Actually after a few days he accepted my result, and I suspect he then no longer doubted my sanity.

This incident illustrates a fact well known to Born's collaborators–that Born had a certain resistance to accept new results obtained by others. Nonetheless, he continued thinking about them, and if they were correct he would eventually apologize for doubting them in the first place.

This trait of Born's personality is very well described by the Polish physicist Leopold Infeld, who collaborated with Born in Cambridge in the 1930s. I will quote shortly some very perceptive observations Infeld made about Born in his biography (Infeld, 1941); but before doing so I would like to mention a small incident relating to this book.

One day I browsed through a bookstore in Edinburgh and found a used copy of Infeld's book. I was astonished to note that the book had Born's signature on its first page. I purchased it and asked Born the next day whether he knew the book. He said, "Yes, I had a copy of it and there is a funny description of me in it; but I lent it to someone and it was never returned. I cannot remember whom I lent it to." The book I had purchased was obviously Born's missing copy, so I gave it to him, much to his delight.

In the book Infeld describes some of his experiences in Cambridge. He started working with Dirac but found him rather uncommunicative. Later Infeld attended some of Born's lectures. During one of them Born gave an account of some results that he had recently obtained. Infeld could not understand one of Born's arguments. He borrowed his notes so that he could study the argument more closely later. Let me now quote from Infeld's biography (Infeld, 1941, p.208 *et seq.*):

> *On the evening of the day I received the paper the point suddenly became clear to me. I knew that the mass of the electron was wrongly evaluated in Born's paper and I knew how to find the right value. My whole argument seemed simple and convincing to me. I could hardly wait to tell it to Born, sure that he would see my point immediately. The next day I went to him after his lecture and said: "I read your paper; the mass of the electron is wrong." Born's face looked even more tense than usual. He said: "This is very interesting. Show me why." Two of his audience*

Fig. 10. Sir Edmund Whittaker. (Reproduced by the courtesy of the University of Edinburgh.

were still present in the lecture room. I took a piece of chalk and wrote a relativistic formula for the mass density. Born interrupted me angrily: "This problem has nothing to do with relativity theory. I don't like such a formal approach. I find nothing wrong with the way I introduced the mass." Then he turned toward the two students who were listening to our stormy discussion. "What do you think of my derivation?" They nodded their heads in full approval. I put down the piece of chalk and did not even try to defend my point. Born felt a little uneasy. Leaving the lecture room, he said, "I shall think it over."

Infeld then goes on to say:

I was annoyed at Born 's behavior as well as at my own and was, for one afternoon, disgusted with Cambridge. I thought: "Here I met two great physicists. One of them does not talk. I could as easily read his papers in Poland as here. The other talks, but he is rude." The next day I went again to Born's lecture. He stood at the door before the lecture room. When I passed him he said to me. "I am waiting for you. You were quite right. We will talk it over after the lecture. You must not mind my being rude. Everyone who has worked with me knows it. I have a resistance against accepting something from outside. I get angry and swear but always accept it after a time if it is right"

Our collaboration had begun with a quarrel, but a day later complete peace and understanding had been restored.

A little further on in his biography, Infeld speaks about Born again, and this is what he says:

I marveled at the way in which he managed his heavy correspondence, answering letters with incredible dispatch, at the same time looking through scientific papers. His tremendous collection of reprints was well ordered; even the reprints from cranks and lunatics were kept, under the heading "Idiots." Born functioned like an entire institution, combining vivid imagination with splendid organization He worked quickly and in a restless mood. As in the case of nearly all scientists, not only the result was important but the fact that he had achieved it.

Infeld later continues:

There was something childish and attractive in Born's eagerness to go ahead quickly, in his restlessness and his moods, which changed suddenly from high enthusiasm to deep depression. Sometimes when I would come with a new idea he would say rudely, "I think it is rubbish," but he never minded if I applied the same phrase to some of his ideas. But the great, the celebrated Born was as happy and as pleased as a young student at words of praise and encouragement. In his enthusiastic attitude, in the vividness of his mind, the impulsiveness with which he grasped and rejected ideas, lay his great charm.

I regard these remarks of Infeld as a true and very perceptive description of Born's mode of work and of Born's personality.

7 Kind and compassionate

In spite of Born's occasional irritation and impatience, he was a person who cared deeply for the well-being of his fellow scientists and collaborators. His wife, Hedwig Born, was likewise a person with deep compassion for others. She too was a remarkable and gifted person. Mrs. Born published a number of books, especially poetry, and around 1938 became a Quaker, remaining active in the Quaker movement for the rest of her life.

I would like to give just one example from my own experience, which illustrates Born's concern for others. A few months after I began working with Born, I was getting married. In those days it was difficult to rent an apartment in Edinburgh. One day during the time when we were searching for a home I received a letter from Mrs. Born, who was then with Professor Born on a visit to Germany. She said that they had heard about our problem and were very concerned that we might have to postpone getting married if we did not find somewhere to live. She then offered to help us, suggesting that we share with them their small house in Edinburgh. In the end we found an apartment elsewhere; but this small episode is an indication of the

"In an Age of mediocrity and moral pygmies, the lives of Albert Einstein and Max Born shine with an intense beauty. Something of this is reflected in their correspondence, and the world is the richer for its publication."

(From Bertrand Russell's Foreword to *The Born-Einstein Letters*, Ref. 4)

Fig. 11.

warmth of their personalities and of their willingness to make a personal sacrifice to help, when help was needed.

I mentioned earlier, that one of Born's great heroes was the mathematician David Hilbert. But there was another, even greater hero in Born's life: Albert Einstein, with whom he and also Mrs. Born maintained close personal friendships for almost half a century. Unfortunately, after Einstein left Europe for America in 1932 they did not see each other again, but they carried on extensive correspondence until Einstein's death in 1955. The letters they exchanged were published in 1971, together with Born's commentary, and the volume (Born, 1971) is a precious contribution to the history of physics and of the times in which they lived.

There is an episode I would like to relate briefly in connection with Born's friendship with Einstein. In the early 1950s, when Sir Edmund Whittaker was preparing the second volume of his classic work *A History of the Theories of Aether and Electricity*, he sent Born the manuscript of a section dealing with the special theory of relativity. Whittaker's treatment placed a heavy emphasis on the work of Poincare and Lorentz and dismissed Einstein's contribution as being of rather minor significance. Born, who himself wrote a book on the theory of relativity, was most unhappy with Whittaker's manuscript and sent him a long report in which he analyzed in detail the various contributions, indicating why he considered Einstein's contribution to be much more fundamental.

Fig. 12. Members of Max Born's department at the time of his retirement (1953) from the Tait Chair of Natural Philosophy at the University of Edinburgh. Standing (from left to right)" E. Wolf, D. J. Hooton, A. Nisbet. Sitting: Mrs Chester (secretary), M. Born, R. Schlapp.

Reprinted from New Scientist, January 13, 1966, Pages 74 78

Dear Emil,
I nominate you as my Chief Prophet
Love
Dennis

HOLOGRAPHY, OR
THE "WHOLE PICTURE"
By Professor Dennis Gabor, FRS
Department of Electrical Engineering Imperial College London

Fig. 13. A dedication from Dennis Gabor.

However, Born did not succeed in changing Whittaker's opinion.[2] In

[2] Born's opinion on this question rather than Whittaker's is generally accepted. See,

Fig. 14. Max Born in Bad Pyrmont feeding pigeons.

September of 1953, around the time Whittaker's book was published, Born wrote to Einstein about this. Let me quote from Born's letter (Born, 1971, p.197): "Many people may now think (even if you do not) that I played a rather ugly role in this business. After all it is common knowledge that you and I do not see eye to eye over the question of determinism."

Einstein was not concerned. This is what he said in his reply to Born (Born, 1971, p.199): "Don't lose any sleep over your friend's book .... If he manages to convince others, that is their own affair. I myself have certainly found satisfaction in my efforts...." and then Einstein added, "After all, I do not need to read the thing."

Born retired that year, in 1953. The accompanying photograph shows Born with the members of his department at the time of his retirement.

---

for example, D. Martin's biographical note about E. T. Whittaker in *Dictionary of Scientific Biography*, C. C. Gillespie, editor-in-chief (Charles Scribner's Sons, New York, 1976), Vol.XIV, p.317; or A. Pais: *Subtle Is the Lord, The Science and the Life of Albert Einstein* (Clarendon Press, Oxford, and Oxford University Press, New York, 1982), p 168.

"Everybody who
had contact with him
remembers him
not only as
a brilliant scientist but
also as
a man of human warmth
and greatness."

Fig. 15.

8 Life in retirement

Soon afterward the Born's left Edinburgh and settled in Bad Pyrmont, a spa in West Germany, not far from Göttingen, where they built a small house. When they left Edinburgh our book was far from finished. We corresponded about it, and I visited Born in his new home several times. Born was hoping that he and Mrs. Born would be able to lead a more quiet life in Bad Pyrmont, but he told me on one of my visits that this proved difficult to achieve. For example, soon after they settled in Bad Pyrmont, Born was invited to address a meeting of a West German physical society. He declined the invitation, saying that he was too old to travel. He received a reply stating that in view of this the meeting would be moved to Bad Pyrmont!

In 1954, the year after his retirement, Born was awarded the Nobel Prize. He was, of course, delighted, but I am quite sure he felt, as many others did, that this great recognition had come somewhat late. The Nobel Prize was awarded to him for contributions that he made almost 30 years earlier. However, as his son Gustav later noted in a postscript to Born's memoirs (Born, 1978, p.296), it came at the right time to add weight to his main retirement occupation, which was to educate thinking people in Germany and elsewhere in the social, economic, and political consequences of science and also of the dangers of nuclear weapons and re-armament.

In 1957 I was a Visiting Scientist at the Courant Institute of New York University, still working on our book. One day I received a letter from Born asking me why the book was not yet finished. I replied that practically the whole manuscript was completed, except for a chapter on partial coherence on which I was still working. Born wrote back almost at once, saying something like, "Who apart from you is interested in partial coherence? Leave

Fig. 16. Max Born in front of his lıbrary at his home in Bad Pyrmont

Fig. 17. Max Born with two of his grandsons, Max and Sebastian (children of Gustav) in Bad Pyrmont. (Credit" AIP Niels Bohr Lıbrary)

that chapter out and send the rest of the manuscript to the printers." Actually I completed that chapter shortly afterward and it was included in the book.

It so happened that within about two years after the publication of our book (in 1959) the laser was invented and optical physicists and engineers then became greatly interested in questions of coherence. Our book was the first that dealt in depth with this subject, and Born was then as pleased as I was that the chapter was included.

Our book was also one of the first textbooks containing an account of holography. Gabor was very happy about it. Later, when holography became popular and useful, he sent me a reprint of one of his papers with a charming dedication (Figure 13).

As I approach the end of my reminiscences about Max Born, I would like to say that I hope my talk conveyed to you the warmth and the affection with which he remains in my memory, not only as a great scientist, but also as a kind and remarkable human being. My feelings about our collaboration are well described by exactly the same words that Born used when he spoke about his association with David Hilbert, quoted earlier, namely that my appointment with him was precious to me beyond description, because it enabled me to see and to talk to him every day.

9 Olivia

Before ending I would like to show you a few pictures taken in Bad Pyrmount during Born's retirement and also to mention one more episode. One shows Professor and Mrs. Born with one of their daughters, Irene. Some years ago I learned that Irene is the mother of a lady who has achieved fame comparable to that of Max Born himself, but in an entirely different field. I am speaking of the pop singer Olivia Newton-John. Shortly after I learned that Olivia Newton-John was Max Born's granddaughter, I was on a sabbatical leave at the University of Toronto. Olivia was scheduled to give a concert in Toronto while I was there. I wrote to her, told her I had collaborated with her grandfather in the writing of a book, and asked her whether we could meet. I received a charming reply in which she invited me to meet her after the concert. We met then and talked mainly about her grandparents. Before I left Olivia gave me two autographed photos of herself. Let me add that to some of my students I am known not so much as the co-author of *Principles of Optics* but rather as the person who knows Olivia Newton-John and who has a picture of her hanging in his office signed "To Emil, Love, Olivia."

I cannot bring you the voice of Max Born, but I will end my presentation with one of the songs that made Olivia famous. (The lectures on which this

Fig. 18. Hedwig Born and Max Born, with their daughter Irene Newton-John in Bad Pyrmont, 1957. (Credit: AIP Niels Bohr Library).

Fig. 19. *Left*: Olivia Newton-John, granddaughter of Max Born *Right*: Max Born (Credit: Lotte Meitner-Graf)

article is based concluded with an excerpt from the song "If You Love Me Let Me Know.")

10 Acknowledgments

In preparing this article for publication I received assistance with obtaining some of the photographs, determining the approximate dates when they were taken and with checking some of the references. I am particularly obliged to G. V. R. Born (London University), R. M. Sillitto and S. D. Fletcher (University of Edinburgh), L. H. Caren (University of Rochester), and D. Dublin (American Institute of Physics) for their help.

References

Born, M. : 1971, *The Born–Einstein Letters*, Walker and Company, New York

Born, M. : 1978, *My Life*, Taylor and Francis, London, and Charles Scribner's Sons, New York, p. 296

Infeld, L.: 1941, *Quest: The Evolution of a Scientist*, Doubleday, Doran and Co., New York

Reid, C.: 1970, *Hilbert*, Springer-Verlag, New York, Heidelbert, Berlin, p. 103

LIST OF FORTHCOMING PAPERS

H.-G. Grothues: What Do Streamers in the Dust Tail Tell about the Cometary Nucleus? (Received 5 January, 1995.)

G. E. Romero and J. A. Combi: Variability in the Southern Blazars PKS 1921–293 and PKS 2155-152. (Received 20 September, 1994; accepted 23 February, 1995.)

A. M. Cherepashchuk, K. V. Bychkov and E. V. Seifina: Colliding Winds Model of X-Ray Radiation in SS 433: Analysis of X-Ray Eclipses. (Received 22 February, 1995; accepted 10 March, 1995.)

P. Hall and D. A. Williams: Polyatomic Molecules in Diffuse Clouds. (Received 24 November, 1994; accepted 20 March, 1995.)

I. Eglītis and M. Eglīte: The Abundance Ratio C/O in the Atmospheres of 343 Carbon Stars. I. The Orion Galactic Arm. (Received 24 October, 1994; accepted 21 March, 1995.)

J.-M. Wang and Y.-Y. Zhou: On the Limit of Brightness Temperature in Nonstationary Radio Sources. (Received 18 January, 1995; accepted 27 March, 1995.)

V. S. Troitskij: A Static Model of the Universe. (Received 14 April, 1993; accepted 29 March, 1995.)

L. Herrera and N. Falcon: Secular Stability Behaviour of Nuclear Burning before Relaxation. (Received 31 October, 1994; accepted 12 April, 1995.)

E. Viñuales, R. Cid and A. Elipe: Visual Binary Star Orbits in Universal Variables. (Received 21 February, 1995; accepted 13 April, 1995.)

GPSR Compliance

The European Union's (EU) General Product Safety Regulation (GPSR) is a set of rules that requires consumer products to be safe and our obligations to ensure this.

If you have any concerns about our products, you can contact us on ProductSafety@springernature.com

In case Publisher is established outside the EU, the EU authorized representative is:

Springer Nature Customer Service Center GmbH
Europaplatz 3
69115 Heidelberg, Germany

Batch number: 10607320

Printed by Printforce, the Netherlands